Towards Sustainable Low-Carbon Concrete

Towards Sustainable Low-Carbon Concrete

Guest Editors

Jiaxiang Lin
Prabir K. Sarker

MDPI

Basel • Beijing • Wuhan • Barcelona • Belgrade • Novi Sad • Cluj • Manchester

Guest Editors
Jiaxiang Lin
School of Civil and Transportation Engineering
Guangdong University of Technology
Guangzhou
China

Prabir K. Sarker
School of Civil and Mechanical Engineering
Curtin University
Perth
Australia

Editorial Office
MDPI AG
Grosspeteranlage 5
4052 Basel, Switzerland

This is a reprint of the Special Issue, published open access by the journal *Materials* (ISSN 1996-1944), freely accessible at: https://www.mdpi.com/journal/materials/special_issues/8R737YC29L.

For citation purposes, cite each article independently as indicated on the article page online and as indicated below:

Lastname, A.A.; Lastname, B.B. Article Title. *Journal Name* **Year**, *Volume Number*, Page Range.

ISBN 978-3-7258-8521-3 (Hbk)
ISBN 978-3-7258-8522-0 (PDF)
https://doi.org/10.3390/books978-3-7258-8522-0

Contents

About the Editors

Jiaxiang Lin

Jiaxiang Lin is an Associate Professor at the School of Civil and Transportation Engineering, Guangdong University of Technology. His research focuses on sustainable and high-performance cementitious composites, low-carbon binder systems, fiber-reinforced and strain-hardening concrete, and durability enhancement technologies for infrastructure in aggressive service environments. He has led and participated in multiple research projects related to green construction materials and advanced structural applications, with emphasis on solid waste utilization, multifunctional concrete, and performance-based durability design. He has received the First Prize of the Science and Technology Progress Award in Higher Education and the First Prize of the Guangdong Provincial Green and Prefabricated Construction Science and Technology Award, and his work has resulted in numerous peer-reviewed publications, patents, and industry collaborations promoting the implementation of low-carbon concrete technologies.

Prabir K. Sarker

Prabir K. Sarker is a Professor at the School of Civil and Mechanical Engineering, Curtin University, Australia, and is well known for his work in sustainable concrete and construction materials. He has held senior academic and research roles and has established expertise in relation to geopolymer and low-carbon concrete, the durability of cementitious materials, and the use of industrial by-products in construction. He is actively involved in international professional and academic communities and serves on the editorial boards of several journals in the fields of concrete and civil engineering. His research activities, publications, and postgraduate supervision reflect a sustained commitment to advancing practical and scientifically grounded solutions for low-carbon and durable concrete infrastructure.

Preface

The Reprint, "Towards Sustainable Low-Carbon Concrete", was developed in response to the accelerating global demand for carbon reduction in the construction sector. Concrete is indispensable to modern infrastructure, yet its production remains a major source of greenhouse gas emissions. The subject of this Reprint therefore centers on innovative material systems, structural solutions, and performance evaluation methods that enable meaningful reductions in embodied carbon without compromising safety, durability, or service life.

The scope of this Reprint spans alternative binders such as alkali-activated and geopolymer systems, the valorization of industrial and agricultural by-products, high-ductility fiber-reinforced composites, and performance-based structural applications in aggressive or seismic environments. Contributions also address durability mechanisms, interfacial behavior, and emerging construction technologies that improve material efficiency. The aim is to present both scientific depth and engineering relevance, bridging laboratory research and practical implementation.

This Reprint is motivated by the urgent need to align concrete technology with climate resilience strategies and sustainable development goals. It is addressed to researchers, practicing engineers, material scientists, policy makers, and industry professionals seeking technically rigorous and application-oriented knowledge to guide the transition toward low-carbon infrastructure.

Jiaxiang Lin and Prabir K. Sarker
Guest Editors

materials

MDPI

Article

Study on the Compressive Strength and Reaction Mechanism of Alkali-Activated Geopolymer Materials Using Coal Gangue and Ground Granulated Blast Furnace Slag

Xiaoping Wang [1,2], Feng Liu [1], Lijuan Li [1], Weizhi Chen [1], Xinhe Cong [1], Ting Yu [3] and Baifa Zhang [3,*]

[1] School of Civil and Transportation Engineering, Guangdong University of Technology, Guangzhou 510006, China; 1122109004@mail2.gdut.edu.cn (X.W.)
[2] School of Architecture and Engineering, Huangshan University, Huangshan 245041, China
[3] School of Environmental Science and Engineering, Guangdong University of Technology, Guangzhou 510006, China
* Correspondence: zhangbaifa@gdut.edu.cn; Tel.: +86-17688480289

Abstract: By reutilizing industrial byproducts, inorganic cementitious alkali-activated materials (AAMs) contribute to reduced energy consumption and carbon dioxide (CO_2) emissions. In this study, coal gangue (CG) blended with ground granulated blast furnace slag (GGBFS) was used to prepare AAMs. The research focused on analyzing the effects of the GGBFS content and alkali activator (i.e., Na_2O mass ratio and alkali modulus [SiO_2/Na_2O]) on the mechanical properties and microstructures of the AAMs. Through a series of spectroscopic and microscopic tests, the results showed that the GGBFS content had a significant influence on AAM compressive strength and paste fluidity; the optimal replacement of CG by GGBFS was 40–50%, and the optimal Na_2O mass ratio and alkali modulus were 7% and 1.3, respectively. AAMs with a 50% GGBFS content exhibited a compact microstructure with a 28 d compressive strength of 54.59 MPa. Increasing the Na_2O mass ratio from 6% to 8% promoted the hardening process and facilitated the formation of AAM gels; however, a 9% Na_2O mass ratio inhibited the condensation of SiO_4 and AlO_4 ions, which decreased the compressive strength. Increasing the alkali modulus facilitated geopolymerization, which increased the compressive strength. Microscopic analysis showed that pore size and volume increased due to lower Na_2O concentrations or alkali modulus. The results provide an experimental and theoretical basis for the large-scale utilization of AAMs in construction.

Keywords: alkali-activated materials; coal gangue; ground granulated blast furnace slag; mechanical properties; microstructure

1. Introduction

Ordinary Portland cement (OPC) is an essential component of concrete. Generally, the production of 1 ton of OPC generates 0.8–1.0 tons of carbon dioxide (CO_2), and as the global population increases, the demand for OPC increases. The cement industry is estimated to account for 7–8% of global CO_2 emissions, which is challenging for human societies [1–3]. It is predicted that the production of OPC will reach 5 billion tons in 2030, and the projected emission of CO_2 could be as high as 4.7 billion tons [4]. The global consensus on CO_2 reduction and emissions has promoted research on low-carbon cementitious materials as alternatives to OPC [5,6].

In recent years, alkali-activated materials (AAMs) have been extensively used in construction [7]. AAMs are the product of polymerization reactions between aluminosilicate precursors and alkali activators [8–10]. They exhibit high compressive strength [11], good thermal properties [12], and corrosion resistance [13,14]. The production of AAMs consumes less energy and produces lower CO_2 emissions [15,16]. As recorded, AAMs show an 80% reduction in CO_2 emissions compared to OPC [17]. Therefore, AAMs are considered an alternative material to OPC due to their attractive physical properties and low CO_2 emissions. Increasing numbers of byproducts have been used as precursor materials for preparing AAMs, such as metakaolin (MK) [18], fly ash (FA) [19], and ground granulated blast furnace slag (GGBFS) [20]. Furthermore, other aluminosilicate-rich byproducts are being explored to develop new AAMs.

Coal gangue (CG), a byproduct of coal mining, accounts for 10–25% of total coal production [21]. CG is one of the largest solid wastes in India, Australia, South Africa, and China [22]. China's stockpile of CG is more than 7 billion tons; it occupies land resources and threatens the environment [23]. Published studies indicate that CG's chemical composition is Al_2O_3 and SiO_2; therefore, it has potential as a precursor for AAMs. Zhang et al. [24] revealed that 800 °C was optimal for forming metakaolin in calcined CG; however, the mechanical properties of the obtained AAMs decreased when the temperature reached 900 °C. Li et al. [21] explored the effects of CG particle size on the compressive strength of AAMs. Their results showed that 200 mesh CG achieved the necessary strength. In addition, Wang et al. [25] investigated alkali activator effects on the microstructures and mechanical properties of AAMs prepared from CG. They concluded that the maximum compressive strength of CG-based AAMs was 24.75 MPa, which required a Na_2SiO_3/NaOH ratio of 2.0 and a liquid/solid ratio of 0.50. Other research [25,26] indicates that the poor mechanical characteristics of CG-based AAMs are due to low activity, suggesting that improving the strength of CG-based AAMs is required.

Combining two or more precursors is an effective method for enhancing the mechanical performance of AAMs [27]. For example, Deb et al. [28] investigated AAMs with varying GGBFS/FA ratios (0%, 10%, and 20%). Their results showed that the maximum compressive strength (51 MPa) was achieved when the precursors were synthesized with 80% FA and 20% GGBFS. Bernal [29] prepared MK/GGBFS-based AAMs with a compressive strength of approximately 45 MPa (50% GGBFS) and 20 MPa (0% GGBFS). Additionally, Venkatesan and Pazhani [30] demonstrated that incorporating 10% rice husk ash (RHA) into GGBFS-based AAMs resulted in a compressive strength greater than 72.3 MPa. According to these studies, GGBFS (Ca-containing) AAM gels composed of calcium silicate hydrate (C-(A)-S-H) enhanced compressive strength [31], whereas Ca-free mineral-reaction gels (i.e., FA, MK, and RHA) composed of N-(A)-S-H decreased autogenous shrinkage [32].

Furthermore, the compressive strength of AAMs is significantly influenced by alkali activators such as the silicate modulus (SiO_2/Na_2O) and alkalinity (Na_2O content) of the solution [33–38]. Pelisser et al. [34] found that a silicate modulus of 1.6 was beneficial for the mechanical properties of MK-based AAMs. However, Thaarrini et al. [35] recommended that a silicate modulus of 1.0 would achieve a higher compressive strength of FA-based AAMs, which differs from Pelisser et al.'s result of 1.6. Wang et al. [37] observed that the gel formation of MK-based AAMs increased as the NaOH solution concentrations (4–12 mol/L) (i.e., alkalinity) increased. Zhang et al. [38] found that the optimum NaOH concentration was 10 mol/L for FA-based AAMs and 12.5 mol/L for MK-based AAMs [33]. These findings highlight that the alkali activators' optimum silicate modulus and alkalinity are distinctive for different precursors.

Therefore, the blending of precursors or alkali activation can promote the mechanical properties of AAMs. This study used a precursor of CG blended with GGBFS, and a mixture of Na_2SiO_3 and NaOH solutions served as the alkali activator. The research investigated the influence of the GGBFS/CG ratio, alkalinity (Na_2O content), and alkali activator modulus (SiO_2/Na_2O) on the mechanical performance and microstructure of CG/GGBFS-based AAMs. The mineral composition, microstructure, and mechanical properties of AAMs were analyzed via X-ray diffraction (XRD), Fourier transform infrared (FTIR) spectroscopy, scanning electron microscopy (SEM), and energy-dispersive X-ray (EDX) spectroscopy. Additionally, mercury intrusion porosimetry (MIP) is employed to detect the pore characteristics and parameters of the AAMs. This study aimed to provide experimental and theoretical bases for the large-scale utilization of AAMs in construction.

2. Materials and Methods

2.1. Materials

The CG used in this study was sourced from Zhengzhou, Henan Province, China. Before use, it was subjected to calcination at 700 °C for 3 h. The GGBFS was procured from Nanyang, Henan Province, China.

Table 1 shows the chemical compositions of the CG and GGBFS, which were determined via X-ray fluorescence. The CG was rich in SiO_2 and Al_2O_3, constituting 86.12 wt% of the total CG composition. The XRD patterns (Figure 1a) showed that quartz, kaolinite, and muscovite were the main minerals in the raw CG. Following the calcination process, the reflections of kaolinite disappeared, and the calcined CG exhibited a broad diffraction band centered at approximately 23.7° (2θ) (Figure 1b), suggesting the transformation of kaolinite into an amorphous phase, while quartz was preserved in the calcined CG due to its thermal stability. The GGBFS primarily contained CaO, SiO_2, and Al_2O_3, which totaled 86.11 wt% of the whole composition. The XRD pattern of GGBFS (Figure 1c) featured a broad diffraction band centered at approximately 31.5° (2θ), indicating the GGBFS was predominantly amorphous. Calcite and dolomite were identified in the GGBFS.

Table 1. Main chemical composition of CG and GGBFS (wt%).

Chemical Composition	SiO_2	Al_2O_3	Fe_2O_3	CaO	Na_2O	MgO	K_2O	TiO_2	Others
CG	62.84	23.28	4.44	1.73	1.04	1.99	3.22	0.98	0.48
GGBFS	28.03	14.65	0.44	43.43	0.51	8.57	0.44	0.99	2.94

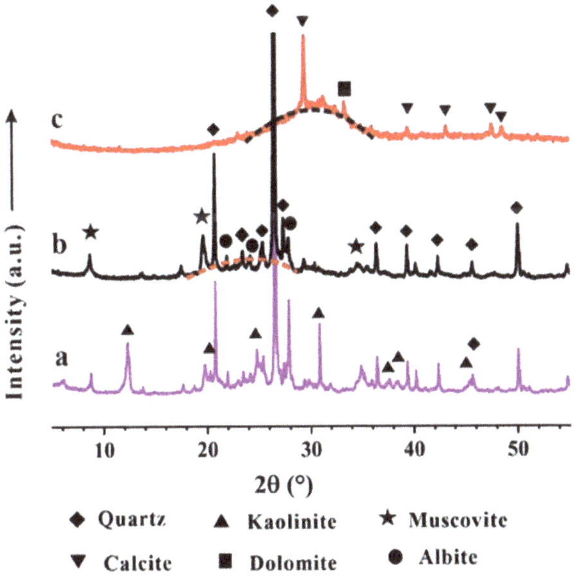

Figure 1. The XRD patterns of (**a**) Raw CG, (**b**) Calcined CG, and (**c**) Raw GGBFS.

Figure 2 shows the SEM images and particle size distributions of the calcined CG and GGBFS. The CG possesses a loose surface morphology and porous structure (Figure 2a), and the GGBFS exhibits a smooth surface morphology and angular shape (Figure 2b). The median particle size (d_{50}) of the CG and GGBFS are 8.59 and 10.29 μm, respectively.

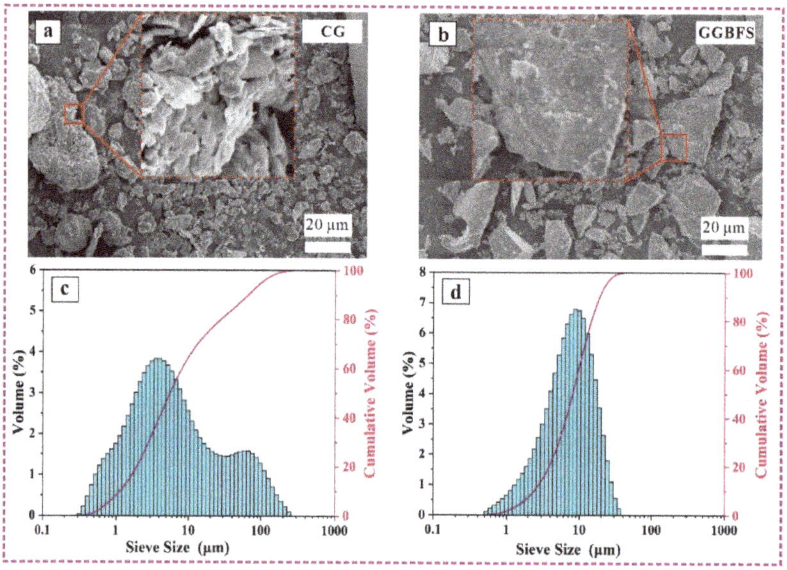

Figure 2. SEM images of CG (**a**) and GGBFS (**b**); and particle size distributions of CG (**c**) and GGBFS (**d**).

The materials used as alkali activators were composed of sodium silicate solution (SiO_2 27.3 wt%, Na_2O 8.5 wt%, and H_2O 64.2 wt%) and sodium hydroxide particles (purity ≥96%). Distilled water was used for adjusting alkalinity. All chemical agents were obtained from laboratory and commercial vendors.

2.2. AAMs Preparation

The preparation process is illustrated in Figure 3, and the specific mixing ratio of CG/GGBFS-based AAMs is detailed in Table 2. The AAMs were prepared by adding NaOH and Na_2SiO_3 to distilled water to formulate the alkali activators, and the precursors (calcined CG and GGBFS) were mixed for 2 min in a planetary mixer. The alkali activators were then blended with the precursors for 5 min to create a uniform slurry paste. The paste was subsequently poured into silica molds and vibrated on an electric vibration table for 90 s to eliminate the remaining bubbles. A thin polyethylene film was applied to cover all samples. Following this, the samples were cured in a controlled environment at 60 °C for 24 h, unmolded, and left at ambient temperature (~25 °C) at 3, 7, and 28 d before conducting the test.

Figure 3. Preparation process of CG-based AAMs.

Table 2. Mixture proportions of CG/GGBFS-based AAMs.

Specimens	CG (wt%)	GGBFS (wt%)	Na_2O (%)	Ms = (SiO_2/Na_2O)	Liquid/Solid	Fluidity (mm)
C-7.0-1.3	0	100				107.6
S_1C_4-7.0-1.3	20	80				118.1
S_2C_3-7.0-1.3	40	60				129.3
S_1C_1-7.0-1.3	50	50	7.0	1.3		135.3
S_3C_2-7.0-1.3	60	40				142.8
S_4C_1-7.0-1.3	80	20				154.4
S-7.0-1.3	100	0			0.55	167.7
S_1C_1-6.0-1.3			6.0			133.2
S_1C_1-8.0-1.3			8.0	1.3		137.1
S_1C_1-9.0-1.3	50	50	9.0			138.2
S_1C_1-7.0-0.9				0.9		127.3
S_1C_1-7.0-1.1			7.0	1.1		131.9
S_1C_1-7.0-1.5				1.3		139.5

The prepared AAMs were labeled S_XC_Y-W-Z, where S and C represent the raw precursors of GGBFS and CG; X and Y represent the mass ratios of GGBFS and CG; W indicates the Na_2O content (mass ratio of Na_2O); and Z indicates the alkali modulus (SiO_2/Na_2O) of the alkali activator. The liquid/solid ratio was 0.55. For example, S_1C_1-7-1.3 signified the specimens were fabricated with a mass ratio of GGBFS/CG of 1:1, and the Na_2O content and alkali modulus were 7 wt% and 1.3, respectively.

2.3. *Methods*

2.3.1. Physical Properties Tests

The fluidity of AAM paste was determined using a shortened conical apparatus (Cangzhou, China). The setting times of AAM paste were determined using a Vicat apparatus (Shanghai, China). The compressive strength of AAM samples was determined using an STS100K testing machine (Xiamen, China). Three samples were tested, and the average results were considered the final compressive strength.

2.3.2. Microscopic Properties Tests

XRD patterns of the samples (raw materials and AAMs) were recorded on a D8 Advance diffractometer (Bruker, Mannheim, Germany) using Cu Kα radiation. The diffractometer was operated at 40 kV and 30 mA, and the scanning range was 3–70° (2θ) with a scanning speed of 3°/min. FTIR spectroscopy of AAM samples was recorded on a Nicolet IS 50 spectrometer (Thermo Fisher, Waltham, MA, USA). The spectra ranged from 400 to 4000 cm^{-1} with a resolution of 4 cm^{-1} and 64 scans. SEM images and EDX results of the samples (raw materials and AAMs) were obtained using an S-3400N-II instrument (Hitachi, Tokyo, Japan). The acceleration voltage was 35 kV, and the current was 10 mA. The MIP of AAM samples was measured using an AutoPore IV 9510 porosimeter (Micromeritics, Norcross, GA, USA). The intrusion pressures ranged from 0.5 to 33,000 psi with a contact angle of 130° and a surface tension of 485 dynes/cm.

3. Results

3.1. *Setting Time and Compressive Strength*

3.1.1. Fluidity

The fluidity of the freshly mixed AAM pastes is shown in Table 2. The paste fluidity results changed from 107.6 to 167.7 mm as the mixture varied and was notably affected by the GGBFS content. For example, compared to C-7-1.3, the fluidity of pastes with GGBFS contents of 20%, 40%, 50%, 60%, 80%, and 100% increased by 9.76%, 20.17%, 25.74%, 32.71%, 43.49%, and 55.86%, respectively. This was attributed to the following: (1) CG has a more porous structure that can absorb free water; and (2) The smaller particle size of CG, inducing a larger specific surface area compared to GGBFS (Figure 2), increases the available water for the AAM paste, resulting in increased fluidity when CG is substituted with GGBFS. Kramar et al. [39] found that the fluidity of slag-based mortar was higher than that of FA (180 vs. 152 mm). However, the paste fluidity of the S_1C_1-W-1.3 samples was not altered (around 135.3 ± 3 mm). With changes in Na_2O content from 6% to 9%, the fluidity increased from 133.2 to 138.2 mm. Similar changes were observed for the S_1C_1-7-Z samples; the fluidity changed from 127.3 to 139.5 mm as the SiO_2/Na_2O modulus increased from 0.9 to 1.5, respectively. This indicates that the Na_2O content and modulus of the alkali activator were insensitive to the AAM paste fluidity.

3.1.2. Setting Time

Figure 4 illustrates the setting times of the AAM pastes. The setting time was closely related to the geopolymerization rate, which was influenced by either the precursor materials or alkali activators. The final setting times of the AAM pastes appeared to be shortened, indicating that curing the samples in the controlled environment (60 °C) accelerated the geopolymerization process.

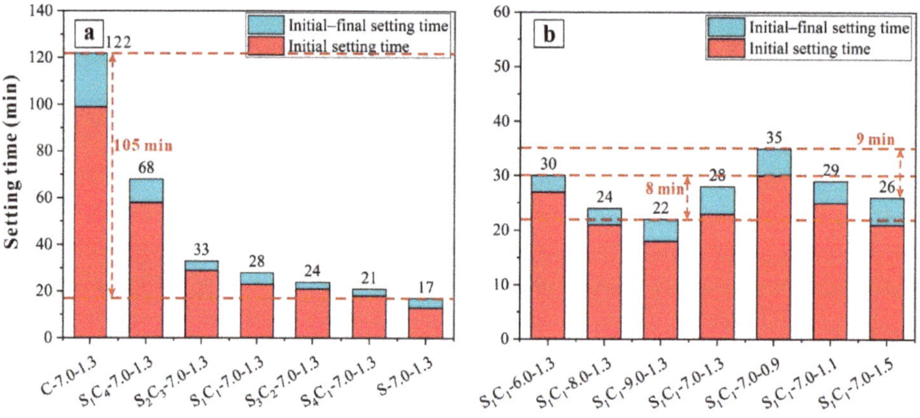

Figure 4. Setting times of AAMs: (**a**) S_XC_Y-7.0-1.3, and (**b**) S_1C_1-W-1.3 and S_1C_1-7.0-Z.

The setting times of the AAM pastes decreased as the GGBFS content increased, impacting both the initial and final setting times. Compared to the C-7.0-1.3 sample, the final setting times of samples with 100% GGBFS content (S-7.0-1.3) were significantly shortened, with an 86.1% reduction. This can be attributed to the higher reactivity of GGBFS; the reaction of the AAM paste increased when CG was replaced with GGBFS.

With increasing Na_2O content, the setting time was slightly reduced (from 30 to 22 min), indicating that higher alkalinity (Na_2O concentration) promoted the dissolution of Al and Si ions, thereby accelerating the geopolymerization process. However, with a higher alkali modulus (SiO_2/Na_2O), both the initial and final setting times of the AAM pastes were shortened. For example, the final setting time of S_1C_1-7.0-0.9 was 35 min, while that of S_1C_1-7.0-1.5 was shortened to 26 min when the alkali modulus increased to 1.5. Studies have reported that the higher alkali modulus in the reaction is conducive to geopolymerization, resulting in more AAM pastes and shorter paste hardening times [40].

3.1.3. Compressive Strength

Figure 5 illustrates the compressive strengths of the AAM specimens at various curing ages (i.e., 3 d, 7 d, and 28 d). The strength of the specimens exhibited a gradual increase throughout the curing period, indicating a continuous ongoing process of geopolymerization. Notably, there was a rapid increase in compressive strength during the early stages, especially by day 3. The compressive strength by day 3 was about 75% of that of the 28 d specimens, and by day 7, the compressive strength of the specimens was about 85% of that of the 28 d specimens. This was attributed to two factors: (1) more soluble silicates participated in the reactions during the initial stages of AAM curing, and (2) an elevation in curing temperature accelerated the geopolymerization rate.

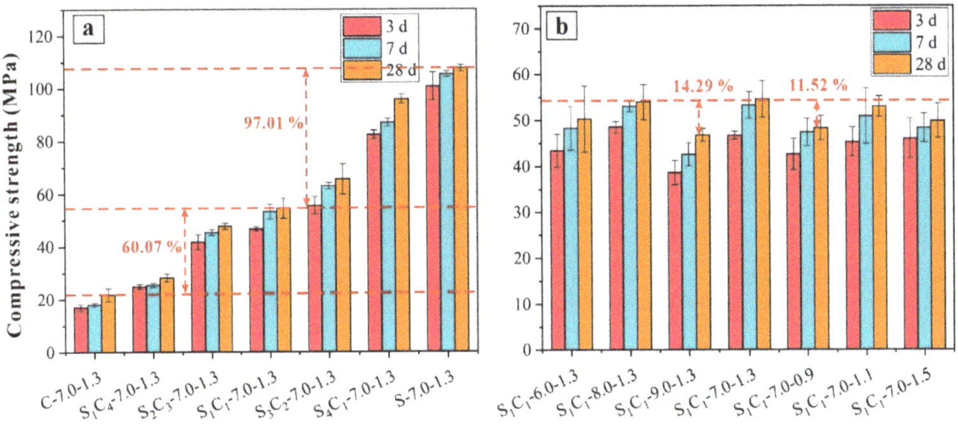

Figure 5. Compressive strength of AAM specimens (**a**) S_XC_Y-7.0-1.3, and (**b**) S_1C_1-W-1.3 and S_1C_1-7.0-Z.

Furthermore, with an increase in GGBFS content, the compressive strength of the AAM showed a significant improvement. The 28 d compressive strength of S-7.0-1.3 (100% GGBFS content) was 393.34% higher than that of C-7.0-1.3 (100% CG content). When the specimens had a CG/GGBFS ratio of 1:1, the compressive strength of S_1C_1-7.0-1.3 reached 54.59 MPa, equivalent to that of grade 52.5 OPC.

Moreover, an increase in the Na_2O mass ratio from 6% to 8% initially improved compressive strength, followed by a decline in strength as the ratio increased to 9% (S_1C_1-9.0-1.3). Compared to S_1C_1-6.0-1.3, the compressive strengths of S_1C_1-7.0-1.3 and S_1C_1-8.0-1.3 increased by 8.44% and 7.21%, respectively, indicating that a higher Na_2O mass ratio was beneficial for generating more paste gels and achieving higher compressive strength through enhanced dissolution of GGBFS and CG particles [41]. Increasing the Na_2O mass ratio to 9% had a notable impact on the compressive strength of the AAMs. This was for two reasons: (1) inadequate alkalinity at low Na_2O concentrations hindered polymerization; and (2) higher Na_2O concentrations led to more precursors rapidly reacting with alkali activators, resulting in geopolymer gels forming on precursor surfaces and limiting their participation in polymerization.

With a rise in the alkali modulus, the compressive strength of the AAM specimens was only slightly affected. As the alkali modulus changed from 0.9 to 1.3, the compressive strength increased by 11.52%. However, the S_1C_1-7.0-1.5 specimens, with a modulus of 1.5, showed a lower compressive strength (49.85 MPa) than that of the S_1C_1-7.0-1.3 specimens (54.59 MPa), a decrease of 8.68%. This finding was contrasted with previous research. Guo et al. found that the compressive strength of 28 d FA-based AAMs significantly increased (from 17.1 to 59.3 MPa) as the SiO_2/Na_2O modulus changed from 1.0 to 1.5 [36].

3.2. *Geopolymer Structure*

3.2.1. XRD Analysis

Figure 6 illustrates the XRD patterns of the AAM specimens. There is a broad reflection within the 18°–40° (2θ), with a prominent peak at ~29° (2θ). Compared to the C-7.0-1.3 specimens, the patterns of the AAM specimens shifted toward a higher angle of ~30° (2θ) due to the inclusion of GGBFS. This indicates the presence of amorphous phases, which primarily consist of sodium aluminate silicate hydrate (N-A-S-H) and calcium aluminate silicate hydrate (C-A-S-H) [42–44]. He et al. [44] identified that the N-A-S-H gel was the product of low-calcium materials, and the C-A-S-H gel was predominant in rich-calcium precursors. Additionally, minerals like quartz and muscovite were detected in the specimens, suggesting that CG was insufficient for the geopolymerization reaction [45].

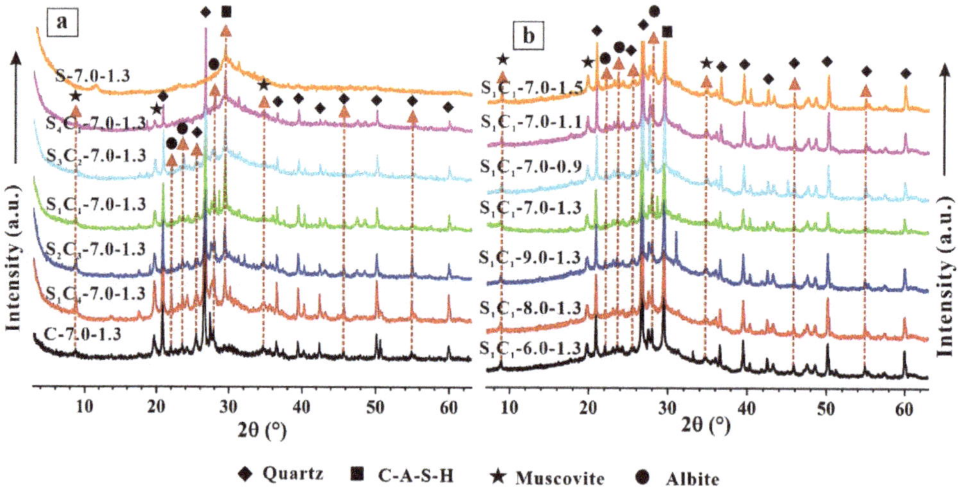

Figure 6. XRD patterns of AAM specimens (**a**) S_XC_Y-7.0-1.3, and (**b**) S_1C_1-W-1.3 and S_1C_1-7.0-Z.

The intensity of the C-A-S-H crystalline phase increased with higher GGBFS content due to the reaction between the GGBFS and alkali activators [46,47]. However, increasing the Na_2O mass ratio from 6% to 8% did not significantly impact mineral composition (Figure 6b). Compared to S_1C_1-6.0-1.3, the C-A-S-H diffraction peak of S_1C_1-8.0-1.3 was enhanced, indicating that more geopolymerization gels generated at higher alkalinity, resulting in the higher compressive strength of S_1C_1-8.0-1.3.

Varying the SiO_2/Na_2O modulus of the alkali solution had minimal effect on geopolymers' compound synthesis. Compared to S_1C_1-7.0-0.9, the XRD patterns of S_1C_1-7.0-1.3 and S_1C_1-7.0-1.5 showed a lower intensity of the quartz diffraction peak, suggesting the greater consumption of quartz crystals in S_1C_1-7.0-1.3 and S_1C_1-7.0-1.5. This confirms that a higher SiO_2/Na_2O modulus promoted geopolymerization.

3.2.2. FTIR Spectroscopy

Figure 7 displays the FTIR spectra of the AAM specimens between 400 and 1800 cm^{-1}, while Table 3 provides the vibrational modes of FTIR band assignments for the selected samples. In Figure 7a, a broad band between 800 and 1200 cm^{-1} was observed and shifted, indicating the presence and alteration of amorphous products [48,49]. With an increase in GGBFS content, the spectra of the specimens shifted towards lower wavenumbers. For example, compared to C-7.0-1.3, the center of the S_1C_1-7.0-1.3 and S-7.0-1.3 broad bands moved from 1027 to 1007 and 971 cm^{-1}, respectively. Research [50] indicated that the N-A-S-H gel is primarily associated with 1030 cm^{-1}, while the C-A-S-H gel is mainly focused on 940 cm^{-1}. This suggests that the components of the AAM specimens were transformed from N-A-S-H to C-A-S-H as the GGBFS content increased.

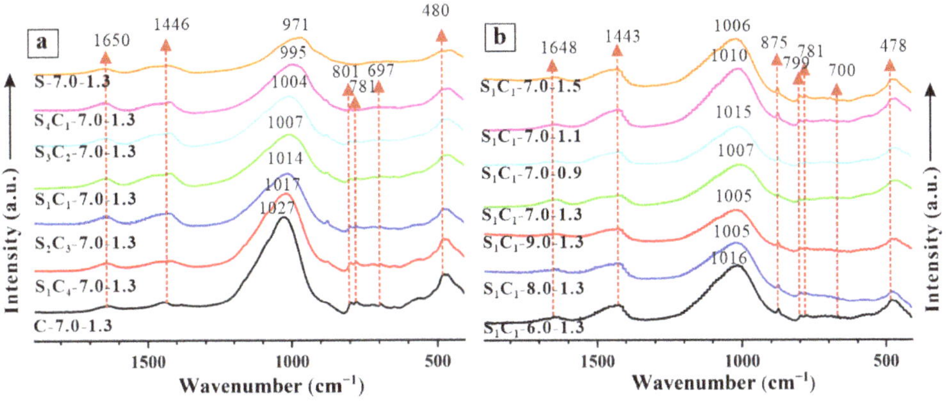

Figure 7. FTIR spectra of AAM specimens (**a**) S_XC_Y-7.0-1.3, and (**b**) S_1C_1-W-1.3 and S_1C_1-7.0-Z.

Table 3. The vibrational modes of FTIR band assignments for the selected samples.

Wavenumber (cm^{-1})	Assignment	References
1650	Stretching vibration of O–H bonds	[51]
1445	Asymmetric stretching vibration of C–O bonds	[52]
1081-1010	In-plane stretching vibrations of Si–O bonds	[53,54]
875	Out-of-plane bending vibrations of C–O bonds	[55]
801	Symmetric stretching vibrations of Si–O bonds	[56]
780	Stretching vibration of Al–O bonds	[57]
697	Internal extension of the Si–O bond	[58]
480	Stretching vibration of Si–O–T bands	[59,60]

In Figure 7b, as the Na_2O mass ratio increased, the characteristic bands decreased from 1016 cm^{-1} (S_1C_1-6.0-1.3) to 1005 cm^{-1} (S_1C_1-9.0-1.3). Additionally, with an increase in SiO_2/Na_2O modulus from 0.9 to 1.5, the center of the broad band decreases from 1015 cm^{-1} (S_1C_1-7.0-0.9) to 1006 cm^{-1} (S_1C_1-7.0-1.5). These minimal changes in FTIR peaks indicated limited impacts on the composition of the AAM specimens within the activated system.

3.2.3. SEM/EDX Results

Figures 8–10 illustrate the SEM/EDX findings of the AAM samples. The microstructures of the specimens with various GGBFS contents (i.e., 0%, 50%, and 100%) are examined in Figure 8a–g. In Figure 8a, the C-7.0-1.3 specimens possess a loose microstructure that coexists with isolated particles in the matrix. As the GGBFS content increased, the microstructure of the S_1C_1-7.0-1.3 and S-7.0-1.3 specimens became more compact (Figure 8c,e), and there were fewer smaller pores in the stone body. This suggested that more AAM gels (C-(N)-A-S-H) were produced as the GGBFS content increased, leading to enhanced compressive strength. The high-magnification SEM images (Figure 8d,f) showed microcracks in the bodies of the S_1C_1-7.0-1.3 and S-7.0-1.3 specimens. Research [61–63] revealed that the shrinkage value of the AAMs increased with the addition of GGBFS, contributing to the formation of more microcracks. The EDX analysis (Figure 8g, spot 1) revealed that the C-7.0-1.3 specimens were primarily composed of N-A-S-H, while the S_1C_1-7.0-1.3 and S-7.0-1.3 specimens (Figure 8g, spots 2 and 3) contained C-(N)-A-S-H, which aligns with the elemental analysis. Additionally, the analyzed AAM samples exhibited varying Ca/Si ratios. For example, the Ca/Si ratio of C-7.0-1.3 was 0.03, considerably lower than that of S_1C_1-7.0-1.3 (0.59) and S-7.0-1.3 (1.27). This result suggests that the formation of more C-A-S-H gel was generated with increasing GGBFS content. Temuujin et al. [31] demonstrated that the more C-A-S-H gel generated, the higher the compressive strength achieved for FA-based AAMs.

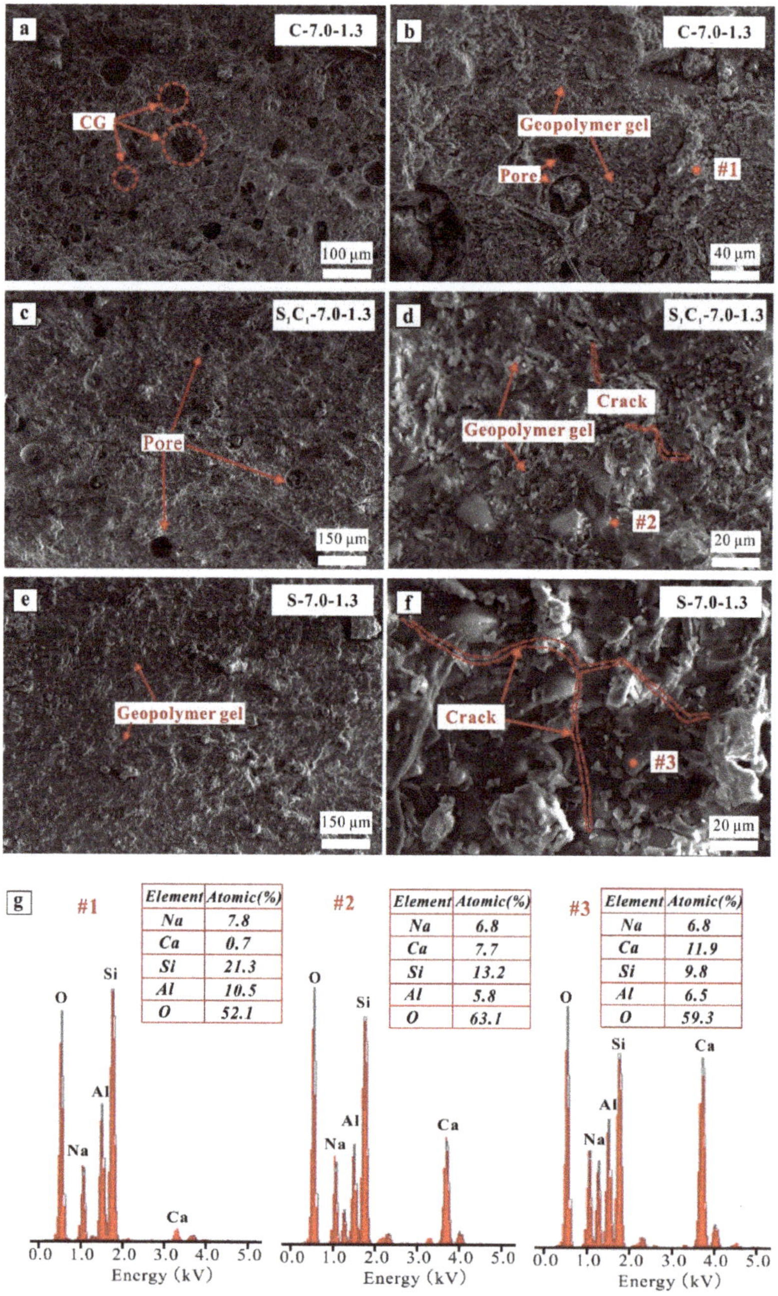

Figure 8. SEM images of AAM specimens: (**a**,**b**) C-7.0-1.3, (**c**,**d**) S_1C_1-7.0-1.3, and (**e**,**f**) S-7.0-1.3; (**g**) EDX results of spots highlighted in images (**b**,**d**,**f**).

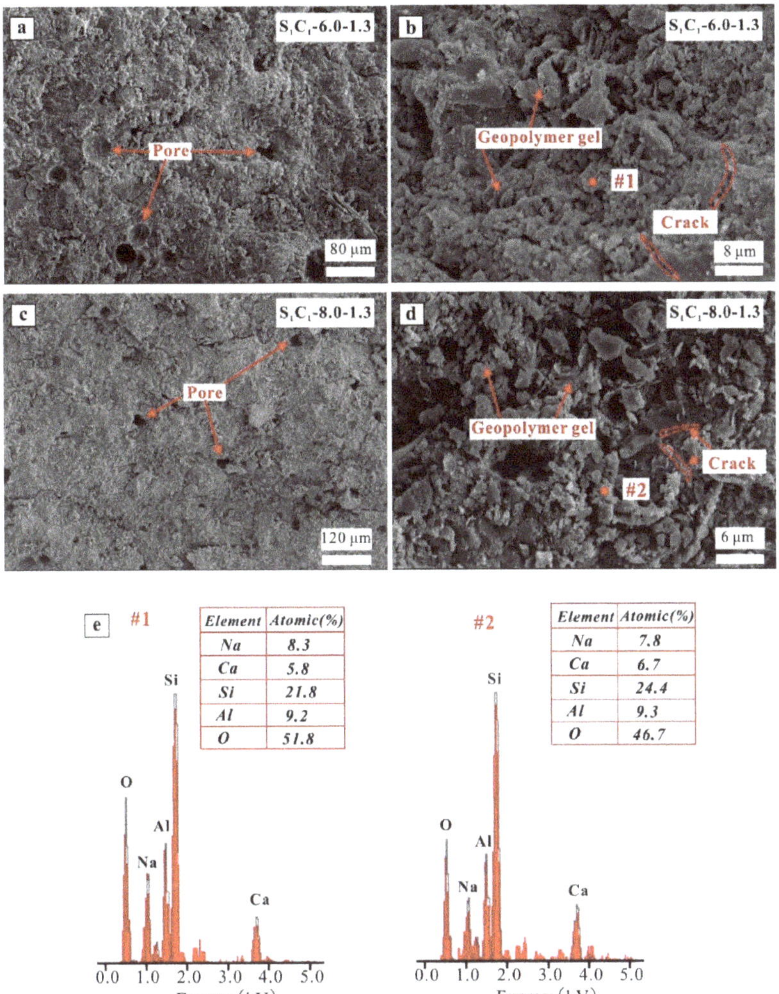

Figure 9. SEM images of AAM specimens: (**a**,**b**) S_1C_1-0.6-1.3, (**c**,**d**) S_1C_1-0.8-1.3, and (**e**) EDX results of spots highlighted in images (**b**,**d**).

As the Na_2O mass ratio increased from 6% to 8%, there was no apparent difference observed between the S_1C_1-0.6-1.3 and S_1C_1-0.8-1.3 specimens, with some pores being evident in the matrix of their compact microstructure (Figure 9a,c). The high-magnification SEM images showed an increase in AAM gels generated with greater Na_2O content; however, some microcracks occurred in the stone body (Figure 9b,d). The EDX results, as depicted in Figure 9e (spots 1, and 2), demonstrated that the Ca/Si ratios were similar between the S_1C_1-0.6-1.3 and S_1C_1-0.8-1.3 (0.26 vs. 0.27); the (Ca + Na)/(Si + Al) ratios of the S_1C_1-0.6-1.3 and S_1C_1-0.8-1.3 were 0.45 and 0.43, respectively. Furthermore, the lower Ca/Si ratio in S_1C_1-0.6-1.3 resulted in a decrease in compressive strength.

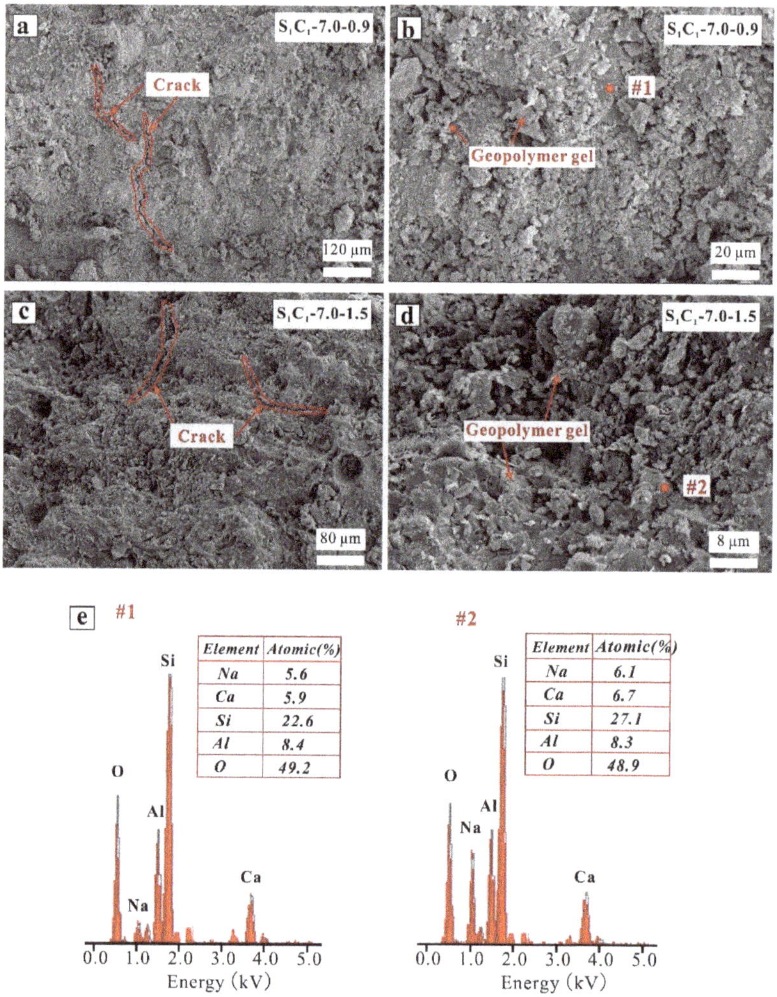

Figure 10. SEM images of AAM specimens: (**a**,**b**) S_1C_1-7.0-0.9, (**c**,**d**) S_1C_1-7.0-Z, and (**e**) EDX results of spots highlighted in images (**b**,**d**).

Figure 10 illustrates the SEM images and EDX findings of the S_1C_1-7.0-0.9 and S_1C_1-7.0-1.5 specimens, with both exhibiting compact morphology (Figure 10a,c). In addition, the high-magnification SEM images showed that less AAM gel was generated in the S_1C_1-7.0-0.9 specimens due to the incomplete reaction of the low soluble silicate content. The EDX results, as depicted in Figure 10e (spots 1 and 2), demonstrate that the Ca/Si ratios of the S_1C_1-7.0-0.9 and S_1C_1-7.0-1.5 (0.26 vs. 0.25) specimens were similar, and the (Ca + Na)/(Si + Al) ratios of the S_1C_1-7.0-0.9 and S_1C_1-7.0-1.5 were 0.37 and 0.36, respectively, suggesting that more Si atoms were involved in the S_1C_1-7.0-1.5 reaction.

3.2.4. MIP Results

Figure 11 displays the distribution of pore size in the AAM specimens, and the pore parameters are shown in Table 4, which were analyzed by MIP. In Figure 11a, the samples exhibited a predominant pore size range of 10 to 200 nm. They were primarily medium capillary pores (10–100 nm) and large capillary pores (100–1000 nm), with large capillary pores having a negative impact on the mechanical properties of the AAMs [64].

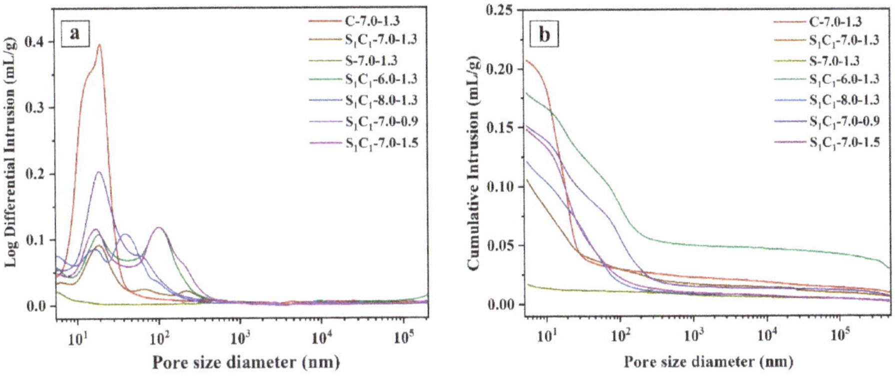

Figure 11. MIP results of AAM specimens: (**a**) log differential intrusion; (**b**) cumulative intrusion.

Table 4. The pore parameters for the AAM samples determined via MIP.

Samples	Total Pore Area (m^2/g)	Average Pore Diameter (nm)	Porosity (%)
C-7.0-1.3	36.58	35.70	31.74
S_1C_1-7.0-1.3	24.89	22.84	18.38
S-7.0-1.3	7.64	17.06	3.36
S_1C_1-6.0-1.3	20.11	25.93	27.19
S_1C_1-8.0-1.3	22.84	21.29	20.58
S_1C_1-7.0-0.9	27.91	29.07	24.58
S_1C_1-7.0-1.5	20.88	21.28	24.42

With 100% CG, the average pore size of C-7.0-1.3 specimens was 35.70 nm, and the porosity was 31.74%. As the GGBFS content increased, the pore size and porosity of the specimens decreased. For example, the average pore sizes of S_1C_1-7.0-1.3 (50% GGBFS content) and S-7.0-1.3 (100% GGBFS content) were 22.84 and 17.06 nm, respectively, and the porosity decreased by 42.1% and 89.4%, respectively. Figure 11b illustrates the decline in the cumulative pore volume of the AAM specimens with added GGBFS, showing a higher cumulative intrusion for C-7.0-1.3 (0.208 mL/g) compared to that for S-7.0-1.3 (0.017 mL/g). This suggests that incorporating GGBFS boosts the reactivity of raw precursors, resulting in the generation of more AAM gels and the compact structures of the specimens.

Additionally, the S_1C_1-6.0-1.3 specimens exhibited a critical pore size of 31.71 mm, which was greater than that of S_1C_1-8.0-1.3. This indicated that increasing Na_2O content promotes hydrolysis of the aluminosilicate precursors and improves the compact structure of the specimens. These findings are aligned with the compressive strengths of the samples. Porosity analysis in Table 4 revealed that S_1C_1-6.0-1.3 had a porosity of 27.19%; therefore, the greater pore size induced the inferior compressive strength of S_1C_1-6.0-1.3 compared to S_1C_1-8.0-1.3 (Figure 5b). Fan et al. [65] emphasized the impact of pore dimensions on the mechanical characteristics of cementitious substances.

Furthermore, the porosity of the AAM specimens remained unchanged (~24.50%) as the SiO_2/Na_2O modulus varied, while the average pore size of S_1C_1-7.0-0.9 was larger (29.07 mm) than that of S_1C_1-7.0-1.5 (21.28 mm). This suggests that an insufficient number of Si atoms participate in the S_1C_1-7.0-0.9 reactions, resulting in a larger pore size (Table 4), consistent with the change in compressive strength.

4. Discussion

The research determined that the compressive strength and microstructure of CG/GGBFS-based AAMs were influenced by GGBFS and alkali activator contents. As the GGBFS content increased from 0% to 50% (or 100%), more AAM gels (C-(A)-S-H) were generated, forming a three-dimensional network and resulting in a higher compressive strength [66]. For example, the compressive strength was 47.72 MPa for S_2C_3-7.0-1.3 (40% GGBFS content), and that of S_1C_1-7.0-1.3 (50% GGBFS content) was 54.59 MPa. This is favorable for the preparation of AAM concrete as an alternative to OPC materials. Additionally, as the Na_2O mass ratio changed from 6% to 9%, the compressive strength increased by 8.44%, which was compared to that of S_1C_1-6.0-1.3. The variation of the SiO_2/Na_2O modulus from 0.9 to 1.5 only slightly affected the strength of the AAM specimens (48.3 vs. 54.59 MPa). This suggests that the compressive strength of CG/GGBFS-based AAMs is mainly dependent on the GGBFS content. A Na_2O mass ratio of 7% and a SiO_2/Na_2O modulus of 1.3 were the most beneficial for generating AAM gels.

Furthermore, the setting times of the CG/GGBFS-based AAMs were faster due to the high reactivity of GGBFS. Several researchers have explored how a retarder admixture prolongs the setting times of AAM systems. For example, Wang et al. [67] found that zinc nitrate and sodium gluconate were excellent retarders for slag-based AAMs. Cong et al. [68] modified the properties of FA-based AAMs by adding borate. Both Kalina [69] and Gong [70] explored how sodium phosphate influences the hydration process of red mud and slag-based AAMs. Their findings indicated that sodium phosphate effectively prolonged the hydration and the setting time for AAMs. In related work, Brough et al. [71] reported that incorporating 0.5 wt% malic acid extended the initial setting time, which ranged from 4 to 20 h. The literature records effective retarders for controlling the setting times of AAMs, but fewer studies have discussed the effects of existing retarders on CG/GGBFS-based AAMs; this remains an investigation for future work.

5. Conclusions

This study investigated the effects of the GGBFS/CG ratio, alkalinity (Na_2O content), and alkali modulus (SiO_2/Na_2O) on the mechanical properties and microstructures of CG/GGBFS-based AAMs. The results of a series of spectroscopic and microscopic tests led to the following conclusions:

The addition of GGBFS significantly influenced the compressive strength and microstructure of CG/GGBFS-based AAMs. Higher GGBFS content led to more AAM gels (C-(A)-S-H) generated, resulting in the fluidity of the AAM paste increasing. Alkali activators (Na_2O and SiO_2/Na_2O contents) had only a slight influence on the chemical composition and microstructure of the AAM.

The compressive strengths of S_2C_3-7.0-1.3 and S_1C_1-7.0-1.3 were 47.72 MPa and 54.59 MPa, respectively, which favors the use of AAM concrete for OPC materials. However, by increasing the Na_2O mass ratio from 6% to 9%, more precursor materials (CG and GGBFS) participated in the depolymerization reactions, resulting in the structure of the AAMs becoming more compact. The porosities of S_2C_3-7.0-1.3 and S_1C_1-7.0-1.3 decreased from 27.19% to 20.58%, and their compressive strengths increased by 8.44% compared to S_1C_1-6.0-1.3. As the alkali modulus changed, specimens with a lower alkali modulus (S_1C_1-7.0-0.9) exhibited a loose microstructure, resulting in decreased compressive strength. The porosity of the AAM specimens was approximately 24.50%, and the compressive strengths of S_1C_1-7.0-0.9 and S_1C_1-7.0-1.5 were 48.3 MPa and 49.85 MPa, respectively.

The findings of this research provide theoretical and technical insights into the reuse of CG and GGBFS waste and the applications of CG- and GGBFS-based AAMs.

Author Contributions: Conceptualization, X.W., F.L. and B.Z.; data curation, X.W. and B.Z.; formal analysis, X.W., X.C., and W.C.; funding acquisition, F.L., B.Z. and L.L.; investigation, X.C. and W.C.; methodology, X.W. and B.Z.; resources, F.L. and L.L.; project administration, F.L. and B.Z.; software

X.C.; supervision, F.L.; writing—original draft preparation, X.W.; writing—review and editing, X.W., T.Y. and B.Z. All authors have read and agreed to the published version of the manuscript.

Funding: This research was funded by the National Natural Science Foundation of China, (Grant No. 12032009, 12072080, and 42302032); and the Basic and Applied Basic Research Foundation of Guangdong Province, (Grant No. 2023A1515012180).

Institutional Review Board Statement: Not applicable.

Informed Consent Statement: Not applicable.

Data Availability Statement: The original data will be made available upon requirement.

Acknowledgments: The authors gratefully thanks all technical personnel from the Structural Laboratory of Guangdong University of Technology for their assistance during the experiment.

Conflicts of Interest: The authors declare that they have no known competing financial interests or personal relationships that could have appeared to influence the work reported in this paper.

References

1. Okoye, F.N.; Durgaprasad, J.; Singh, N.B. Mechanical Properties of Alkali Activated Fly ash/Kaolin Based Geopolymer Concrete. *Constr. Build. Mater.* **2015**, *98*, 685–691. [CrossRef]
2. Liu, Q.; Hu, Z.; Wang, X.; Zhao, H.; Qian, K.; Li, L.; Meng, Z. Numerical study on cracking and its effect on chloride transport in concrete subjected to external load. *Constr. Build. Mater.* **2022**, *325*, 126797. [CrossRef]
3. Tan, Y.; He, Y.; Cui, X.; Liu, L. Design and performance optimization of alkali-activated waste coal bottom ash/slag porous concrete. *Constr. Build. Mater.* **2022**, *359*, 129413. [CrossRef]
4. Dong, K.; Jiang, H.; Sun, R.; Dong, X. Driving Forces and Mitigation Potential of Global CO_2 Emissions from 1980 through 2030: Evidence from Countries with Different Income Levels. *Sci. Total Environ.* **2019**, *649*, 335–343. [CrossRef] [PubMed]
5. Xi, F.; Davis, S.J.; Ciais, P.; Crawford-Brown, D.; Guan, D.; Pade, C.; Shi, T.; Syddall, M.; Lv, J.; Ji, L.; et al. Substantial global carbon uptake by cement carbonation. *Nat. Geosci.* **2016**, *9*, 880. [CrossRef]
6. Tran, T.; Kim, Y.; Kang, G.; Dinh, B.; Do, T. Feasibility of reusing marine dredged clay stabilized by a combination of by-products in coastal road construction. *Transp. Res. Record.* **2019**, *2673*, 519–528. [CrossRef]
7. Pobłocki, K.; Pawlak, M.; Drzeżdżon, J.; Gawdzik, B.; Jacewicz, D. Clean production of geopolymers as an opportunity for sustainable development of the construction industry. *Sci. Total Environ.* **2024**, *928*, 172579. [CrossRef] [PubMed]
8. Longhi, M.A.; Zhang, Z.; Rodriguez, E.D.; Kirchheim, A.P.; Wang, H. Efflorescence of alkali-activated cements (geopolymers) and the impacts on material structures: A critical analysis. *Front. Mater.* **2019**, *6*, 89. [CrossRef]
9. Lee, N.K.; Lee, H.K. Setting and mechanical properties of alkali-activated fly ash/slag concrete manufactured at room temperature. *Constr. Build. Mater.* **2013**, *47*, 1201–1209. [CrossRef]
10. Abdollahnejad, Z.; Luukkonen, T.; Mastali, M.; Giosue, C.; Favoni, O.; Ruello, M.L.; Kinnunen, P.; Illikainen, M. Microstructural analysis and strength development of one-part alkali-activated slag/ceramic binders under different curing regimes. *Waste Biomass Valorization* **2020**, *11*, 3081–3096. [CrossRef]
11. Zurinskas, D.; Vaiciukyniene, D.; Stelmokaitis, G.; Dorosevas, V. Clayey soil strength improvement by using alkali activated slag reinforcing. *Minerals* **2020**, *10*, 1076. [CrossRef]
12. Zhang, H.; Kodur, V.; Qi, S.; Cao, L.; Wu, B. Development of metakaolin–fly ash based geopolymers for fire resistance applications. *Constr. Build. Mater.* **2014**, *55*, 38–45. [CrossRef]
13. Kong, L.; Zhao, W.; Xuan, D.; Wang, X.; Liu, Y. Application potential of alkali-activated concrete for antimicrobial induced corrosion: A review. *Constr. Build. Mater.* **2022**, *317*, 126169. [CrossRef]
14. Aiken, T.A.; Gu, L.; Kwasny, J.; Huseien, G.F.; McPolin, D.; Sha, W. Acid resistance of alkali-activated binders: A review of performance, mechanisms of deterioration and testing procedures. *Constr. Build. Mater.* **2022**, *342*, 128057. [CrossRef]
15. Puertas, F.; Fernández-Jiménez, A. Mineralogical and microstructural characterisation of alkali-activated fly ash/slag pastes. *Cem. Concr. Compos.* **2003**, *25*, 287–292. [CrossRef]
16. Luukkonen, T.; Abdollahnejad, Z.; Yliniemi, J.; Kinnunen, P.; Illikainen, M. One-part alkali-activated materials: A review. *Cem. Concr. Res.* **2018**, *103*, 21–34. [CrossRef]
17. Cloete, S.; Giuffrida, A.; Romano, M.C.; Zaabout, A. The swing adsorption reactor cluster for post-combustion CO_2 capture from cement plants. *J. Clean. Prod.* **2019**, *223*, 692–703. [CrossRef]
18. Chen, L.; Wang, Z.; Wang, Y.; Feng, J. Preparation and properties of alkali activated metakaolin-based geopolymer. *Materials* **2016**, *9*, 767. [CrossRef] [PubMed]
19. Li, X.; Zhao, Y.; Hu, Y.; Wang, G.; Xia, M.; Luo, B.; Luo, Z. Influence of multiple factors on the workability and early strength development of alkali-activated fly ash and slag-based geopolymer-stabilized soil. *Materials* **2022**, *15*, 2682. [CrossRef] [PubMed]
20. Zhang, Q.; Ji, T.; Yang, Z.; Wang, C.; Wu, H. Influence of different activators on microstructure and strength of alkali-activated nickel slag cementitious materials. *Constr. Build. Mater.* **2020**, *235*, 117449. [CrossRef]

21. Li, Z.; Gao, Y.; Zhang, J.; Zhang, C.; Chen, J.; Liu, C. Effect of particle size and thermal activation on the coal gangue based geopolymer. *Mater. Chem. Phys.* **2021**, *267*, 124657. [CrossRef]
22. Jablonska, B.; Kityk, A.V.; Busch, M.; Huber, P. The structural and surface properties of natural and modified coal gangue. *J. Environ. Manag.* **2017**, *190*, 80–90. [CrossRef] [PubMed]
23. Yang, L.; Song, J.; Bai, X.; Song, B.; Wang, R.; Zhou, T.; Jia, J.; Pu, H. Leaching behavior and potential environmental effects of trace elements in coal gangue of an open-cast coal mine area, inner mongolia, China. *Minerals* **2016**, *6*, 50. [CrossRef]
24. Zhang, W.; Dong, C.; Huang, P.; Sun, Q.; Li, M.; Chai, J. Experimental study on the characteristics of activated coal gangue and coal gangue-based geopolymer. *Energies* **2020**, *13*, 2504. [CrossRef]
25. Wang, X.; Liu, F.; Pan, Z.; Chen, W.; Muhammad, F.; Zhang, B.; Li, L. Geopolymerization of coal gangue via alkali-activation: Dependence of mechanical properties on alkali activators. *Buildings* **2024**, *14*, 787. [CrossRef]
26. Yi, C.; Ma, H.; Chen, H.; Wang, J.; Jing, S.; Li, Z.; Yu, M. Preparation and characterization of coal gangue geopolymers. *Constr. Build. Mater.* **2018**, *187*, 318–326. [CrossRef]
27. Kizilkanat, A.B. Experimental evaluation of mechanical properties and fracture behavior of carbon fiber reinforced high strength concrete. period. *Polytech. Civil Eng.* **2016**, *60*, 289–296. [CrossRef]
28. Deb, P.S.; Nath, P.; Sarker, P.K. The effects of ground granulated blast-furnace slag blending with fly ash and activator content on the workability and strength properties of geopolymer concrete cured at ambient temperature. *Mater. Des.* **2014**, *62*, 32–39. [CrossRef]
29. Buchwald, A.; Tatarin, R.; Stephan, D. Reaction progress of alkaline-activated metakaolin-ground granulated blast furnace slag blends. *J. Mater. Sci.* **2009**, *44*, 5609–5617. [CrossRef]
30. Venkatesan, R.P.; Pazhani, K.C. Strength and durability properties of geopolymer concrete made with Ground Granulated Blast Furnace Slag and Black Rice Husk Ash. *KSCE J. Civ. Eng.* **2016**, *20*, 2384–2391. [CrossRef]
31. Temuujin, J.; van Riessen, A. Effect of fly ash preliminary calcination on the properties of geopolymer. *J. Hazard. Mater.* **2009**, *164*, 634–639. [CrossRef] [PubMed]
32. Khalifa, A.Z.; Cizer, Ö.; Pontikes, Y.; Heath, A.; Patureau, P.; Bernal, S.A.; Marsh, A.T.M. Advances in alkali-activation of clay minerals. *Cem. Concr. Res.* **2020**, *132*, 106050. [CrossRef]
33. Song, M.; Qian, J.; Zhong, L.J.; Liang, S. *From a View of Alkali Solution: Alkali Concentration to Determine Hydration Process of Alkali Activating Metakaolin*; Springer: Dordrecht, The Netherlands, 2015; Volume 10, pp. 305–313. [CrossRef]
34. Pelisser, F.; Guerrino, E.L.; Menger, M.; Michel, M.D.; Labrincha, J.A. Micromechanical characterization of metakaolin-based geopolymers. *Constr. Build. Mater.* **2013**, *49*, 547–553. [CrossRef]
35. Thaarrini, J.; Venkatasubramani, R. Feasibility Studies on Compressive Strength of Ground Coal Ash Geopolymer Mortar (Retracted Paper). *Period. Polytech. Civ. Eng.* **2015**, *59*, 373–379. [CrossRef]
36. Guo, X.; Shi, H.; Dick, W.A. Compressive strength and microstructural characteristics of class C fly ash geopolymer. *Cem. Concr. Compos.* **2010**, *32*, 142–147. [CrossRef]
37. Wang, H.; Li, H.; Yan, F. Synthesis and mechanical properties of metakaolinite-based geopolymer. *Colloids Surf. A Physicochem. Eng. Asp.* **2005**, *268*, 1–6. [CrossRef]
38. Zhang, Z.; Huo, Z.; Yang, Z.; Wang, X.; Xia, J. Curing condition and NaOH concentration on the mechanical properties of fly ash based geopolymer mmortars. *Int. J. Sustain. Constr. Eng Technol.* **2023**, *14*, 279–286. [CrossRef]
39. Kramar, S.; Šajna, A.; Ducman, V. Assessment of alkali activated mortars based on different precursors with regard to their suitability for concrete repair. *Constr. Build. Mater.* **2016**, *124*, 937–944. [CrossRef]
40. Gebregziabiher, B.S.; Thomas, R.; Peethamparan, S. Very early-age reaction kinetics and microstructural development in alkali-activated slag. *Cem. Concr. Compos.* **2015**, *55*, 91–102. [CrossRef]
41. Longhi, M.A.; Walkley, B.; Rodríguez, E.D.; Kirchheim, A.P.; Zhang, Z.; Wang, H. New selective dissolution process to quantify reaction extent and product stability in metakaolin-based geopolymers. *Compos. Part B Eng.* **2019**, *176*, 107172. [CrossRef]
42. Chen, Z.; Li, J.-S.; Zhan, B.-J.; Sharma, U.; Poon, C.S. Compressive strength and microstructural properties of dry-mixed geopolymer pastes synthesized from GGBS and sewage sludge ash. *Constr. Build. Mater.* **2018**, *182*, 597–607. [CrossRef]
43. Zhang, B.; Guo, H.; Yuan, P.; Li, Y.; Wang, Q.; Deng, L.; Liu, D. Geopolymerization of halloysite via alkali-activation: Dependence of microstructures on precalcination. *Appl. Clay Sci.* **2020**, *185*, 105375. [CrossRef]
44. He, J.; Jie, Y.; Zhang, J.; Yu, Y.; Zhang, G. Synthesis and characterization of red mud and rice husk ash-based geopolymer composites. *Cem. Concr. Compos.* **2013**, *37*, 108–118. [CrossRef]
45. van Jaarsveld, J.G.S.; van Deventer, J.S.J. Effect of the alkali metal activator on the properties of fly ash-based geopolymers. *Ind. Eng. Chem. Res.* **1999**, *38*, 3932–3941. [CrossRef]
46. Liew, Y.M.; Kamarudin, H.; Al Bakri, A.M.M.; Bnhussain, M.; Luqman, M.; Nizar, I.K.; Ruzaidi, C.M.; Heah, C.Y. Optimization of solids-to-liquid and alkali activator ratios of calcined kaolin geopolymeric powder. *Constr. Build. Mater.* **2012**, *37*, 440–451. [CrossRef]
47. Zuhua, Z.; Xiao, Y.; Huajun, Z.; Yue, C. Role of water in the synthesis of calcined kaolin-based geopolymer. *Appl. Clay Sci.* **2009**, *43*, 218–223. [CrossRef]
48. Aiken, T.A.; Kwasny, J.; Sha, W.; Soutsos, M.N. Effect of slag content and activator dosage on the resistance of fly ash geopolymer binders to sulfuric acid attack. *Cem. Concr. Res.* **2018**, *111*, 23–40. [CrossRef]

49. Yaseri, S.; Masoomi Verki, V.; Mahdikhani, M. Utilization of high volume cement kiln dust and rice husk ash in the production of sustainable geopolymer. *J. Clean. Prod.* **2019**, *230*, 592–602. [CrossRef]
50. Zhang, M.; Zhao, M.; Zhang, G.; El-Korchi, T.; Tao, M. A multiscale investigation of reaction kinetics, phase formation, and mechanical properties of metakaolin geopolymers. *Cem. Concr. Compos.* **2017**, *78*, 21–32. [CrossRef]
51. Fernández-Jiménez, A.; Palomo, A. Mid-infrared spectroscopic studies of alkali-activated fly ash structure. *Micropor. Mesopor. Mat.* **2005**, *86*, 207–214. [CrossRef]
52. Li, N.; Farzadnia, N.; Shi, C. Microstructural changes in alkali-activated slag mortars induced by accelerated carbonation. *Cem. Concr. Res.* **2017**, *100*, 214–226. [CrossRef]
53. El Alouani, M.; Alehyen, S.; El Achouri, M.; Taibi, M. Preparation, Characterization, and Application of Metakaolin-Based Geopolymer for Removal of Methylene Blue from Aqueous Solution. *J. Chem.* **2019**, *2019*, 4212901. [CrossRef]
54. Dehghani, A.; Aslani, F.; Ghaebi Panah, N. Effects of initial SiO_2/Al_2O_3 molar ratio and slag on fly ash-based ambient cured geopolymer properties. *Constr. Build. Mater.* **2021**, *293*, 123527. [CrossRef]
55. Wu, C. Comparative assessment of the interface between poly(3-hydroxybutyrate-Co-3-hydroxyvalerate) and Fish Scales in composites: Preparation, characterization, and applications. *Mater. Sci. Eng. C Mater. Biol. Appl.* **2019**, *104*, 109878. [CrossRef] [PubMed]
56. Yu, Q.; Li, S.; Li, H.; Chai, X.; Bi, X.; Liu, J.; Ohnuki, T. Synthesis and characterization of Mn-slag based geopolymer for immobilization of Co. *J. Clean. Prod.* **2019**, *234*, 97–104. [CrossRef]
57. Yuan, P.; Tan, D.; Annabi-Bergaya, F.; Yan, W.; Fan, M.; Liu, D.; He, H. Changes in structure, morphology, porosity, and surface activity of mesoporous halloysite nanotubes under heating. *Clay Clay Min.* **2012**, *60*, 561–573. [CrossRef]
58. Ryu, G.S.; Lee, Y.B.; Koh, K.T.; Chung, Y.S. The mechanical properties of fly ash-based geopolymer concrete with alkaline activators. *Constr. Build. Mater.* **2013**, *47*, 409–418. [CrossRef]
59. Sun, Z.; Vollpracht, A. One year geopolymerisation of sodium silicate activated fly ash and metakaolin geopolymers. *Cem. Concr. Compos.* **2019**, *95*, 98–110. [CrossRef]
60. Koshy, N.; Dondrob, K.; Hu, L.; Wen, Q.; Meegoda, J.N. Synthesis and characterization of geopolymers derived from coal gangue, fly ash and red mud. *Constr. Build. Mater.* **2019**, *206*, 287–296. [CrossRef]
61. Li, Z.; Lu, T.; Liang, X.; Dong, H.; Ye, G. Mechanisms of autogenous shrinkage of alkali-activated slag and fly ash pastes. *Cem. Concr. Res.* **2020**, *135*, 106107. [CrossRef]
62. Mastali, M.; Kinnunen, P.; Dalvand, A.; Mohammadi Firouz, R.; Illikainen, M. Drying shrinkage in alkali-activated binders—A critical review. *Constr. Build. Mater.* **2018**, *190*, 533–550. [CrossRef]
63. Melo Neto, A.A.; Cincotto, M.A.; Repette, W. Drying and autogenous shrinkage of pastes and mortars with activated slag cement. *Cem. Concr. Res.* **2008**, *38*, 565–574. [CrossRef]
64. Wang, Y.-S.; Alrefaei, Y.; Dai, J.-G. Silico-aluminophosphate and alkali-aluminosilicate geopolymers: A comparative review. *Front. Mater.* **2019**, *6*, 106. [CrossRef]
65. Fan, Y.; Luan, H. Pore structure in concrete exposed to acid deposit. *Constr. Build. Mater.* **2013**, *49*, 407–416. [CrossRef]
66. Rashad, A.M. Alkali-activated metakaolin: A short guide for civil Engineer—An overview. *Constr. Build. Mater.* **2013**, *41*, 751–765. [CrossRef]
67. Wang, J.; Han, L.; Liu, Z.; Wang, D. Setting controlling of lithium slag-based geopolymer by activator and sodium tetraborate as a retarder and its effects on mortar properties. *Cem. Concr. Compos.* **2020**, *110*, 103598. [CrossRef]
68. Cong, X.; Zhou, W.; Geng, X.; Elchalakani, M. Low field NMR relaxation as a probe to study the effect of activators and retarders on the alkali-activated GGBFS setting process. *Cem. Concr. Compos.* **2019**, *104*, 103399. [CrossRef]
69. Kalina, L.; Bilek, V.; Novotny, R.; Moncekova, M.; Masilko, J.; Koplik, J. Effect of Na_3PO_4 on the hydration process of alkali-activated blast furnace slag. *Materials* **2016**, *9*, 395. [CrossRef]
70. Gong, C.M.; Yang, N.R. Effect of phosphate on the hydration of alkali-activated red mud-slag cementitious material. *Cem. Concr. Res.* **2000**, *30*, 1013–1016. [CrossRef]
71. Brough, A.R.; Holloway, M.; Sykes, J.; Atkinson, A. Sodium silicate-based alkali-activated slag mortars Part II. The retarding effect of additions of sodium chloride or malic acid. *Cem. Concr. Res.* **2000**, *30*, 1375–1379. [CrossRef]

materials

MDPI

Article

Evaluation of Bonding Behavior between Engineered Geopolymer Composites with Hybrid PE/PVA Fibers and Concrete Substrate

Yu Ling [1], Xiafei Zhang [1], Yanwei Wu [1], Weiyu Zou [1], Chuang Wang [1], Chaosen Li [2] and Wen Li [3,*]

[1] Guangzhou Power Supply Bureau, Guangdong Power Grid Co., Ltd., China Southern Power Grid Co., Ltd., Guangzhou 510620, China; 15915714058@163.com (Y.L.); 18102278577@189.cn (X.Z.); wyw1400@163.com (Y.W.); 13822112199@139.com (W.Z.); 18022327865@163.com (C.W.)
[2] School of Civil and Transportation Engineering, Guangdong University of Technology, Guangzhou 510006, China; 2112109088@mail2.gdut.edu.cn
[3] School of Civil Engineering and Transportation, South China University of Technology, Guangzhou 510640, China
* Correspondence: liiween@scut.edu.cn; Tel.: +86-159-2092-4878

Abstract: Engineered geopolymer composites (EGCs) exhibit excellent tensile ductility and crack control ability, making them promising for concrete structure repair. However, their widespread use is limited by high costs of reinforcement fiber and a lack of an EGC–concrete interface bonding mechanism. This study investigated a hybrid PE/PVA fiber-reinforced EGC using domestically produced unoiled PVA fibers to replace commonly used PE fibers. The bond performance of the EGC–concrete interface was evaluated through direct tensile and slant shear tests, focusing on the effects of PE fiber content (1%, 2%, and 3%), fiber hybrid ratios (2.0:0.0, 1.5:0.5, 1.0:1.0, 0.5:1.5, and 0.0:2.0), concrete substrate strength (C30, C50, and C70), and the ratio of fly ash (FA) to ground granulated blast furnace slag (GGBS) (6:4, 7:3, and 8:2) on interface bond strength. Results showed that the EGCs' compressive strength ranged from 77.1 to 108.9 MPa, with increased GGBS content significantly enhancing the compressive strength and elastic modulus. Most of the specimens exhibited strain-hardening behavior after initial cracking. Interface bonding tests revealed that a PE/PVA ratio of 1.0 increased tensile bond strength by 8.5% compared with using 2.0% PE fiber alone. Increasing the PE fiber content, PVA/PE ratio, GGBS content, and concrete substrate strength all improved the shear bond strength. This improvement was attributed to the flexible fibers' ability to restrict thermo–hydro damage and deflect and blunt microcracks, enhancing the interface's failure resistance. Cost analysis showed that replacing 50% of the PE fiber in EGC with unoiled PVA fiber reduced costs by 44.2% compared with PE fiber alone, offering the best cost–performance ratio. In summary, hybrid PE/PVA fiber EGC has promising prospects for improving economic efficiency while maintaining tensile ductility and crack-control ability. Future optimization of fiber ratios and interface design could further enhance its potential for concrete repair applications.

Keywords: EGC; unoiled PVA fiber; hybrid fiber; bond behavior; cost analysis

1. Introduction

Concrete is the cornerstone of modern infrastructure, used extensively in buildings, bridges, dams, cement concrete pavement, and other critical structures due to its versatility, strength, and durability [1–4]. However, concrete structures degrade over time, experiencing issues like cracking, rebar corrosion, and spalling, which compromise their integrity and safety [5,6]. This premature aging of reinforced concrete structures poses significant challenges, threatening public safety, increasing maintenance costs, and potentially straining future economies [7]. To mitigate these problems while preserving the function and integrity of these structures, it is crucial to minimize the frequency and extent of repair interventions. Traditional concrete repair and reinforcement methods typically use supplementary materials to restore or enhance deteriorated structures. These materials include FRP, steel plates, and various concrete repair compounds, applied through bonding, overlaying, spraying, and pressure grouting techniques [8–10]. However, the durability of these repair methods is often questioned, with concerns about thermal degradation of epoxy resins and corrosion of steel [11]. As a result, concrete repairs are frequently criticized for their lack of early performance and long-term durability.

Cement-based repair materials are inherently brittle, exhibiting low tensile strength, low strain capacity, and poor ductility. Engineered cementitious composites (ECCs), known for their excellent tensile properties and crack control, offer a solution to these issues [12–15]. Research has shown the feasibility of using ECCs for repair and reinforcement [16–20]. However, ECCs require significant amounts of ordinary Portland cement, contributing to global CO_2 emissions, which increases their carbon footprint by at least 5–8% [21]. To address the global challenge of climate change, researchers have proposed using industrial by-products, such as fly ash (FA) and ground granulated blast furnace slag (GGBS), to produce engineered geopolymer composites (EGCs). This approach avoids high-carbon cement materials, reducing the carbon footprint of repair materials [22–25]. The utilization of EGCs can be adopted as repair materials for the rehabilitation of existing building structures, concrete pavements, etc.

Controlling the cost of repair materials is a crucial issue for the application of EGCs. In EGC-like materials, polyvinyl alcohol (PVA) fibers, polyethylene (PE) fibers, and other polymer fibers like aramid fibers or polyethylene terephthalate (PET) fibers have been adopted for their various benefits [26–29]. However, frequently used fiber types like oiled PVA fibers or PE fibers are expensive. For example, Kuraray Co., Ltd. in Japan monopolizes the surface-coating technology for PVA fibers, making them accountable for up to 70% of the total material costs. To promote widespread application, reducing preparation costs is essential. One effective method is to partially replace high-performance fibers with more affordable ones such as unoiled domestic PVA fibers, at only 10–20% of the cost of PE fibers.

It was reported that oiled PVA fibers reduced excessive bonding between hydrophilic PVA fibers and the surrounding matrix, achieving the desired tensile performance for PVA-ECC [30]. Nevertheless, although prone to premature rupture due to excessive bonding, unoiled PVA fibers can be suitable for ECC materials when the matrix is customized to compensate for these changes [28,31]. Fortunately, it was reported that the fracture toughness of the EGC matrix was so low that the fiber's mechanical performance requirements as well as the fiber content could be reduced. In the study by Wang et al. [32], it was found that using 0.1–1.5% PE fibers maintained the high tensile strain capacity of EGC. Thus, it can be inferred that incorporating unoiled PVA fibers can significantly reduce costs while maintaining the high tensile strain capacity of EGC, compared with using PE fibers or oiled PVA fibers alone.

Despite the promising properties of EGCs with hybrid PVA/PE fibers, the bond performance of the EGC–concrete interface deserves attention when it comes to the repairing of concrete structures. The interface bond strength is crucial to the performance of repaired structures. Factors affecting bond strength include the strength grade of the concrete substrate and repair material [18,33], interface roughness, and casting methods [17,34,35]. Existing studies have shown a positive correlation between the strength of the concrete substrate, the repair material, and the interface bond strength [17]. For instance, an improvement in interface shear strength of up to three-fold was reported when the compressive strength of ECC increased from 21.7 MPa to 40.8 MPa [17]. Properly treating the interface to increase roughness can improve bond strength, but overly rough surfaces may be detrimental [33]. Increasing the sand-to-cement ratio can reduce microcracks at the EGC–concrete interface, improving bond strength [36]. However, comprehensive studies on bonding between hybrid fiber EGC and concrete substrates are limited.

Systematic research is urgently required to unravel the unique bonding mechanisms between hybrid fiber EGC and concrete for advanced engineering applications. This study introduces a new approach by utilizing domestically sourced unoiled PVA fibers as a replacement for conventional PE fibers, leading to the development of an innovative hybrid PE/PVA fiber EGC. The bonding performance of the EGC–concrete interface was investigated through direct tensile and slant shear tests, exploring the effects of different fiber hybrid ratios, concrete substrate strength, and FA/GGBS ratios on interface bond strength. Additionally, a cost analysis compared the EGC developed in this study with others reported in the existing studies, discussing the feasibility of replacing PE fibers with domestic unoiled PVA fibers. This research not only advances the understanding of bonding behavior between hybrid PE/PVA fiber EGC and concrete substrate, but also paves the way for its practical application in repair and reinforcement, thereby promoting its widespread adoption in the industry.

2. Experimental Program

2.1. Materials and Mix Proportion

2.1.1. EGC with Hybrid PE/PVA Fibers

Figure 1 shows the raw materials used to prepare the EGC in this study, including Class F FA, S105 GGBS, quartz powder (QP), sodium hydroxide (SH), sodium silicate (SS), polyethylene fibers (PE fibers), domestic unoiled polyvinyl alcohol fibers (PVA fibers), barium chloride (BC), defoamer, and water. Class F FA and S105 GGBS conformed to the requirements of GB/T 51003-2014 [37]. Figure 2 displays the particle size distribution of FA, GGBS, and QP, determined through laser particle size analysis. Table 1 presents the chemical composition of FA and GGBS using X-ray fluorescence (XRF). X-ray fluorescence (XRF) is an analytical technique used to determine the elemental composition of materials, which is based on the principle that individual elements emit characteristic "secondary" (or fluorescent) X-rays when they are excited by high-energy X-rays or gamma rays.

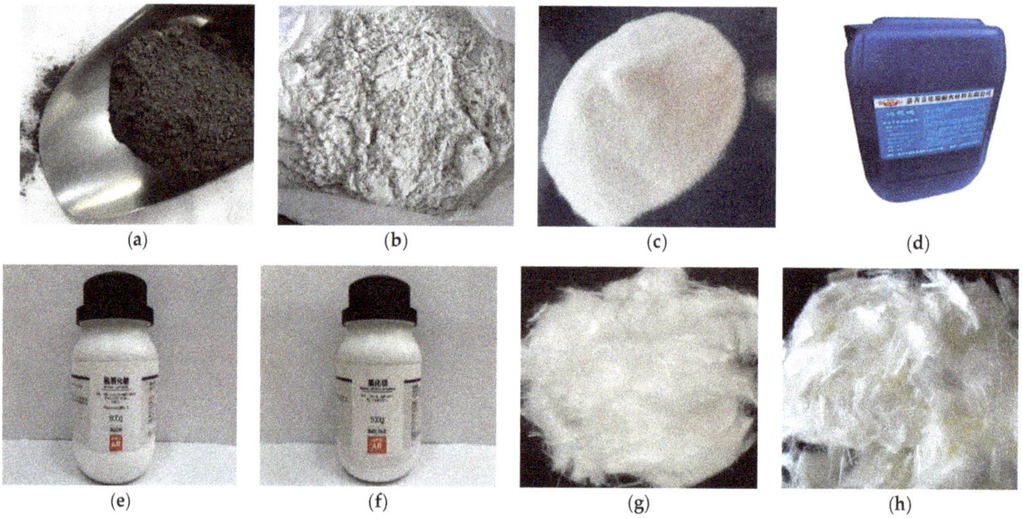

Figure 1. Raw materials of EGC: (**a**) fly ash; (**b**) ground granulated blast furnace slag; (**c**) quartz powder; (**d**) sodium silicate; (**e**) sodium hydroxide; (**f**) barium chloride; (**g**) polyethylene fiber; (**h**) polyvinyl alcohol fiber.

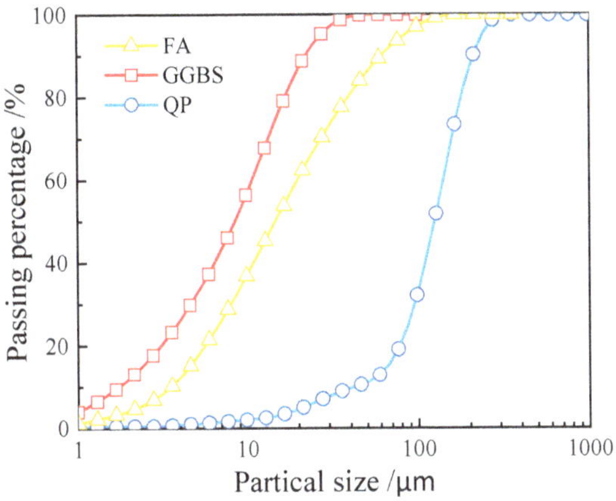

Figure 2. Particle size distributions of raw materials.

Table 1. Chemical composition of GGBS and FA (unit: wt%).

	CaO	SiO_2	Al_2O_3	Fe_2O_3	MgO	TiO_2	SO_3	Others	LOI [1]
GGBS	34.00	34.50	17.70	1.03	6.01	/	1.64	5.12	0.84
FA	4.01	53.97	31.15	4.16	1.01	1.13	2.20	2.37	4.60

[1] LOI means loss on ignition.

SH and SS were used to prepare the alkali activator. The SS solution had a modulus of 2.25, with a mass ratio of $m(SiO_2)$: $m(Na_2O)$: $m(H_2O)$ = 29.99: 13.75: 52.26. The modulus in sodium silicate is given by the molar ratio of SiO_2 to Na_2O, indicating the composition of sodium silicate. The SH solution had a concentration of 10.00 mol/L. The alkali activator was made by mixing SS solution and SH solution in a 2:1 ratio. To prepare the 10 mol/L SH solution, 400 g of SH flakes were dissolved in water to make 1 L of solution, which was then mixed with the SS solution, stirred, sealed, and stored for later use.

BC was used as a retarder to delay the rapid hardening of EGC. Table 2 lists the relevant parameters for PVA and PE fibers. Defoamer was also used to reduce bubbles during mixing. The mix proportions of the hybrid fiber EGC are shown in Table 3. The naming convention for the mix proportions is Mx-UyPz, where x represents FA/GGBS ratios of 7:3, 6:4, and 5:5 (for x = 1, 2, 3, respectively), and y and z represent the volume content of PE fibers and PVA fibers, respectively.

Table 2. Parameters of PVA and PE fiber.

Fiber Type	Length (mm)	Diameter (mm)	Tensile Strength (MPa)	Elastic Modulus (GPa)	Density (g/cm^3)	Elongation (%)
PVA	12	20	2500	120	0.97	3.7
PE	12	15	1830	40	1.29	6.9

Table 3. Mix proportions of EGC.

Mix IDs	Mix Proportion							Volume Fraction/%	
	FA [1]	GGBS [2]	QP [3]	AAS [4]	Water	BC [5]	Deformer	PE [6]	PVA [7]
M1-U1.0P0.0	0.70	0.30	0.20	0.40	0.06	0.01	0.001	1.0	0.0
M1-U1.5P0.0								1.5	0.0

Table 3. *Cont.*

Mix IDs	Mix Proportion							Volume Fraction/%	
	FA [1]	GGBS [2]	QP [3]	AAS [4]	Water	BC [5]	Deformer	PE [6]	PVA [7]
M1-U2.0P0.0	0.70	0.30	0.20	0.40	0.06	0.01	0.001	2.0	0.0
M1-U1.5P0.5								1.5	0.5
M1-U1.0P1.0								1.0	1.0
M1-U0.5P1.5								0.5	1.5
M1-U0.0P2.0								0.0	2.0.
M2-U2.0P0.0	0.60	0.40	0.20	0.40	0.06	0.01	0.001	2.0	0.0
M3-U2.0P0.0	0.50	0.50	0.20	0.40	0.06	0.01	0.001	2.0	0.0

[1] fly ash; [2] ground granulated blast furnace slag; [3] quartz powder; [4] alkali-activated solution; [5] barium chloride; [6] polyethylene fibers; [7] polyvinyl alcohol fibers.

2.1.2. Concrete Substrate

In this study, ordinary concretes with compressive strengths of 30, 50, and 70 MPa were used as the concrete substrate. The mix proportions are detailed in Table 4. The raw materials included water, cement, coarse aggregate (gravel), fine aggregate (river sand), and a water-reducing agent. The cement was 42.5R ordinary Portland cement. The coarse aggregate consisted of 5–10 mm granite gravel, and the fine aggregate was dried river sand. The water-reducing agent, produced by Guangdong Strong Building Materials Co., Ltd., mainly consisted of β-naphthalene sulfonate condensate, with a water reduction rate of approximately 15–20%. The compressive strength of the concrete substrate was measured using 150 mm cubic specimens, based on the national standard GB50081-2019 [38]. The results showed compressive strengths of 35.3 MPa for C30, 51.3 MPa for C50, and 70.6 MPa for C70.

Table 4. Mix proportions of concrete substrate.

Mix IDs	Cement	Aggregate	Sand	Water	Water Reducer
C30	1.00	3.56	2.37	0.66	0.000
C50	1.00	3.41	1.92	0.63	0.000
C70	1.00	1.75	0.98	0.28	0.005

2.2. *Specimens and Preparation*

Figure 3 illustrates the specimen dimensions for testing the mechanical properties of hybrid fiber EGC and the concrete substrate. Figure 4 shows the dimensions for the bond performance tests between hybrid fiber EGC and the concrete substrate. In accordance with ASTM C469 [39], a cylindrical specimen with a diameter of 50 mm and a height of 100 mm (Figure 3a) was used to test the axial compression performance of the EGC. A dumbbell-shaped specimen (Figure 3b), per JSCE regulations [40], was used to test the axial tensile performance of the EGC. For bond performance tests, the specimens for direct tension test and slant shear test were designed according to T/CBMF37-2018 [41] (Figure 4a) and ASTM-C882 [42], respectively. Specifically, dumbbell-shaped specimens as shown in Figure 4a were used for direct tension testing, and cylindrical specimens with dimensions of Φ75 mm × 150 mm and an interface inclination angle of 30° were adopted for slant shear tests (Figure 4b). Note that three samples were prepared in each series.

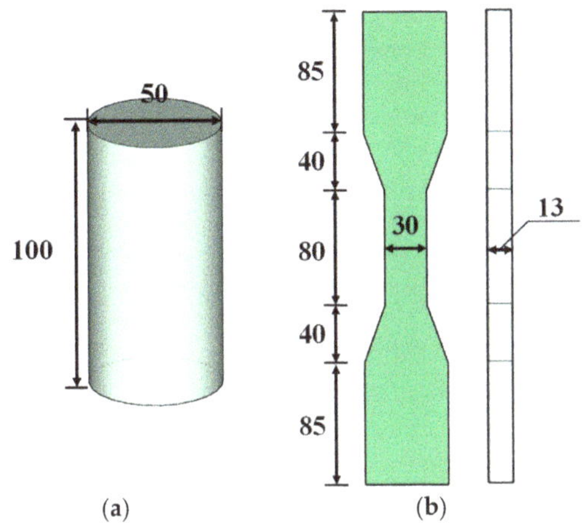

Figure 3. Specimens used for testing the basic mechanical properties of EGC: (**a**) axial compression test; (**b**) axial tension test (dimensions in mm).

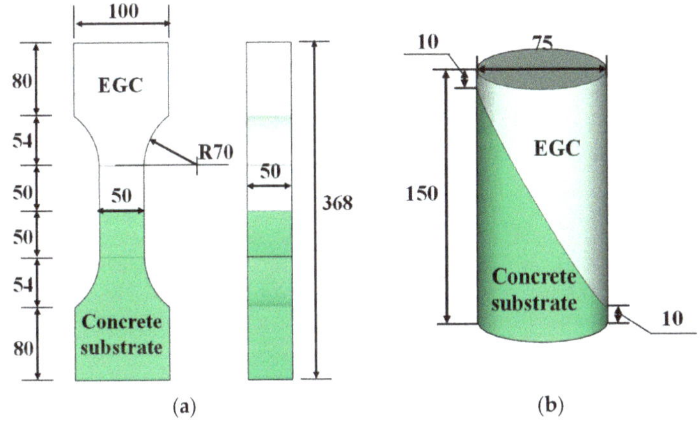

Figure 4. Specimens adopted for bonding behavior between EGC and concrete substrate: (**a**) direct tension test; (**b**) compressive slant shear test (dimensions in mm).

The preparation process of hybrid fiber EGC was as follows: FA, GGBS, BC, and QP were mixed at 76 rpm for two minutes. The prepared solution and water were then added during slow mixing, followed by one minute of medium-speed mixing at 135 rpm. Finally, the fibers were added during slow mixing, which continued for another minute to ensure even distribution. The prepared EGC was placed in molds on a vibrating table for compaction and leveling, covered with plastic film, cured indoors for 24 h, demolded, and cured in water for 28 days.

Table 5 details the specimens used for testing the bond performance between hybrid fiber EGC and the concrete substrate. Specifically, direct tensile tests were conducted to examine the effects of different PE/PVA fiber hybrid ratios on bond performance, keeping the concrete substrate strength grade and FA/GGBS ratio constant. Compressive shear tests investigated the effects of the concrete substrate strength grade, FA/GGBS ratio, and PE/PVA fiber hybrid ratio on bond performance.

Table 5. Details of test specimens for assessment of bonding behavior between EGC and concrete substrate.

Specimen IDs	Testing Method	Substrate Concrete Strength Grade	Repairing Material	FA/GGBS Ratio of EGC	Volume Fraction of EGC/%	
					PE	PVA
T7 [1]-M1-U2.0P0.0	direct tension	C70	M1-U2.0P0.0	7:3	2.0	0.0
T7-M1-U1.5P0.5	direct tension	C70	M1-U1.5P0.5	7:3	1.5	0.5
T7-M1-U1.0P1.0	direct tension	C70	M1-U1.0P1.0	7:3	1.0	1.0
T7-M1-U0.5P1.5	direct tension	C70	M1-U0.5P1.5	7:3	0.5	1.5
T7-M1-U0.0P2.0	direct tension	C70	M1-U0.0P2.0	7:3	0.0	2.0
S7 [2]-M1-U1.0P0.0	compressive slant shear	C70	M1-U1.0P0.0	7:3	1.0	0.0
S7-M1-U1.5P0.0	compressive slant shear	C70	M1-U1.5P0.0	7:3	1.5	0.0
S7-M1-U2.0P0.0	compressive slant shear	C70	M1-U2.0P0.0	7:3	2.0	0.0
S7-M1-U1.5P0.5	compressive slant shear	C70	M1-U1.5P0.5	7:3	1.5	0.5
S7-M1-U1.0P1.0	compressive slant shear	C70	M1-U1.0P1.0	7:3	1.0	1.0
S7-M1-U0.5P1.5	compressive slant shear	C70	M1-U0.5P1.5	7:3	0.5	1.5
S7-M1-U0.0P2.0	compressive slant shear	C70	M1-U0.0P2.0	7:3	0.0	2.0
S7-M2-U2.0P0.0	compressive slant shear	C70	M2-U2.0P0.0	6:4	2.0	0.0
S7-M3-U2.0P0.0	compressive slant shear	C70	M3-U2.0P0.0	5:5	2.0	0.0
S3 [2]-M1-U2.0P0.0	compressive slant shear	C30	M1-U2.0P0.0	7:3	2.0	0.0
S5 [2]-M1-U2.0P0.0	compressive slant shear	C50	M1-U2.0P0.0	7:3	2.0	0.0

[1] T7 denotes direct tension test and that the substrate concrete strength grade was C70; [2] S3, S5 and S7 denote that the substrate concrete strength grades were C30, C50, and C70, respectively, in the compressive slant shear test.

Figure 5 depicts the specimen preparation process for bond performance tests. Dumbbell-shaped specimens were prepared by first preparing the concrete side specimens. The mixed concrete was poured into large dumbbell-shaped molds, vibrated, covered with plastic film, demolded after 24 h, and cured in water for 28 days. The concrete specimens were then cut in half, and one half was placed back into the mold, followed by the casting of the EGC. Specimens for slant shear tests required pre-prepared molds. PVC pipes with an inner diameter of 75 mm were cut to a length of 150 mm, and detachable inner cores were 3D printed and placed inside the PVC pipes. The concrete substrate was mixed, cast into the mold, covered with plastic film, and demolded after 24 h of indoor curing. The concrete substrate part of the shear specimen was cured in water for 28 days. Subsequently, the floating slurry at the concrete interface was removed, the concrete part was placed back into the PVC pipe, and the other half of the EGC was cast. After curing indoors for 24 h, the specimens were demolded and cured in water for 28 days until formal testing.

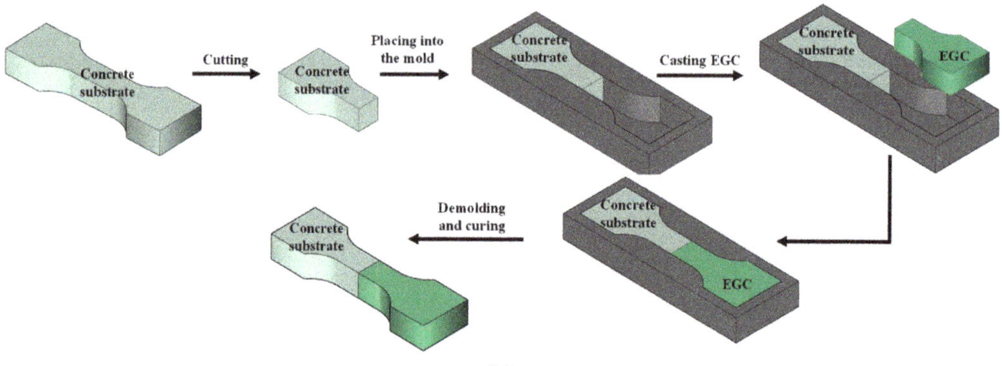

(a)

Figure 5. *Cont.*

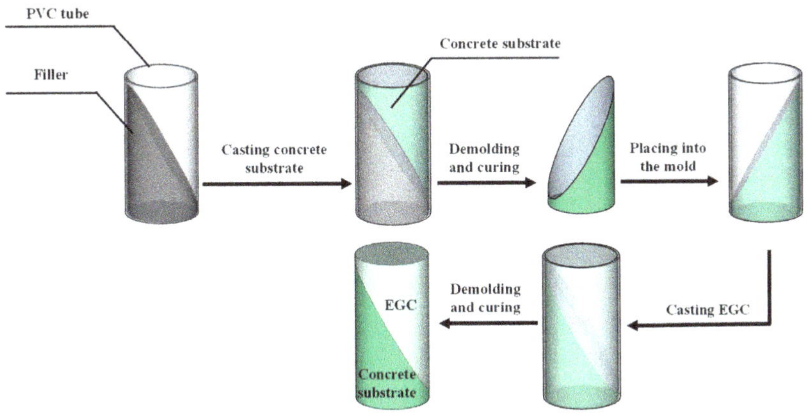

(**b**)

Figure 5. Preparation of specimens adopted for bonding behavior between EGC and concrete substrate: (**a**) direct tension test; (**b**) compressive slant shear test.

3. Experimental Setup and Procedure

3.1. Axial Compressive Test

To ensure accurate and reliable measurements, specimens were leveled with high-strength gypsum to avoid eccentric loading. Tests were conducted in displacement control mode with a loading rate of 0.2 mm/min. Axial load was measured using the load cell of the testing machine, while axial strain was measured with strain gauges symmetrically attached to the middle section of each specimen. Axial deformation was monitored by two symmetrically distributed linear variable differential transformers (LVDTs). Data on axial load and deformation were synchronously collected at a frequency of 1 Hz using a static acquisition system.

3.2. Axial Tensile Test

The displacement control mode was used with a loading rate of 0.5 mm/min. During the loading process, axial deformation in the central 80 mm region of the specimen was monitored by two symmetrically placed LVDTs.

3.3. Test for Assessing Bonding Behavior

For the bond performance tests between concrete and EGC, both direct tensile and shear tests were performed. In the direct tensile tests, universal joints were installed on both sides of the fixtures to prevent eccentric tension, with the loading rate set at 0.2 mm/min, following T/CBMF37-2018 guidelines [41].

In the shear tests, gypsum was applied to the upper and lower surfaces of the specimens to prevent eccentric compression. The loading rate was maintained at 0.2 mm/min in displacement control mode. Load data were collected via the pressure sensor built into the testing machine.

4. Results

4.1. Compressive Properties of EGC

Figure 6 illustrates the variations in compressive strength and elastic modulus of hybrid fiber EGCs across different FA/GGBS ratios and hybrid fiber ratios. Overall, the compressive strength and elastic modulus of the prepared hybrid fiber EGC in this study ranged from 77.1 to 108.9 MPa and 16.0 to 20.3 GPa. Specifically, incorporating PE fibers alone resulted in a reduction in compressive strength as the PE fiber content increased. For example, increasing PE fiber content from 1.0% to 2.0% decreased compressive strength by

17.9%. Conversely, substituting PE fibers with PVA fibers led to a consistent increase in the compressive strength of the EGC. In particular, compared with M1-U2.0P0.0 (2.0% PE fibers), M1-U2.0P0.0 (2.0% PVA fibers) exhibited a 21.1% increase in compressive strength. Yet, it was observed that the elastic modulus of the EGC was influenced by PE fiber content and FA/GGBS ratio rather than the fiber hybrid ratio. Furthermore, Figure 6 shows that with identical fiber content, both the compressive strength and elastic modulus of the EGC increased with higher GGBS content. For example, M3-U2.0P0.0 (FA:GGBS = 5:5) exhibited a 16.6% increase in compressive strength and a 24.5% increase in elastic modulus compared with M1-U2.0P0.0 (FA:GGBS = 7:3).

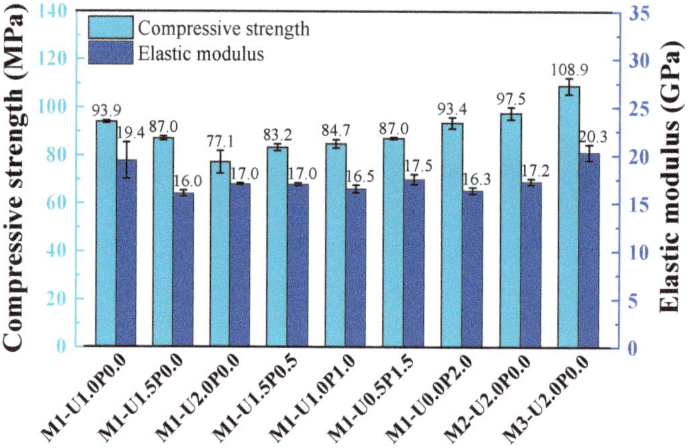

Figure 6. Compressive properties of EGC with hybrid PE/PVA fibers.

4.2. Tensile Properties of EGC

Figure 7 illustrates representative tensile stress–strain curves of EGC under different variations, highlighting the tensile performance of hybrid fiber EGC under various hybrid PVA/PE fiber ratios and FA/GGBS ratios. Most of specimens demonstrated significant strain-hardening behavior, capable of sustaining loads even after matrix cracking initiated. This behavior is essential for applications requiring high durability and resilience. However, notable exceptions were observed in specimens with higher PVA/PE fiber ratios. These specimens lost their crack-control ability, which is not acceptable in EGC-like materials, transitioning into a strain-softening stage soon after the initial cracking. This loss of tensile performance is crucial as it suggests limitations in the use of high PVA fiber content for certain structural applications. As depicted in Figure 7a, increasing the PE fiber content enhanced the tensile strength and ultimate tensile strain of the EGC, indicating that PE fibers contributed significantly to the material's load-bearing capacity and tensile ductility. Conversely, it was observed with increasing PVA fiber content, the tensile performance of EGC gradually declined, with the material losing its characteristic tensile properties. Figure 7c indicates that reducing the FA/GGBS (increasing GGBS content) improved tensile performance, suggesting that GGBS played a vital role in improving the overall tensile behavior of the EGC. Overall, hybrid fiber ratios and FA/GGBS ratios were both demonstrated to affect the tensile properties of the EGC, implying that these two essential parameters should be chosen cautiously to preserve excellent tensile properties while saving costs.

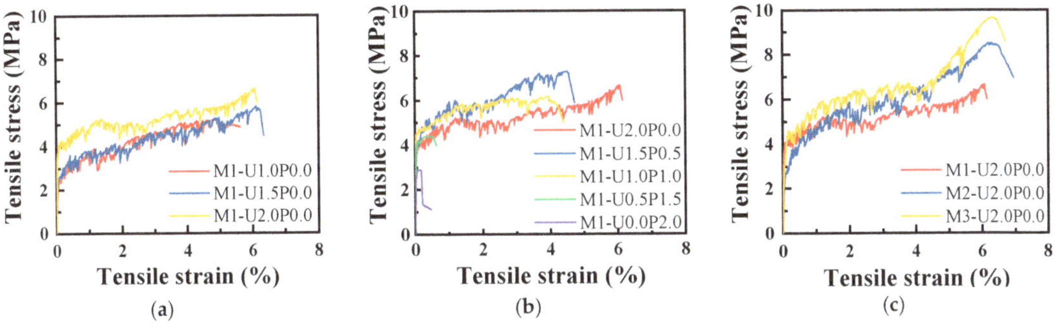

Figure 7. Representative stress–strain curves of EGC with hybrid PE/PVA fibers under axial tension: effects of (**a**) PE fiber content; (**b**) fiber hybrid ratio; (**c**) FA/GGBS.

4.3. Bonding Behavior

4.3.1. Failure Mode of Specimens in Direct Tension Tests

Figure 8 depicts the failure modes of specimens in the direct tensile tests on the bond performance of the EGC–concrete interface, with concrete substrate on the left and EGC on the right. It is evident from Figure 8 that the main cracks after failure appeared alternately on both sides of the interface, resulting in the pulling out of both the concrete substrate and the EGC. This phenomenon is commonly referred to in the literature as interface alternating failure [43]. Bonding of EGC to the concrete side was consistently observed across all groups, as shown in Figure 8. Moreover, an increase in the proportion of unoiled PVA fibers correlated with worsened damage on the concrete substrate side, evidenced by an increasing area of exposed aggregate. For instance, in group I7-M1-U0.0P2.0, where the EGC was reinforced with PVA fiber alone, the concrete side exhibited the most prominent exposed aggregate. This observation may be attributed to the relatively minor strength difference between C70 concrete and EGC repair materials. As the proportion of PVA fibers increased from 0% to 100%, the compressive strength of EGC gradually improved, thereby enhancing to some extent the bond strength at the interface.

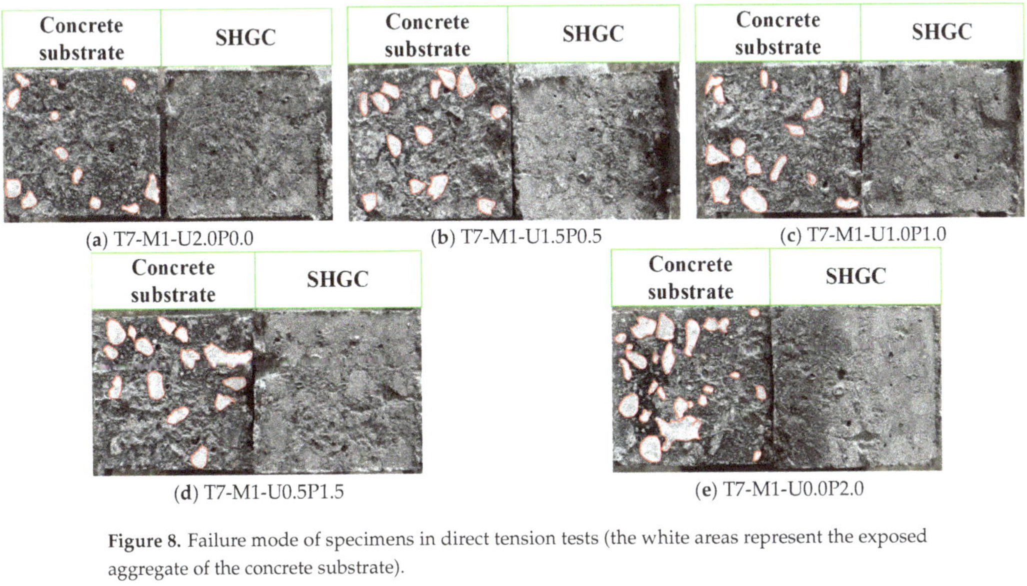

Figure 8. Failure mode of specimens in direct tension tests (the white areas represent the exposed aggregate of the concrete substrate).

4.3.2. Direct Tensile Bond Strength

After the failure of the specimens, the peak load at failure was recorded, and the direct tensile bond strength f_s was calculated using the formula:

$$f_s = \frac{p}{A} \quad (1)$$

where f_s is the direct tensile bond strength (MPa), P is the peak load (N), and A is the bonding area (50 × 50 = 2500 mm^2). The results of the bond strength between the EGC repair material and ordinary concrete substrate are presented in Table 6.

Table 6. Results of bonding behavior between EGC and concrete substrate in direct tension tests.

Specimen IDs	Compressive Strength (MPa)		Peak Load (kN)	Bonding Area (mm^2)	Average Tensile Bond Strength (MPa)
	EGC	Concrete Substrate			
T7-M1-U2.0P0.0	77.1	70.6	1.165	2500	0.466
T7-M1-U1.5P0.5	83.2	70.6	1.171	2500	0.468
T7-M1-U1.0P1.0	84.7	70.6	1.270	2500	0.508
T7-M1-U0.5P1.5	87.0	70.6	1.371	2500	0.549
T7-M1-U0.0P2.0	93.4	70.6	1.390	2500	0.556

Table 6 presents the results of the direct tensile tests for the bond performance of the EGC–concrete interface. It can be observed that substituting PVA fibers for PE fibers slightly increased the tensile bond strength. When PE/PVA = 1.0, the direct tensile bond strength was 8.5% higher than with 2.0% PE fibers alone, indicating that using cost-effective unoiled PVA fibers did not negatively affect the bond performance of the EGC–concrete interface. Instead, it slightly improved the bond strength while saving costs. Since the material lost its characteristic tensile properties when PE/PVA was below 1.0, it is recommended to use M1-U1.0P1.0 (PE/PVA = 1.0) as the repair material for the EGC prepared in this study. This approach slightly enhanced the direct tensile bond strength while maintaining the tensile properties of the EGC and reducing costs. Note that in this study, the concrete substrate surface was not roughened. Given the significant impact of interface roughness, it is suggested to increase the surface roughness to further enhance bond strength.

4.3.3. Failure Mode of Specimens in Compressive Slant Shear Test

Figure 9 illustrates the failure modes of the interface specimens on the concrete substrate side in the compression shear test, with dark-colored areas representing the concrete substrate. Despite showing only the concrete substrate side after specimen failure, it was evident that the concrete substrate side and EGC side of specimens maintained fundamentally good integrity. However, fractures in the concrete substrate occurred when there was a significant strength difference between the concrete substrate and the repair EGC. Integrity is defined as the condition of the concrete substrate and repair material; cracks or fractures in either indicate poor integrity. The integrity of both sides also reflects the sequence of failure at the bonding interface. Good integrity suggests that damage first occurs at the interface. Thus, as the bonding performance improves, the integrity of the substrate and repair mortar decreases. In this study, the fractures observed on the concrete substrate side of S7-M2-U2.0P0.0, S7-M3-U2.0P0.0, and S3-M1-U2.0P0.0 indicated relatively good bonding performance, consistent with the calculated results. Therefore, it was concluded that the EGC showed better repair effects in the process of concrete rehabilitation, which was likely to have been due to the increased tensile strength of the EGC, leading to improved bonding performance.

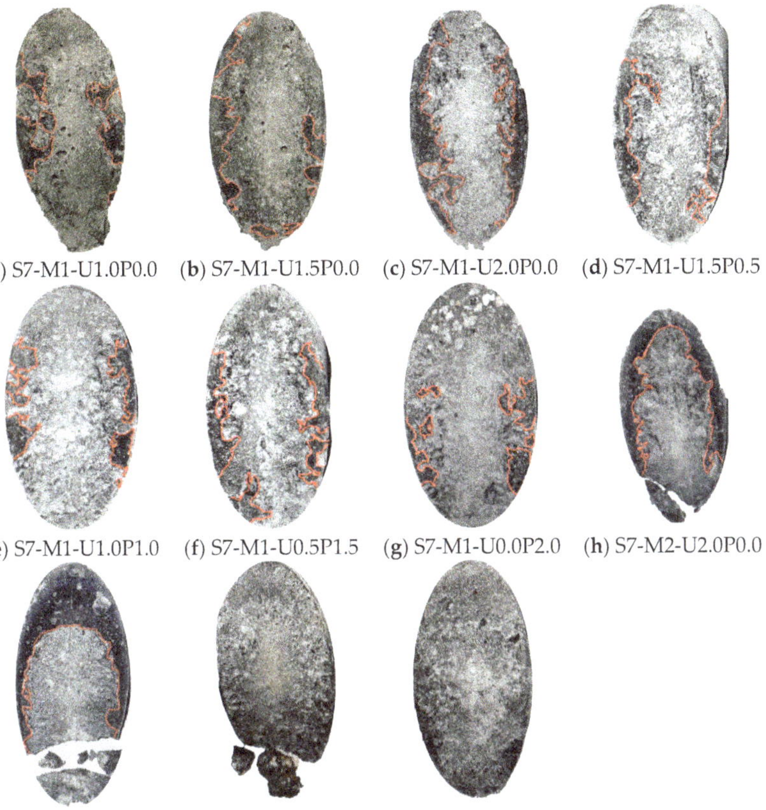

(**a**) S7-M1-U1.0P0.0 (**b**) S7-M1-U1.5P0.0 (**c**) S7-M1-U2.0P0.0 (**d**) S7-M1-U1.5P0.5

(**e**) S7-M1-U1.0P1.0 (**f**) S7-M1-U0.5P1.5 (**g**) S7-M1-U0.0P2.0 (**h**) S7-M2-U2.0P0.0

(**i**) S7-M3-U2.0P0.0 (**j**) S3-M1-U2.0P0.0 (**k**) S5-M1-U2.0P0.0

Figure 9. Failure mode of specimens in the compressive slant shear test (concrete substrate side with the shaded area representing EGC).

4.3.4. Shear Bond Strength

As per ASTM C882 [42], failure loads in the slant shear test were recorded. The shear bond strength was calculated by dividing the load at failure by the bonded area. The specific calculation formula was as follows:

$$\tau_n = \frac{P}{A} \tag{2}$$

where P is the peak load (N), and A is the elliptical bonded area of the cylindrical specimen as specified by ASTM C882 (9116 mm^2) [42]. To eliminate the impact of size, this study ensured that the repair interface area of each sample was the same. The shear bond strength of the specimens was calculated to analyze the effects of PE/PVA fiber hybrid ratios, FA/GGBS ratios, and PE fiber content on the EGC–concrete shear interface performance. The calculated results are shown in Table 7.

Figure 10 presents the effects of various variables on the shear bond strength. From Figure 10, it is evident that PE fiber contents, fiber hybrid ratios, FA/GGBS ratios, and the concrete substrate strength all influenced the shear bond strength to varying degrees. The shear bond strength generally met or exceeded the range recommended by ACI 546.3R [44], a guide to materials selection for concrete repair reported by the American Concrete Institute (ACI) committee 546.

Table 7. Test results of slant shear test.

Specimen IDs	Compressive Strength (MPa)		Applied Stress σ_0 (MPa)	Interfacial Shear Strength τ_n (MPa)
	EGC	Concrete Substrate		
S7-M1-U1.0P0.0	93.9	70.6	31.4	15.5
S7-M1-U1.5P0.0	87.0	70.6	32.4	16.0
S7-M1-U2.0P0.0	77.1	70.6	35.6	17.5
S7-M1-U1.5P0.5	83.2	70.6	38.5	18.9
S7-M1-U1.0P1.0	84.7	70.6	41.4	20.5
S7-M1-U0.5P1.5	87.0	70.6	41.7	20.7
S7-M1-U0.0P2.0	93.4	70.6	43.9	21.0
S7-M2-U2.0P0.0	97.5	70.6	50.0	24.5
S7-M3-U2.0P0.0	108.9	70.6	53.5	26.1
S3-M1-U2.0P0.0	77.1	35.3	46.6	23.1
S5-M1-U2.0P0.0	77.1	51.3	50.1	24.5

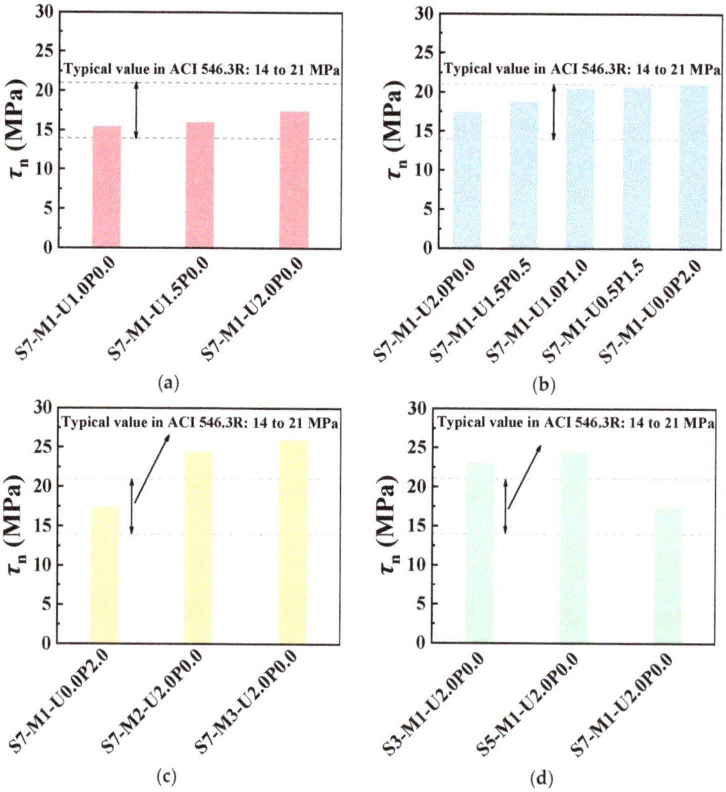

Figure 10. Interfacial shear bond strength between EGC and concrete substrate based on compressive slant shear test. Effects of (**a**) PE fiber content; (**b**) PE/PVA fiber hybrid ratio; (**c**) FA/GGBS; (**d**) concrete substrate strength grade.

As the PE fiber content increased, the shear bond strength increased as well. For instance, with PE fiber contents of 1% (S7-M1-U1.0P0.0), 1.5% (S7-M1-U1.5P0.0), and 2% (S7-M1-U2.0P0.0), the shear bond strengths were 15.5 MPa, 16.0 MPa, and 17.5 MPa, respectively, which were 10.7%, 14.3%, and 25.0% higher than the minimum standard of ACI 546.3R [44]. Additionally, when the PE fiber content of EGC was 2.0%, the shear bond strength increased by 3.2% and 12.9% compared with that with 1.0% and 1.5% PE fiber content, respectively. Similarly, increasing the PVA fiber hybrid ratio, GGBS ratio, and concrete substrate strength grade all exerted positive influence on the shear bond strength. However, for the C70 concrete substrate, the shear bond strength was lower than that of specimens with C30 and C50 concrete substrates.

4.4. Cost Analysis

Table 8 shows the costs of raw materials used in hybrid fiber EGCs, based on bulk commodity prices for laboratory-purchased raw materials. Note that the prices for laboratory-purchased raw materials are generally higher in terms of purchasing amount and shipping costs. Table 9 compares the hybrid EGC mix proportions studied here, focusing on key axial compression and tension performance parameters (compressive strength σ_c, tensile strength σ_{tu}, ultimate tensile strain ε_{tu}) as economic indicators. The impact of different PVA/PE ratios on the economic efficiency of EGCs with hybrid fibers was analyzed, with the price of oiled PVA fibers sourced from reference [45]. Cost comparisons with EGCs from other studies were also conducted [27,46,47]. However, shear bond strength data are not included in the reported comparison because variations in casting methods, interface roughness, and test methods significantly affected the bonding performance of the EGC–concrete interface.

Table 8. Cost of raw materials.

Materials	FA [1]	GGBS [2]	QP [3]	NaOH	Na_2SiO_3	BC [4]	Deformer	HRWR [5]	PE [6]	PP [7]	SF [8]	PVA [9]	PVA@O [10]
Cost (USD/kg)	0.27	0.34	0.06	2.80	0.42	3.36	3.08	2.66	56.00	1.89	1.68	6.30	56.00 [45]

[1] fly ash; [2] ground granulated blast furnace slag; [3] quartz powder; [4] barium chloride; [5] high-range water reducer; [6] polyethylene fibers; [7] polypropylene fiber; [8] steel fiber; [9] polyvinyl alcohol fibers; [10] oiled-coating polyvinyl alcohol fibers.

Table 9. Mix proportions of EGC with hybrid fiber.

Mix IDs	Mix Proportion								Volume Fraction/%			
	FA	GGBS	QP	NaOH	Na_2SiO_3	Water	BC	HRWR	PE	PP	SF	PVA@O
A1 [46]	0.50	0.50	0.00	0.00	0.12	0.45	0.00	0.00	1.50	0.00	0.50	0.00
A2 [46]	0.50	0.50	0.00	0.00	0.12	0.45	0.00	0.00	1.00	0.00	1.00	0.00
A3 [46]	0.50	0.50	0.00	0.00	0.12	0.45	0.00	0.00	0.50	0.00	1.50	0.00
B1 [47]	0.50	0.50	0.20	0.13	0.32	0.02	0.00	0.00	0.00	0.00	0.00	2.00
B2 [47]	0.50	0.50	0.20	0.13	0.32	0.02	0.00	0.00	0.00	1.00	0.00	1.00
B3 [47]	0.50	0.50	0.20	0.13	0.32	0.02	0.00	0.00	0.00	2.00	0.00	0.00
C1 [27]	0.60	0.40	0.23	0.04	0.27	0.08	0.00	0.00	2.00	0.00	0.00	0.00
C2 [27]	0.60	0.40	0.23	0.04	0.27	0.08	0.00	0.00	1.00	0.00	1.00	0.00
C3 [27]	0.60	0.40	0.23	0.04	0.27	0.08	0.00	0.00	0.00	0.00	2.00	0.00

Table 10 displays the mechanical performance of EGCs in this study and other existing research, along with the results of the cost analysis. Figure 11 contrasts these cost analyses, with the cost per cubic meter of M1-U2.0P0.0 set as the baseline (1.0). Note that A3, B3, and C3 were excluded from Figure 11 because of their low tensile capacity. Notably, using PP fibers, domestically produced unoiled PVA fibers, or SFs effectively reduced the cost, but a high substitution rate compromised the material's excellent tensile ductility, which is crucial for EGCs. From Table 10, it is evident that M1-U2.0P0.0, M2-U2.0P0.0, and M3-U2.0P0.0 incurred the highest costs. The influence of different FA/GGBS ratios on costs was minor but these ratios did affect mechanical performance. From a cost-effectiveness standpoint, increasing the PE fiber content was found to be economically inefficient in this study.

Table 10. Relevant mechanical properties and cost analysis of EGCs with hybrid fibers.

Mix IDs	Cost	σ_c (MPa)	ε_{tu} (%)	σ_{tu} (MPa)	$\frac{\sigma_c}{Cost}$	$\frac{\sigma_{tu}}{Cost}$	$\frac{\varepsilon_{tu}}{Cost}$
M1-U1.0P0.0	0.503	93.9	4.74	4.92	186.87	9.43	9.79
M1-U1.5P0.0	0.751	87.0	6.16	5.74	115.81	8.20	7.64
M1-U2.0P0.0	1.000	77.1	6.07	6.26	77.10	6.07	6.26
M1-U1.5P0.5	0.779	83.2	4.70	7.04	106.77	6.03	9.03
M1-U1.0P1.0	0.558	84.7	3.80	5.64	151.67	6.80	10.10
M1-U0.5P1.5	0.338	87.0	0.80	4.19	257.65	2.37	12.41
M1-U0.0P2.0	0.117	93.4	0.52	3.10	799.02	4.45	26.52
M2-U2.0P0.0	1.000	97.5	6.62	7.83	97.49	6.62	7.83
M3-U2.0P0.0	1.000	108.90	6.31	9.29	108.89	6.31	9.29
A1 [46]	0.757	62.00	3.18	2.27	81.92	4.20	3.00
A2 [46]	0.516	62.00	1.24	2.61	120.25	2.41	5.06
A3 [46]	0.274	62.00	0.38	2.97	226.05	1.39	10.83
B1 [47]	1.003	37.24	2.41	3.12	37.13	2.40	3.11
B2 [47]	0.522	36.43	2.44	3.07	69.77	4.67	5.88
B3 [47]	0.041	33.27	3.70	1.96	804.10	89.43	47.37
C1 [27]	1.000	96.63	5.54	6.19	96.65	5.54	6.19
C2 [27]	0.517	110.34	2.12	6.00	213.35	4.10	11.60
C3 [27]	0.035	111.82	0.54	6.26	3233.62	15.62	181.03

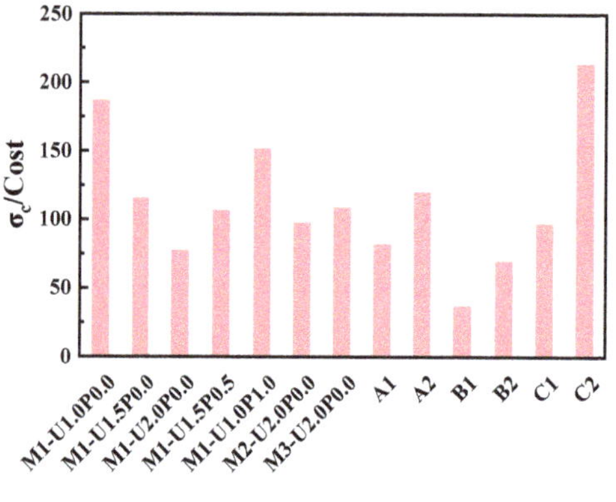

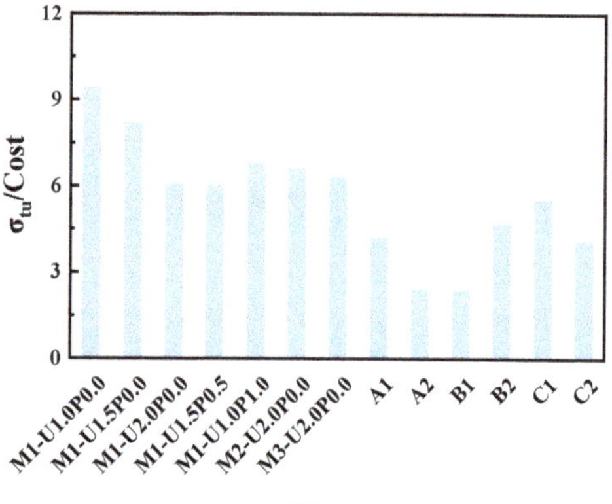

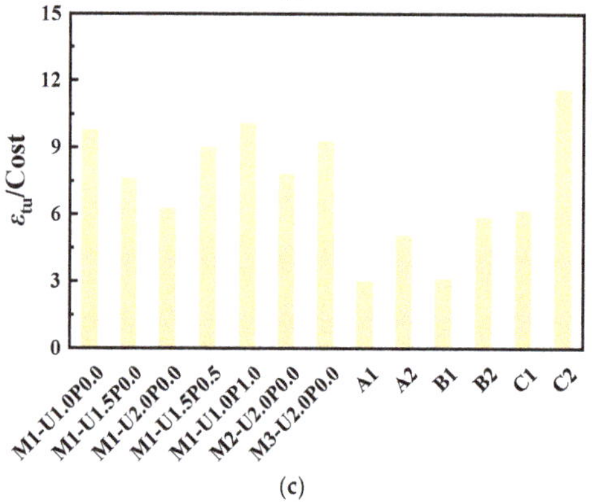

Figure 11. Cost analysis: (**a**) $\frac{\sigma_c}{Cost}$; (**b**) $\frac{\sigma_{tu}}{Cost}$; (**c**) $\frac{\varepsilon_{tu}}{Cost}$.

5. Discussion

5.1. Effects of Hybrid Fiber Ratios and FA/GGBS Ratios on Compressive Properties of EGCs

Single incorporation of PE fibers in EGCs was demonstrated to decrease the compressive properties, different from when PVA fibers were utilized to replace PE fibers. In effect, this finding aligned with Yun [48], who reported that uneven dispersion of high-aspect-ratio PE fibers in mortar weakened fiber–aggregate interfacial sections, resulting in premature compressive failure [48]. Additionally, hydrophobic PE fibers may increase entrapped air content and localized porosity during mixing, which could affect the compressive strength and elastic modulus under varying PE fiber contents. Regarding the FA/GGBS ratio, it was found that higher compressive properties could be acquired with lower FA/GGBS (i.e., higher GGBS content). This enhanced performance with increased GGBS content was attributed to the formation of different reaction products. With higher GGBS content, the elastic modulus of C-A-S-H gel ranged from 12 to 47 GPa [49,50], surpassing the N-A-S-H gel formed by FA activation (4.44–20 GPa) [51]. The presence of C-A-S-H gel significantly enhanced both compressive strength and elastic modulus, as reported in previous studies [52]. Still, it is recommended to be cautious when it comes to the selection of the hybrid fiber ratio and FA/GGBS ratio of EGCs, in that flowability and characteristic tensile behavior should be taken into account as well.

5.2. Effects of Hybrid Fiber Ratios and FA/GGBS Ratios on Tensile Properties of EGCs

Test results showed that EGCs reinforced with various fibers presented diverse tensile properties. Higher single PE fiber content and lower hybrid PVA/PE ratios and FA/GGBS ratios exerted positive influences on the tensile properties of EGC. Optimal hybrid fiber and FA/GGBS ratios are essential for maintaining excellent tensile properties and cost efficiency. The improvement in tensile properties was ascribed to changes in the fiber-bridging strength, fiber/matrix interface performance, and mechanical properties of the matrix under different conditions. Within a certain range, higher PE fiber content, whether hybridized with PVA fibers or not, enhanced fiber-bridging strength, thereby boosting the EGC's tensile performance. The domestically produced unoiled PVA fibers used in this study possessed mechanical properties inferior to PE fibers. Consequently, PVA fibers forming a reliable bond with the matrix (hydrophilic fibers) may experience premature fracture in high-strength matrices, providing lower fiber-bridging strength compared with PE fibers, thus negatively impacting the EGC's tensile performance. Moreover, increasing GGBS content results in higher production of C-A-S-H and C-S-H products, facilitating the formation of a dense matrix and improving fiber/matrix interface bonding, thereby enhancing tensile performance [53,54].

5.3. Effects of Hybrid Fiber Ratios and FA/GGBS Ratios on Bond Behavior

Since the bond behavior between the EGC and the concrete substrate was assessed through both direct tensile tests and slant shear tests, the variations in bonding behavior and the bonding hypothesis are separately discussed in this section. From the results of the direct tensile tests for bond behavior, it was observed that adopting cost-effective unoiled PVA fibers did not harm the bond behavior. Conversely, the tensile bond strength increased modestly while decreasing the cost of the repair material. In effect, with higher hybrid PVA/PE fiber ratios, the compressive strength of the EGC was gradually enhanced, which has been demonstrated to enhance bond strength [55].

As far as shear bond strength is concerned, results showed that higher PE content, PVA/PE ratio, and GGBS/FA ratio were helpful. According to existing studies, the resistance mechanisms of interface shear include interlocking, overriding, and fracturing [56]. Interlocking and overriding are activated by creating a rough interface. Since a smooth interface was utilized in this study, these two mechanisms can be disregarded while focusing primarily on resistance to interface fracture. Additionally, the dowel effect observed in concrete repair materials with steel fibers (SFs) is not applicable to the flexible fibers (PE and PVA fibers) used in this study [56,57]. Therefore, the increase in shear bond strength due to the PE fiber content and the PVA fiber hybrid ratio was attributed to the restriction of thermal and moisture damage by the flexible fibers, and the micro cracks' deflection and blunting. The former was attributed to repair of the cracks and delamination caused by restrained shrinkage, enhancing damage tolerance and promoting interface integrity and thus improving interface failure resistance. This phenomenon has been documented in existing studies.

Cristina Zanotti [58] observed that instability of the interface between plain concrete and fiber-reinforced concrete occurred after the failure of plain concrete. Apart from the contribution of interface roughness to crack propagation, this phenomenon was ascribed to the repair reinforcement of PVA fibers that facilitated stress redistribution around the main crack and the formation of chemical bonds between PVA fibers and the cement paste. Similarly, previous studies indicated that if the repair material had excellent fracture performance, the interface cracks could be prevented from propagating into the repair material, especially when the repair material was fiber-reinforced [59]. Since the excellent tensile ductility and fracture performance of EGC-like materials have been mentioned in existing studies [25,60–63], it can be inferred that the improved tensile ductility and fracture performance of the hybrid EGCs, due to the increased PE fiber content, PVA fiber hybrid ratio, and GGBS ratio, led to enhanced bond strength at the EGC–concrete interface, benefiting from the aforementioned mechanisms.

5.4. Cost Analysis

As mentioned above, the utilization of unoiled PVA fibers to replace commonly used oiled PVA and PE fibers can maintain the mechanical properties of EGC while saving costs to a greater or lesser degree. In practice, this method can be extended to other fiber types (like steel fiber, aramid fiber, PET fiber, etc.). Different fiber types will result in various fiber-bridging capacities due to their mechanical properties themselves and intrinsic hydrophobic or hydrophilic natures, which is of great importance to EGC based on its design principle [10,15,23,48]. In summary, the selection of fiber type should be cautious. It is necessary to comprehensively take the mechanical properties and cost of EGC into consideration.

6. Conclusions

This study developed hybrid PE/PVA fiber-reinforced EGCs using domestically produced PVA fibers without oil coating as a replacement for commonly used PE fibers. The research focused on investigating the bond performance between the EGC and the concrete through direct tensile tests and slant shear bond tests. Key aspects included exploring the influence of different fiber hybrid ratios, concrete substrate strength grades, and FA/GGBS ratios on interfacial bond strength. Additionally, the study compared and analyzed the cost of the EGCs prepared in this research with the selected published literature, discussing the feasibility of replacing PE fibers with domestically produced PVA fibers without oil coating. The conclusions drawn from this study can be summarized as follows:

(1) The compressive strength of EGC ranged from 77.1 to 108.9 MPa. Increasing the PE fiber content decreased the compressive strength, while substitution with PVA fibers led to improvements of up to 21.1%. The elastic modulus was influenced by the PE fiber content and FA/GGBS ratio rather than the fiber hybrid ratio, ranging from 16.0 to 20.3 GPa. When the FA/GGBS was 5:5, the compressive strength and elastic modulus of EGC increased by 16.6% and 24.5%, respectively, compared with when FA/GGBS was 7:3. This improvement was attributed to the increased formation of C-A-S-H gel, which positively affected axial compression performance.

(2) Most specimens exhibited strain-hardening behavior after initial crack formation. Increasing the PE fiber content improved the tensile strength and ultimate tensile strain of EGC. However, as the PVA/PE fiber ratio increased, the tensile performance gradually decreased, with the material completely losing its tensile performance characteristics at a PVA/PE ratio of 1:1. Increasing the GGBS content also contributed to improved tensile

performance by altering the fiber-bridging strength, fiber/matrix interface properties, and mechanical properties of the matrix. PVA fibers exhibited lower mechanical performance than PE fibers, potentially leading to premature fracture in high-strength matrices. Conversely, the C-A-S-H and C-S-H gel resulting from GGBS enhanced the matrix densification and fiber/matrix interface bonding, thereby improving tensile performance.

(3) Substituting PVA fibers for PE fibers slightly increased the direct tensile bond strength between EGC and concrete substrates, particularly at a PE/PVA ratio of 1.0, which increased bond strength by 8.5% compared with the inclusion of 2.0% PE fibers only. Slant shear test results showed that increasing the PE fiber content, PVA fiber hybrid ratio, GGBS ratio, and concrete substrate strength all enhanced the shear bond strength. The contributions of PE fiber content and PVA fiber hybrid ratio to the increase in shear bond strength were due to the restrictive effect of flexible fibers on thermal and moisture damage, as well as the enhancement of interface fracture resistance through microcrack deflection and blunting, thereby increasing the interface bond strength.

(4) The costs of M1-U2.0P0.0, M2-U2.0P0.0, and M3-U2.0P0.0 were the highest, with negligible effects of different FA/GGBS ratios on costs but significant impacts on mechanical performance. From a cost-effectiveness perspective, increasing the PE fiber content was not economical. When the PVA/PE ratio was 1.0, costs decreased by 44.2%, with good cost-effectiveness. However, when the PVA/PE ratio was higher than 1.0, tensile performance significantly decreased despite the decreasing costs. The cost-effectiveness of the EGC developed in this study was relatively superior under the same fiber substitution ratio conditions. Considering both axial compression performance, tensile performance, and cost, M1-U1.0P1.0 is recommended as the preferred repair material.

(5) EGCs with hybrid PE/PVA fiber have promising application prospects as a material for repairing concrete, particularly in enhancing economic efficiency while maintaining excellent mechanical performance. Optimizing the substitution ratio of PE/PVA fibers significantly reduced costs while preserving tensile performance and interface bond strength. Future research can explore the micro-mechanisms of different fiber hybrid ratios on mechanical performance to optimize fiber hybrid ratios and enhance interfacial bond strength. Additionally, studying the effects of interface roughness and casting methods on material performance will promote the practical application and dissemination of EGCs in engineering. Further experimental and simulation studies will also aid in understanding their long-term performance and durability, providing more reliable solutions for concrete repair.

Author Contributions: Conceptualization, Y.L.; methodology, Y.L.; validation, W.L.; formal analysis, X.Z. and Y.W.; investigation, Y.W. and C.L.; resources, W.L.; data curation, X.Z. and W.Z.; writing—original draft preparation, Y.L.; writing—review and editing, W.Z. and C.L.; visualization, C.W.; supervision, W.L.; funding acquisition, W.L. All authors have read and agreed to the published version of the manuscript.

Funding: This research was funded by Guangzhou Power Supply Bureau, Guangdong Power Grid Co., Ltd., China Southern Power Grid Co., Ltd., grant number 0301002024030301XG00030.

Data Availability Statement: The data presented in this study are available on request from the corresponding author. The data are not publicly available due to confidentiality issues.

Conflicts of Interest: Authors Yu Ling, Xiafei Zhang, Yanwei Wu, Weiyu Zou, and Chuang Wang were employed by the company Guangzhou Power Supply Bureau, Guangdong Power Grid Co., Ltd., China Southern Power Grid Co., Ltd. The remaining authors declare that the research was conducted in the absence of any commercial or financial relationships that could be construed as potential conflicts of interest. The authors declare that this study received funding from Guangzhou Power Supply Bureau, Guangdong Power Grid Co., Ltd., China Southern Power Grid Co., Ltd. The funders were not involved in the study design, collection, analysis, interpretation of data, the writing of this article, or the decision to submit it for publication.

References

1. Li, W.; Tang, S.; Huang, Z.; Yang, X.; Shi, T.; Xing, F. Shear behavior of concrete beam reinforced in shear with carbon fiber-reinforced polymer mesh fabric (CFRP-MF) configuration. *Eng. Struct.* **2020**, *218*, 110828. [CrossRef]
2. Wang, Z.; Wang, J.; Chen, K.; Zhang, J. Feasible region of post-tensioning force for precast segmental post-tensioned UHPC bridge columns. *Eng. Struct.* **2019**, *200*, 109685. [CrossRef]
3. Lin, J.-X.; Su, J.-Y.; Pan, H.-S.; Peng, Y.-Q.; Guo, Y.-C.; Chen, W.-S.; Sun, X.-L.; Yuan, B.-X.; Liu, G.-T.; Lan, X.-W. Dynamic compression behavior of ultra-high performance concrete with hybrid polyoxymethylene fiber and steel fiber. *J. Mater. Res. Technol.* **2022**, *20*, 4473–4486. [CrossRef]
4. Decky, M.; Papanova, Z.; Juhas, M.; Kudelcikova, M. Evaluation of the Effect of Average Annual Temperatures in Slovakia between 1971 and 2020 on Stresses in Rigid Pavements. *Land* **2022**, *11*, 764. [CrossRef]
5. Ali, A.H.; Mohamed, H.M.; Benmokrane, B. Bar size effect on long-term durability of sand-coated basalt-FRP composite bars. *Compos. Part B Eng.* **2020**, *195*, 108059. [CrossRef]
6. Sadrinejad, I.; Madandoust, R.; Ranjbar, M.M. The mechanical and durability properties of concrete containing hybrid synthetic fibers. *Constr. Build. Mater.* **2018**, *178*, 72–82. [CrossRef]
7. Wang, P.; Jiao, M.; Hu, C.; Tian, L.; Zhao, T.; Lei, D.; Fu, H. Research on Bonding and Shrinkage Properties of SHCC-Repaired Concrete Beams. *Materials* **2020**, *13*, 1757. [CrossRef] [PubMed]
8. Chen, Z.; Huang, L.; Huang, P.; Xie, J. Axial-Impact Resistance of CFRP-Confined Ultrahigh-Performance Concrete. *J. Compos. Constr.* **2022**, *26*, 04022059. [CrossRef]
9. Lin, J.-X.; Huang, P.-Y.; Guo, Y.-C.; Guo, X.-Y.; Zeng, J.-J.; Zhao, C.; Chen, Z.-B. Fatigue behavior of RC beams strengthened with CFRP laminate under hot-wet environments and vehicle random loads coupling. *Int. J. Fatigue* **2020**, *131*, 105329. [CrossRef]
10. Lin, J.-X.; Luo, R.-H.; Su, J.-Y.; Guo, Y.-C.; Chen, W.-S. Coarse synthetic fibers (PP and POM) as a replacement to steel fibers in UHPC: Tensile behavior, environmental and economic assessment. *Constr. Build. Mater.* **2024**, *412*, 134654. [CrossRef]
11. Ali, A.H.; Mohamed, H.M.; Benmokrane, B.; ElSafty, A.; Chaallal, O. Durability performance and long-term prediction models of sand-coated basalt FRP bars. *Compos. Part B Eng.* **2019**, *157*, 248–258. [CrossRef]
12. Su, J.-Y.; Chen, G.; Pan, H.-S.; Lin, J.-X.; Zhang, J.; Zhuo, K.-X.; Chen, Z.-B.; Guo, Y.-C. Rubber modified high strength-high ductility concrete: Effect of rubber replacement ratio and fiber length. *Constr. Build. Mater.* **2023**, *404*, 133243. [CrossRef]
13. Deng, M.; Yang, S. Experimental and numerical evaluation of confined masonry walls retrofitted with engineered cementitious composites. *Eng. Struct.* **2020**, *207*, 110249. [CrossRef]
14. Ding, Y.; Yu, J.-t.; Yu, K.-Q.; Xu, S.-l. Basic mechanical properties of ultra-high ductility cementitious composites: From 40 MPa to 120 MPa. *Compos. Struct.* **2018**, *185*, 634–645. [CrossRef]
15. Lin, J.-X.; Song, Y.; Xie, Z.-H.; Guo, Y.-C.; Yuan, B.; Zeng, J.-J.; Wei, X. Static and dynamic mechanical behavior of engineered cementitious composites with PP and PVA fibers. *J. Build. Eng.* **2020**, *29*, 101097. [CrossRef]
16. Yuan, F.; Chen, M.; Pan, J. Experimental study on seismic behaviours of hybrid FRP–steel-reinforced ECC–concrete composite columns. *Compos. Part B Eng.* **2019**, *176*, 107272. [CrossRef]
17. Tian, J.; Wu, X.; Zheng, Y.; Hu, S.; Du, Y.; Wang, W.; Sun, C.; Zhang, L. Investigation of interface shear properties and mechanical model between ECC and concrete. *Constr. Build. Mater.* **2019**, *223*, 12–27. [CrossRef]
18. Tian, J.; Wu, X.; Zheng, Y.; Hu, S.; Ren, W.; Du, Y.; Wang, W.; Sun, C.; Ma, J.; Ye, Y. Investigation of damage behaviors of ECC-to-concrete interface and damage prediction model under salt freeze-thaw cycles. *Constr. Build. Mater.* **2019**, *226*, 238–249. [CrossRef]
19. Wu, X.; Tian, J.; Ma, H.; Zheng, Y.; Hu, S.; Wang, W.; Du, Y.; Huang, W.; Sun, C.; Zhu, Z. Investigation on interface fracture properties and nonlinear fracture model between ECC and concrete subjected to salt freeze-thaw cycles. *Constr. Build. Mater.* **2020**, *259*, 119785. [CrossRef]
20. Yang, X.; Gao, W.-Y.; Dai, J.-G.; Lu, Z.-D. Shear strengthening of RC beams with FRP grid-reinforced ECC matrix. *Compos. Struct.* **2020**, *241*, 112120. [CrossRef]
21. Jindal, B.B. Investigations on the properties of geopolymer mortar and concrete with mineral admixtures: A review. *Constr. Build. Mater.* **2019**, *227*, 116644. [CrossRef]
22. Lao, J.-C.; Huang, B.-T.; Fang, Y.; Xu, L.-Y.; Dai, J.-G.; Shah, S.P. Strain-hardening alkali-activated fly ash/slag composites with ultra-high compressive strength and ultra-high tensile ductility. *Cem. Concr. Res.* **2023**, *165*, 107075. [CrossRef]
23. Lin, J.-X.; Chen, G.; Pan, H.-S.; Wang, Y.-C.; Guo, Y.-C.; Jiang, Z.-X. Analysis of Stress-Strain Behavior in Engineered Geopolymer Composites Reinforced with Hybrid PE-PP Fibers: A Focus on Cracking Characteristics. *Compos. Struct.* **2023**, *323*, 117437. [CrossRef]
24. Pan, H.; Xie, Z.; Chen, G.; Su, J.; Zhuo, K.; Chen, Z.; Lin, J.; Feng, C.; Guo, Y. Dynamic compressive behavior of high-strength engineered geopolymer composites. *J. Build. Eng.* **2023**, *80*, 108036. [CrossRef]
25. Peng, Y.-Q.; Zheng, D.-P.; Pan, H.-S.; Yang, J.-L.; Lin, J.-X.; Lai, H.-M.; Wu, P.-Z.; Zhu, H.-Y. Strain hardening geopolymer composites with hybrid POM and UHMWPE fibers: Analysis of static mechanical properties, economic benefits, and environmental impact. *J. Build. Eng.* **2023**, *76*, 107315. [CrossRef]
26. Korniejenko, K.; Lin, W.-T.; Šimonová, H. Mechanical Properties of Short Polymer Fiber-Reinforced Geopolymer Composites. *J. Compos. Sci.* **2020**, *4*, 128. [CrossRef]

27. Zhitong, C. Experimental Study on Fracture Properties of Steel-PE Fiber High Ductile Geopolymer Composites. Master's Thesis, Guangdong University of Technology, Guangzhou, China, 2023.
28. Zhang, Z.; Zhang, Q. Matrix tailoring of Engineered Cementitious Composites (ECC) with non-oil-coated, low tensile strength PVA fiber. *Constr. Build. Mater.* **2018**, *161*, 420–431. [CrossRef]
29. Yang, Z.-M.; Pan, H.-S.; Jiang, Z.-X.; Lv, J.-H.; Ruan, G.-W.; Lai, H.-M.; Lin, J.-X. Pseudo strain-hardening alkali-activated composites with up to 100% rubber aggregate: Static mechanical properties analysis and constitutive model development. *Constr. Build. Mater.* **2024**, *439*, 137338. [CrossRef]
30. Li, V.C.; Wang, S.X.; Wu, C. Tensile strain-hardening behavior of polyvinyl alcohol engineered cementitious composite (PVA-ECC). *ACI Mater. J.* **2001**, *98*, 483–492.
31. Zhu, M.; Chen, B.; Wu, M.; Han, J. Effects of different mixing ratio parameters on mechanical properties of cost-effective green engineered cementitious composites (ECC). *Constr. Build. Mater.* **2022**, *328*, 127093. [CrossRef]
32. Wang, F.; Ma, J.; Ding, Y.; Yu, J.; Yu, K. Engineered geopolymer composite (EGC) with ultra-low fiber content of 0.2%. *Constr. Build. Mater.* **2024**, *411*, 134626. [CrossRef]
33. Wang, B.; Xu, S.; Liu, F. Evaluation of tensile bonding strength between UHTCC repair materials and concrete substrate. *Constr. Build. Mater.* **2016**, *112*, 595–606. [CrossRef]
34. Santos, P.M.D.; Júlio, E.N.B.S.; Silva, V.t.D. Correlation between concrete-to-concrete bond strength and the roughness of the substrate surface. *Constr. Build. Mater.* **2007**, *21*, 1688–1695. [CrossRef]
35. Zhang, X.; Zhang, W.; Luo, Y.; Wang, L.; Peng, J.; Zhang, J. Interface Shear Strength between Self-Compacting Concrete and Carbonated Concrete. *J. Mater. Civ. Eng.* **2020**, *32*, 04020113. [CrossRef]
36. Kumar, S.; Sekhar Das, C.; Lao, J.; Alrefaei, Y.; Dai, J.-G. Effect of sand content on bond performance of engineered geopolymer composites (EGC) repair material. *Constr. Build. Mater.* **2022**, *328*, 127080. [CrossRef]
37. *GB/T 51003-2014*; Technical Code for Application of Mineral Admixture. Standardization Administration of China: Beijing, China, 2014.
38. *GB/T 50081-2019*; Standard for Test Methods of Concrete Physical and Mechanical Properties. Standardization Administration of China: Beijing, China, 2019.
39. *ASTM C469/C469M-22*; Standard Test Method for Static Modulus of Elasticity and Poisson's Ratio of Concrete in Compression. ASTM International: West Conshohocken, PA, USA, 2022.
40. Yokota, H.; Rokugo, K.; Sakata, N. *JSCE Recommendations for Design and Construction of High Performance Fiber Reinforced Cement Composite with Multiple Fine Cracks*; Japan Society of Civil Engineers: Tokyo, Japan, 2007.
41. *T/CBMF37-2018*; Fundamental Characteristics and Test Methods of Ultra-High Performance Concrete. China Building Materials Federation: Beijing, China, 2018.
42. *ASTM C882/C882M-20*; Standard Test Method for Bond Strength of Epoxy-Resin Systems Used with Concrete by Slant Shear. ASTM International: West Conshohocken, PA, USA, 2020.
43. Daneshvar, D.; Deix, K.; Robisson, A. Effect of casting and curing temperature on the interfacial bond strength of epoxy bonded concretes. *Constr. Build. Mater.* **2021**, *307*, 124328. [CrossRef]
44. *ACI 546.3R-14*; Guide to Materials Selection for Concrete Repair; ACI: Farmington Hills, MI, USA, 2014.
45. Wang, H.; Xu, J.; Song, Y.; Hou, M.; Xu, Y. Development of low-cost engineered cementitious composites using Yellow River silt and unoiled PVA fiber. *Constr. Build. Mater.* **2024**, *425*, 136063. [CrossRef]
46. Alrefaei, Y.; Dai, J.-G. Tensile behavior and microstructure of hybrid fiber ambient cured one-part engineered geopolymer composites. *Constr. Build. Mater.* **2018**, *184*, 419–431. [CrossRef]
47. Humur, G.; Çevik, A. Effects of hybrid fibers and nanosilica on mechanical and durability properties of lightweight engineered geopolymer composites subjected to cyclic loading and heating–cooling cycles. *Constr. Build. Mater.* **2022**, *326*, 126846. [CrossRef]
48. Yun, H.-D. Effect of accelerated freeze–thaw cycling on mechanical properties of hybrid PVA and PE fiber-reinforced strain-hardening cement-based composites (SHCCs). *Compos. Part B Eng.* **2013**, *52*, 11–20. [CrossRef]
49. Puertas, F.; Palacios, M.; Manzano, H.; Dolado, J.S.; Rico, A.; Rodríguez, J. A model for the C-A-S-H gel formed in alkali-activated slag cements. *J. Eur. Ceram. Soc.* **2011**, *31*, 2043–2056. [CrossRef]
50. Thomas, R.J.; Gebregziabiher, B.S.; Giffin, A.; Peethamparan, S. Micromechanical properties of alkali-activated slag cement binders. *Cem. Concr. Compos.* **2018**, *90*, 241–256. [CrossRef]
51. Němeček, J.; Šmilauer, V.; Kopecký, L. Nanoindentation characteristics of alkali-activated aluminosilicate materials. *Cem. Concr. Compos.* **2011**, *33*, 163–170. [CrossRef]
52. Chen, G.; Zheng, D.-P.; Chen, Y.-W.; Lin, J.-X.; Lao, W.-J.; Guo, Y.-C.; Chen, Z.-B.; Lan, X.-W. Development of high performance geopolymer concrete with waste rubber and recycle steel fiber: A study on compressive behavior, carbon emissions and economical performance. *Constr. Build. Mater.* **2023**, *393*, 131988. [CrossRef]
53. Zhong, H.; Zhang, M. Engineered geopolymer composites: A state-of-the-art review. *Cem. Concr. Compos.* **2023**, *135*, 104850. [CrossRef]
54. Meng, F.; Luo, S.; Sun, J.; Zhang, C.; Xu, L.; Du, Y.; Zeng, J.; Guo, Y. Compressive and Tensile Behavior of High-Ductility Alkali-Activated Composites with Polyethylene Terephthalate Powder. *Buildings* **2024**, *14*, 1399. [CrossRef]
55. Diab, A.M.; Abd Elmoaty, A.E.M.; Tag Eldin, M.R. Slant shear bond strength between self compacting concrete and old concrete. *Constr. Build. Mater.* **2017**, *130*, 73–82. [CrossRef]

56. Zanotti, C.; Rostagno, G.; Tingley, B. Further evidence of interfacial adhesive bond strength enhancement through fiber reinforcement in repairs. *Constr. Build. Mater.* **2018**, *160*, 775–785. [CrossRef]
57. Gao, D.; Chen, X.; Chen, G.; Zhang, L.; Zhan, Z. Shear-bond behaviour between concrete and hybrid fibre-reinforced cementitious composites for repairing: Experimental and modelling. *J. Build. Eng.* **2023**, *64*, 105636. [CrossRef]
58. Cristina Zanotti, N.B. Towards sustainable repairs: Substrate-repair interface mode-I fracture analysis. *Int. J. Sustain. Mater. Struct. Syst.* **2014**, *1*, 265–281. [CrossRef]
59. Lim, Y.M.; Li, V.C. Durable repair of aged infrastructures using trapping mechanism of engineered cementitious composites. *Cem. Concr. Compos.* **1997**, *19*, 373–385. [CrossRef]
60. Su, J.-Y.; Luo, R.-H.; Chen, Z.-B.; Lin, J.-X.; Huang, P.-Y.; Guo, Y.-C. Experimental study on the fracture performance of rubberized high strength-high ductility concrete with J-integral method. *Constr. Build. Mater.* **2024**, *421*, 135668. [CrossRef]
61. Lin, J.-X.; Liu, R.-A.; Liu, L.-Y.; Zhuo, K.-Y.; Chen, Z.-B.; Guo, Y.-C. High-strength and high-toughness alkali-activated composite materials: Optimizing mechanical properties through synergistic utilization of steel slag, ground granulated blast furnace slag, and fly ash. *Constr. Build. Mater.* **2024**, *422*, 135811. [CrossRef]
62. Xu, S.; Lyu, Y.; Xu, S.; Li, Q. Enhancing the initial cracking fracture toughness of steel-polyvinyl alcohol hybrid fibers ultra high toughness cementitious composites by incorporating multi-walled carbon nanotubes. *Constr. Build. Mater.* **2019**, *195*, 269–282. [CrossRef]
63. Gao, S.; Qi, L.; Zhu, Y.; Wang, W. Effect of notch depth ratio on mode I and mixed mode I-II fracture properties of engineered cementitious composites (ECC). *Int. J. Solids Struct.* **2022**, *236–237*, 111363. [CrossRef]

materials

MDPI

Article

Study on High-Ductility Geopolymer Concrete: The Influence of Oven Heat Curing Conditions on Mechanical Properties and Microstructural Development

Ruihao Luo [1], Runan Liu [1], Guang Qin [2], Minyang Jiang [3], Yixian Wu [4] and Yongchang Guo [1,*]

[1] School of Civil and Transportation Engineering, Guangdong University of Technology, Guangzhou 510006, China; luoryann@outlook.com (R.L.); 2112209125@mail2.gdut.edu.cn (R.L.)
[2] Foshan Road and Bridge Supervision Station Corporation Limited, Foshan 528313, China; qinguang1984@163.com
[3] Fujian Xingyuan Construction Engineering Development Corporation Limited, Fuzhou 350001, China; fjxyjsgc@fjxyjsgc.cn
[4] Fujian Chengshuo Construction Engineering Corporation Limited, Fuzhou 350800, China; rang1128@126.com
* Correspondence: guoyc@gdut.edu.cn

Abstract: Low carbon and high performance have become key trends in the development of construction materials. Understanding the mechanism by which curing conditions affect the mechanical properties of high-ductility geopolymer concrete (HDGC) is of significant importance. This study investigated three sealing curing temperatures (room temperature, 45 °C, and 60 °C) and four curing durations (1 day, 3 days, 5 days, and 7 days), while considering two final curing ages (7 days and 28 days) to explore their effects on the axial tensile and compressive properties of HDGC. The results showed that both 45 °C and 60 °C could improve the brittle failure of HDGC under axial compressive loading. However, curing at 60 °C and for durations longer than 1 day in an oven would catalyze the formation of eight-faced zeolite crystals within the slag–fly ash geopolymer matrix, and it could weaken the matrix's pore structure and subsequently affect the material's later strength development. Nevertheless, oven heat curing enhanced the bridging effect between the fibers and the matrix, partially compensating for the reduction in the initial tensile strength of HDGC. This follows the pseudo-strain-hardening material's saturation cracking criterion to enhance the strain-hardening behavior of HDGC and improve its tensile energy absorption capacity. A curing condition of 45 °C for 5 days is recommended to maximize the ductility of HDGC. This study provides important theoretical support for the design and promotion of green, low-carbon, high-ductility composite materials.

Keywords: high-ductility geopolymer concrete; oven heat curing; curing duration; mechanical properties; bridging effect

1. Introduction

In the modern construction industry, the development of high-performance materials and the overall reduction in carbon emissions have become hotspots. Traditional concrete materials are gradually revealing their limitations in durability, strength, and environmental friendliness [1–4]. Although Engineered Cementitious Composites (ECCs) are considered highly promising due to their high ductility and toughness, their high demand for cement does not align with the construction industry's pursuit of low-carbon and environmentally friendly development [5–11]. Geopolymer alkali-activated materials, typically produced by activating aluminosilicates with alkaline activators through a geopolymerization process that involves dissolution, rearrangement, condensation, and solidification [12–15], are now considered excellent alternatives to Ordinary Portland Cement (OPC). The use of geopolymer alkali-activated materials in concrete is associated with a lower carbon footprint, excellent thermal resistance, and durability [1,13,14,16]. Ohno et al. designed Engineered Geopolymer Composites (EGCs) based on the ECC design methodology, resulting in the creation of environmentally friendly high-ductility composite materials [17].

Existing research indicates that the mechanical properties of EGCs are influenced by various variables, including the type of precursor [18], particle size distribution of the materials [19], alkalinity and modulus of the alkali activator [20], water-to-binder ratio [21], and curing conditions [22–24]. Among these, curing conditions are one of the most critical factors in designing high-performance geopolymer composites, as they significantly affect the development of the geopolymer matrix and consequently impact the mechanical properties. In addition to standard water curing as per ASTM-C511 [25], environmental/room temperature curing (25 °C), oven heat curing, steam curing [26], and intermittent/cyclic heat curing [27] have been identified as catalytic methods that can influence the formation of geopolymer binders. Alexandre et al. [28] found that when the binder products are aluminosilicates, curing temperature and age significantly impact the material's compressive strength. Additionally, the use of fibers and mineral materials is believed to reduce common microcracks in alkali-activated binder matrices. Mo et al. [29] pointed out that elevated curing temperatures could promote the dissolution of geopolymer precursor particles and accelerate the polymerization process. Prolonged curing times favor the early formation of a hardened structure in the matrix.

Changing the curing temperature is widely studied due to its ease of implementation and significant effects. Commonly used geopolymer precursors are typically cured at temperatures ranging from 40 °C to 80 °C for 4–48 h [30–33]. Al Bakri et al. [34] reported that oven heat curing accelerates the development of compressive strength in fly ash-based geopolymers. Similarly, Singh et al. [24] discussed the impact of curing temperature on the mechanical properties of fly ash-based geopolymers, noting that higher temperatures increase the dissolution rate of the glass phase in fly ash, allowing materials cured at 85 °C to achieve the same compressive strength in a shorter period as those cured at 40 °C. The optimal curing temperature for enhancing mechanical properties is not absolute and varies between different precursor compositions. Manvendra Verma [35] found that slag–fly ash-based geopolymer concrete significantly improved in compressive strength after being oven-cured at 80 °C for 24 h, with further gradual enhancement after 28 days compared to ambient-cured samples. Kürklü [36] reported that 60 °C oven curing yielded the highest compressive strength for slag-based geopolymers. Türker et al. [37] evaluated the effects of two curing regimes (ambient curing at 21 ± 2 °C and 60% humidity, and high-temperature curing at 60 °C for 6 h) on the high-temperature performance of alkali-activated mortars. The results indicated that high-temperature curing resulted in a notably high compressive strength, though the flexural strength significantly decreased. Additionally, compared to ambient curing and OPC mortars, high-temperature curing reduced mass loss under high-temperature conditions.

Oven heat curing duration also significantly impacts the mechanical properties of geopolymers. Existing studies often set the heat curing duration within 48 h, with the optimal curing duration defined as 6–24 h [26,36,38–47]. Guo et al. [48] found that the compressive strength of fly ash-based geopolymers cured at 75 °C began to decrease after 8 h due to matrix structure damage and dehydration shrinkage, attributing the decline primarily to prolonged curing. Rovnanik et al. [49] analyzed the mechanism by which prolonged curing led to pore accumulation, forming macropores in the matrix structure and negatively affecting the material's mechanical properties. While extended heat curing has evident adverse effects on fly ash-based geopolymers, this also depends on the precursor composition. The most favorable curing conditions for slag–alkali-activated products are considered to be within a curing temperature range of 30 °C to 60 °C and a curing duration of 2 h to 14 days [50–53].

The superior tensile deformation capacity and tensile strength of EGC compared to OPC are largely attributed to the effective bridging performance between the fibers and the matrix, which is closely linked to the microstructure of the EGC matrix. Lv [54] and Idawati [55] observed similar phenomena, noting that some of the Ca released from slag dissolution is incorporated into N-A-S-H, forming N-(C)-A-S-H-type gels. Prolonged heat curing promotes the formation of more cross-linked binding products and a denser microstructure. Puertas [56] found that at 25 °C ambient curing, the alkali activation of slag was sufficient, with only a partial dissolution of fly ash participating in the reaction, resulting in the absence of "geopolymer gel" in the matrix. Fly ash requires temperatures between 40 °C and 85 °C to accelerate the reaction. S.K. Nath [57] compared the effects of curing conditions and alkali solution concentration on the microstructure and morphological evolution of alkali-activated fly ash, pointed that with increased alkali concentration and curing temperature, the morphology of geopolymers changed and formed fibrous and plate-like structures. The prevailing view is that a dense microstructure in the matrix enhances the fiber bridging capability, which is beneficial for improving deformation capacity and energy absorption [58].

Existing research generally suggests that the optimal oven heat curing conditions for EGC are primarily based on single-component precursors, with a recommendation of approximately 60 °C and a curing duration not exceeding 48 h. However, the effects of more commonly used two-component precursors and longer oven curing times on the performance of these materials have rarely been studied. Moreover, the fiber–matrix interface performance profoundly affects the ductility of EGC, and the mechanism of how curing conditions influence this microscopic behavior has not yet been fully explored. To further determine the optimal curing conditions for the mechanical properties of slag–fly ash-based high-ductility geopolymer concrete (HDGC), this study set three sealed curing temperatures (room temperature, 45 °C, and 60 °C) and four curing durations (1 day, 3 days, 5 days, and 7 days), while also considering two total curing ages (7 days and 28 days). This study explores the effects of curing conditions on the axial tensile and compressive mechanical properties of HDGC. Additionally, by examining the impact of curing temperatures on the microstructure of HDGC and the matrix–fiber interface, this research further investigates the fundamental mechanisms by which these curing conditions regulate the mechanical properties. Through this study, we aim to provide theoretical support for the design of low-carbon, high-performance concrete in terms of curing conditions, thus promoting innovation and development in sustainable building materials.

2. Materials and Methods

2.1. Materials

The composition of HDGC studied in this paper includes precursors, aggregates, alkali activators, water, fibers, and admixtures. The precursors include Class F fly ash [59] (FA), Grade S105 ground granulated blast furnace slag (GGBS) [60], and silica fume (SF), all sourced from Longze Water Purification Materials Co., Ltd., in Gongyi City, China. The aggregates consist of quartz sand of two particle sizes: medium sand (MS, 0.15~0.40 mm) and fine sand (FS, 0.076~0.150 mm). The alkali activators are composed of a mixture of sodium hydroxide solution and sodium silicate solution, obtained from Xilong Scientific Co., Ltd., (Shantou, China) and Yourui Refractories Co., Ltd., (Jiaxing, China) in Jiashan County, China, respectively. The fibers used in HDGC are ultra-high molecular weight polyethylene fibers (UHMWPE fibers, hereafter referred to as PE fibers), sourced from Sovet Special Threads Co., Ltd., in Dongguan, China. The admixture is barium chloride dihydrate (abbreviated as $BaCl_2$), used as a retarder to delay the initial setting time of HDGC, at a dosage of 1% by mass of the precursors, sourced from Xilong Scientific Co., Ltd. The SEM images of the precursor materials are shown in Figure 1. The particle size distribution curves of the precursor materials and the two aggregates are shown in Figure 2. From Figure 2, it can be seen that the particle size distribution of the components in HDGC is relatively uniform in this study. The physical and chemical parameters of all materials are listed in Tables 1 and 2.

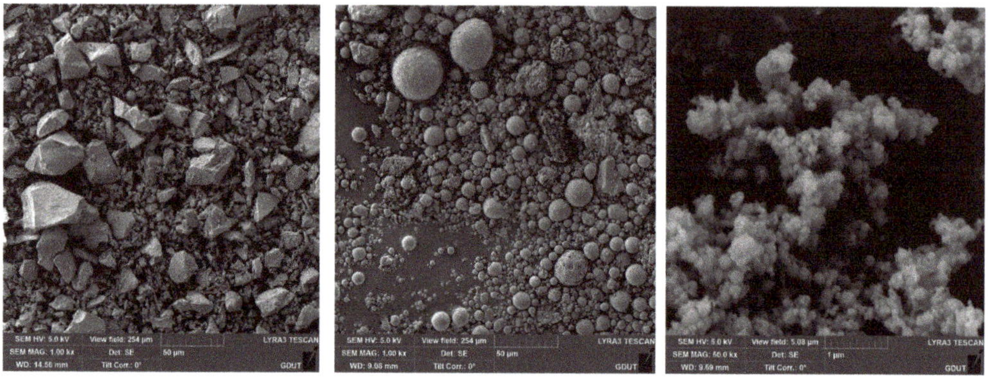

(a) (b) (c)

Figure 1. SEM of precursor materials: (**a**) GGBS, (**b**) FA, and (**c**) SF.

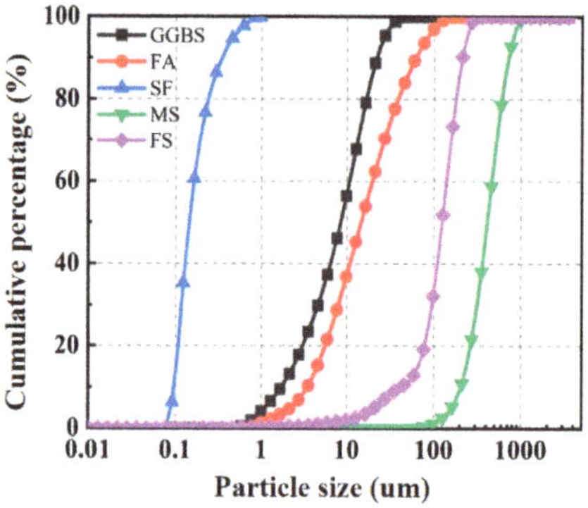

Figure 2. Cumulative particle size distribution curve.

Table 1. Property parameters of precursor materials.

Property Parameters.	CaO	SiO_2	Al_2O_3	SO_3	Fe_2O_3	MgO	TiO_2	Other	Loss on Ignition (%)	Density (g/cm^3)
	wt %									
GGBS	34.0	34.5	17.7	1.64	1.03	6.01	/	5.12	0.840	2.90
FA	4.01	54.8	31.2	2.20	4.16	1.01	1.13	2.37	4.60	2.30
SF	/	94.7	/	0.20	/	/	/	5.10	3.00	0.25

Table 2. Physical and mechanical properties of PE fibers.

Diameter d_f (μm)	Length L_f (mm)	Strength (MPa)	Elastic Modulus (GPa)	Elongation (%)	Density (g/cm^3)	Melting Temperature (°C)
20.0	6/12/18	>3000	>110	3.5–4	0.970	135–145

The HDGC mix used in this study is detailed in Table 3. The alkali activator consists of a 14 mol/L sodium hydroxide solution and a sodium silicate solution with a modulus of 2.25 (i.e., Na_2O_2 = 2.25), which are mixed to prepare a solution with a modulus of 1.5. Specifically, solid sodium hydroxide is first used to prepare a 14 mol/L sodium hydroxide solution, which is allowed to cool for 24 h. Once the solution has cooled to room temperature, it is mixed with the sodium silicate solution, stirred thoroughly, and allowed to stand until the temperature of the mixed solution returns to room temperature, ready for use.

Table 3. Mix ratio of HDGC (unit: kg/m^3).

GGBS	FA	SF	MS	FS	NaOH Solution	Sodium Silicate Solution	Extra Water	PE Fiber	$BaCl_2$
924	280	76.0	68.0	385	105	439	78.8	21.9	12.7

The specific process for preparing HDGC in this study is illustrated in Figure 3. First, the precursors, aggregates, and retarder are added to a planetary mixer and mixed for 4 min to ensure the dry materials are uniformly blended. In the second step, the cooled alkali activator is mixed with the required additional water and then added to the mixer, which is stirred for 3 min to complete the preparation of the HDGC matrix. In the final step, the pre-prepared 2 vol% PE fibers are evenly sprinkled into the mixer within 3 min while the mixer continues to operate. After the fibers are added, the mixer performs an additional 2 min of mixing to fully utilize the slurry's lateral shear force to disperse the PE fibers, ensuring their even distribution throughout the HDGC matrix. Previous studies have shown that the uniformity of fiber distribution significantly affects both the workability and mechanical properties of fiber-reinforced concrete [61].

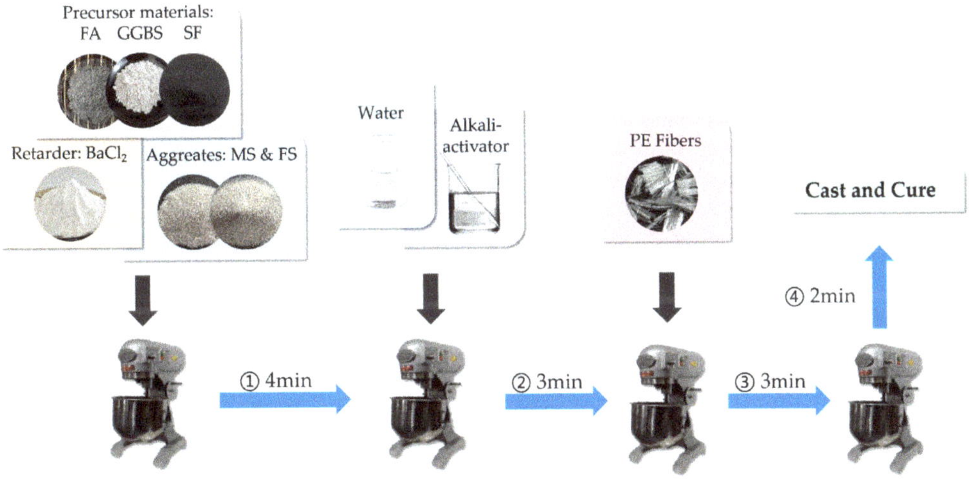

Figure 3. HDGC preparation procedure.

2.2. Curing Methods and Curing Time Design

The different curing conditions and durations for the specimens in this study constitute the primary focus of the research. The curing conditions are divided into four types: sealed indoor curing at room temperature (D-Y, room temperature 20 ± 3 °C), curing in a drying oven at 45 °C (T45-X-Y), and curing in a drying oven at 60 °C (T60-X-Y). Under the two different elevated temperature curing conditions, four curing durations of 1 day, 3 days, 5 days, and 7 days were designed, recorded as X = 1/3/5/7, to further investigate the impact of different temperatures and curing durations on the mechanical properties of HDGC. Additionally, specimens cured in the heated environment were removed from the drying oven after the corresponding curing time X and placed indoors to continue curing until 7 days and 28 days, recorded as Y = 7/28. In summary, the numbering of all experimental groups and their corresponding curing environments and durations in this study are shown in Figure 4.

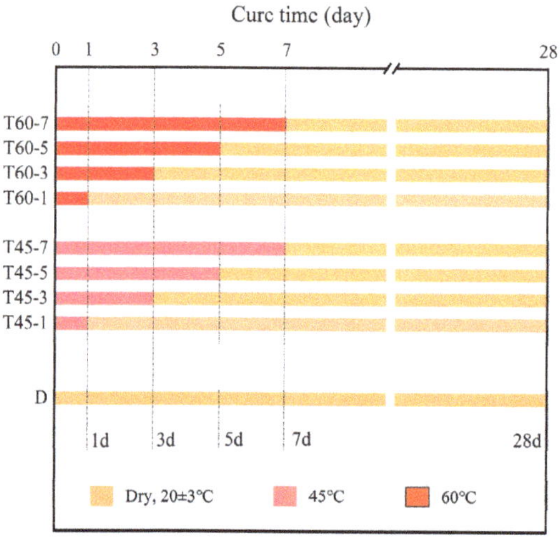

Figure 4. Test group setup.

2.3. Test Setup

2.3.1. Compressive Test

The axial compression tests in this study were conducted following the ASTM-C469/C469M standards [62]. The tests utilized cylindrical specimens with a diameter of 50 mm and a height of 100 mm. Four symmetrically distributed strain gauges, each 30 mm in length, were attached around the middle of the specimens to measure axial and circumferential strains. Additionally, a pair of LVDTs (Linear Variable Differential Transformers) was installed in the axial position to measure the axial strain of the specimens. The tests were performed using a displacement control mode at a rate of 0.2 mm/min. The specific test setup and device configuration are illustrated in Figure 5.

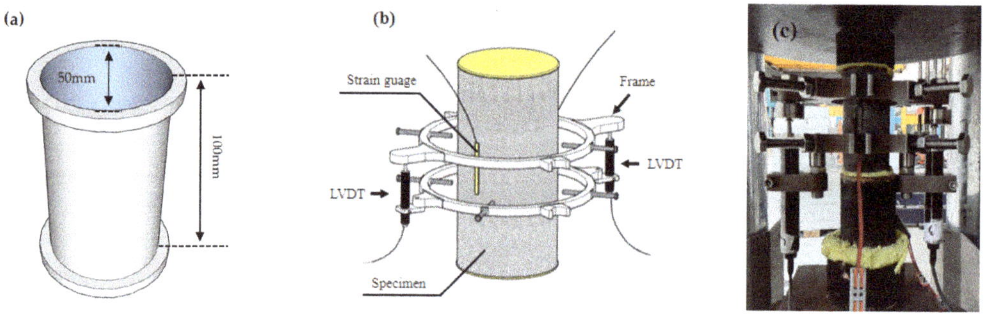

Figure 5. Compressive test setup: (**a**) the size of the mold, (**b**) schematic diagram of the test, (**c**) schematic diagram of the actual test.

2.3.2. Tensile Test

In this study, HDGC dumbbell-shaped specimens were prepared according to JC/T2461-2018 [63]. The specimens were loaded in a tensile testing machine at a displacement rate of 0.5 mm/min. A pair of LVDTs were placed in the effective gauge length area in the middle of the dumbbell-shaped specimens to measure axial deformation. The specific specimen dimensions and test setup are shown in Figure 6.

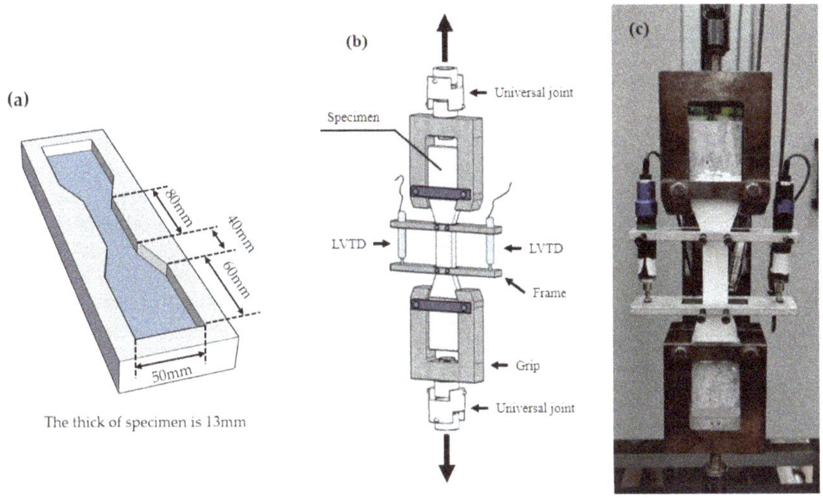

Figure 6. Tensile test setup: (**a**) the size of the mold, (**b**) schematic diagram of the test, (**c**) schematic diagram of the actual test.

2.3.3. SEM and EDS Analysis

The scanning electron microscopy and energy-dispersive X-ray spectroscopy test were conducted using Pharos G2 scanning electron microscope, produced by PhenomScientific (Shanghai, China).

3. Results and Discussion

3.1. Compressive Test

3.1.1. Stress–Strain Relationship of Uniaxial Compressive Test

Figure 7a presents the axial compressive stress–strain curves of HDGC under different curing conditions after a 28-day curing age. The curves generally exhibit three stages: an elastic stage where stress increases steadily, a crack propagation stage where stress decreases post-cracking, and a failure stage where the slope of the stress–strain curve approaches zero. Notably, under the 45 °C curing condition, the longer the curing duration, the steeper and longer the elastic stage of the HDGC curve. In contrast, the results for the 60 °C curing condition showed the opposite trend; the elastic stage seemed to become gradually shorter. Additionally, the stress–strain curves of the specimens cured at room temperature (groups D) displayed a vertical line in the second stage, indicating that the specimens transition directly to the failure stage after cracking. However, the T45 and T60 series exhibited a descending branch post-cracking, known as the ductile failure descending section of concrete [2], with T45-5-28 being particularly prominent. Interestingly, Figure 7 reveals that the axial compressive strength of the T60 series is significantly lower than that of the specimens cured at room temperature, especially for T60-3-28.

Considering that T45-5-28 showed the high compressive strength and a relatively excellent ductility, T60-3-28 showed the opposite, and the compressive strength was even lower than that of D-28. Therefore, Figure 7b compares the axial compressive stress–strain curves of T45-5-7 and T60-3-7. The stress–strain curve of T45-5-28 appears to be the stress–strain curve following the development of compressive strength in T45-5-7. However, the stress–strain development of T60-3-28 and T60-3-7 differ significantly, with the former exhibiting a substantial strength reduction and the latter showing a brittle failure descending section post-cracking.

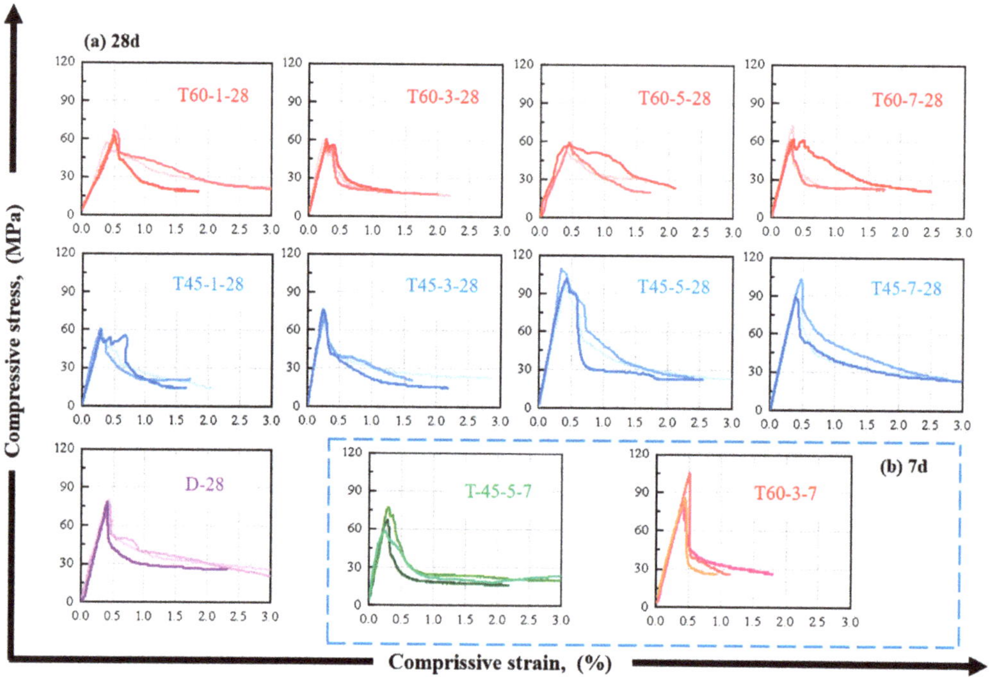

Figure 7. Compressive stress–strain curve of HDGC: (**a**) after aging to 28 d, (**b**) after aging to 7 d.

3.1.2. Compressive Properties Parameters

Figure 8 illustrates the compressive strength of HDGC under different curing conditions, and Table 4 summarizes the compressive performance data of HDGC for each condition. In Figure 8a, it could be observed that only T45-5-28 and T45-7-28 exhibited a higher axial compressive strength than D-28, which showed a value of 78.5 MPa. This suggests that a curing temperature of 60 °C and curing times exceeding 3 days are not conducive to the 28-day strength development of the material. The highest compressive strength of 103 MPa was observed in T45-5-28, while the lowest compressive strength of 60.1 MPa was found in T60-3-28.

Table 4. Compressive properties parameters under different curing conditions.

Specimen	Compressive Strength (MPa)	Peak Compressive Strain (%)	Elastic Modulus (GPa)
D-7	49.9 (0.9)	0.005 (0.001)	14.4 (0.9)
D-28	78.5 (1.5)	0.004 (0.002)	21.5 (1.8)
T45-1-28	60.4 (0.5)	0.003 (0.001)	22.9 (2.7)
T45-3-28	75.9 (1.3)	0.002 (0.001)	34.1 (0.7)
T45-5-28	104 (2.9)	0.004 (0.003)	28.1 (1.4)
T45-7-28	97.8 (4.3)	0.004 (0.001)	25.0 (0.5)
T60-1-28	62.5 (2.8)	0.005 (0.002)	13.1 (0.7)
T60-3-28	60.2 (0.8)	0.003 (0.001)	27.8 (3.4)
T60-5-28	65.2 (3.8)	0.003 (0.002)	22.3 (0.6)
T60-7-28	56.4 (2.4)	0.004 (0.001)	15.3 (1.2)
T45-5-7	67.7 (5.3)	0.003 (0.001)	34.3 (5.5)
T60-3-7	91.979 (6.575)	0.005 (0.001)	22.1 (1.1)

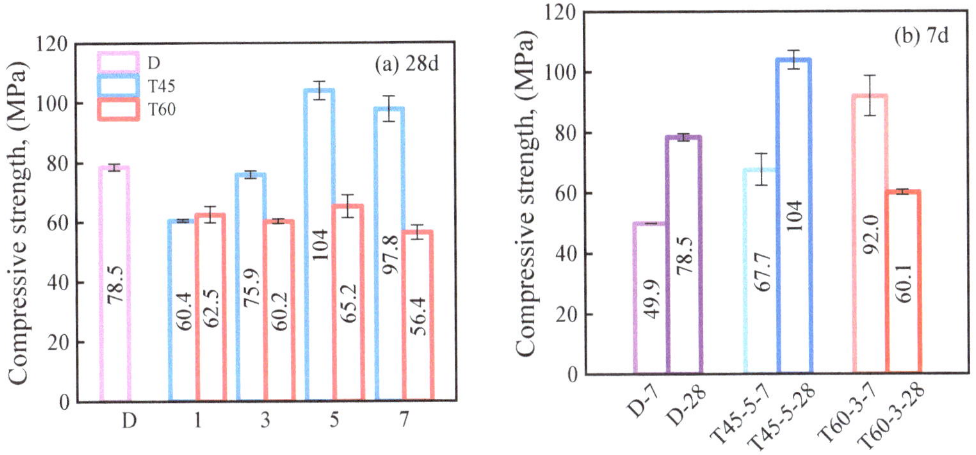

Figure 8. Compressive strength of HDGC: (**a**) after aging to 28 d, (**b**) after aging to 7 d.

Figure 8b shows the 7-day compressive strength of certain test groups. Notably, the T45-5 and T60-3 series reveal that 60 °C significantly promotes the early strength development of HDGC. This observation is consistent with previous studies [34,64]. However, after 7 days of oven curing, the final strength at 28 days decreased by 34.7%. As other researchers have reported [65,66], prolonged oven curing can cause shrinkage and cracking of the precursor gel products, disrupting the original particle size distribution matrix and hindering later strength development. Additionally, Zhang [67] reported that at 60 °C or higher, zeolite crystals start to form in the matrix of alkali-activated fly ash-based materials, leading to changes in the pore structure of the matrix [68]. This explains the significant reduction in compressive strength for T60-3-28 after the removal from the thermal curing environment and subsequent aging. The performance differences demonstrated by the stress–strain curves of T60-3-28 and T45-5-28 in 3.1.1 are explained. Moreover, based on the test results of the T60 series in Figure 8a, it can be observed that the impact of oven curing time on compressive strength is not as direct as that of the oven curing temperature. Therefore, curing temperature plays a more decisive role in adjusting the compressive strength of HDGC.

3.2. Tensile Test

3.2.1. Failure Mode

Figure 9 presents the typical tensile failure modes of specimens at 7 days and 28 days under three different curing temperatures: D, T45, and T60. The images focus on the central 80 mm effective gauge length area. Visually, it can be observed that HDGC exhibited a characteristic multi-cracking failure mode under all three curing conditions. There is no significant difference in the number of cracks for specimens at the same age under different temperatures, but a noticeable change in the number of cracks is observed as the curing age increases from 7 days to 28 days. Kenda et al. [69] pointed out in their study on PE-fiber-reinforced composites that the development of σ_{cu}^{n} (σ means strength) should meet the condition that $\sigma_{cu}^{n} > \sigma_{cu}^{i}$ $(i = 1,\ 2,\ \cdots n-1)$ at that time, which means that a greater number of developed cracks indicates an improvement in the fiber bridging capacity.

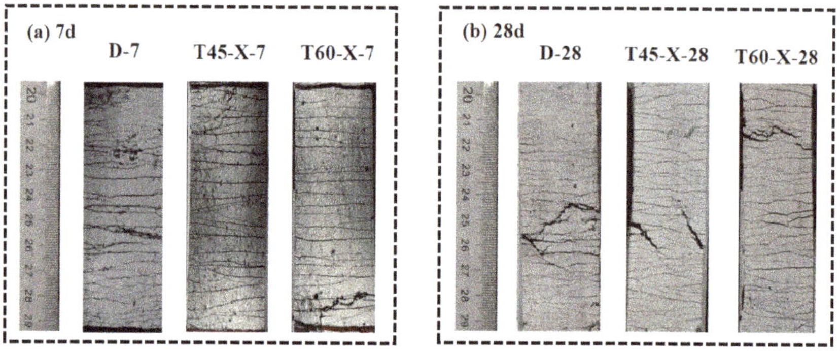

Figure 9. Failure mode of uniaxial tensile test: (**a**) after aging to 7 d, (**b**) after aging to 28 d.

3.2.2. Tensile Stress–Strain Curves

Figure 10 shows the tensile stress–strain curves of HDGC at 7 days and 28 days under different curing conditions. From Figure 10a, it can be seen that all test groups exhibited three stages in tensile stress development: elastic development, pseudo-strain-hardening, and failure softening. Notably, within the 7-day curing period, a curing temperature of 60 °C is more conducive to the development of tensile strain during the pseudo-strain-hardening stage of HDGC, and the highest ultimate tensile strain is visually observed in T60-7-7. Additionally, the pseudo-strain-hardening stage is significantly longer in the T60 series compared to the T45 series. When the curing age reached 28 days, the pseudo-strain-hardening stage shortened in some test groups, such as T45-1-28, T45-3-28, T60-1-28, and T60-5-28. Specifically, the pseudo-strain-hardening stage extended in T45-7-28, with a significant increase in tensile strength during this stage, following the same pattern as seen in the D-series test group.

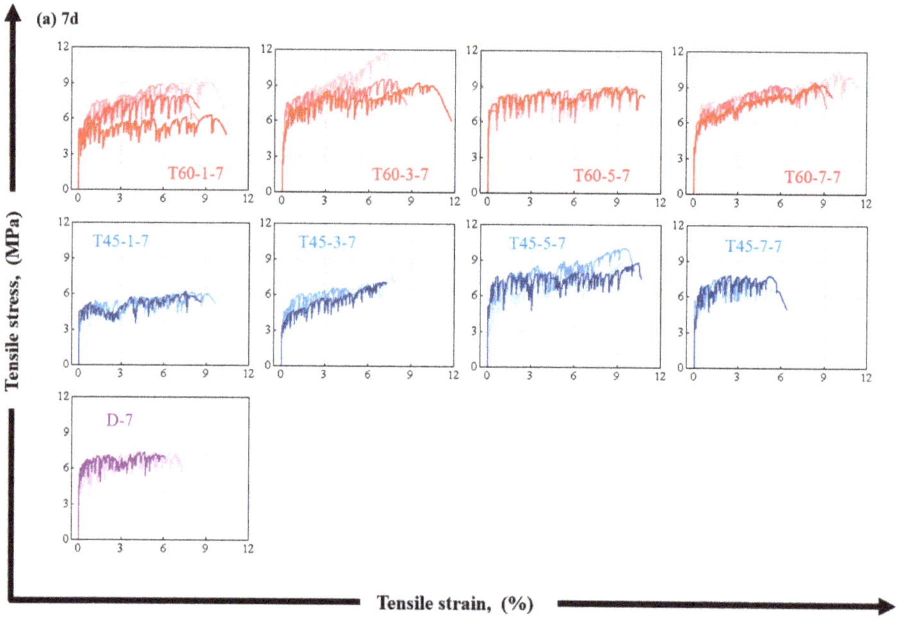

Figure 10. *Cont.*

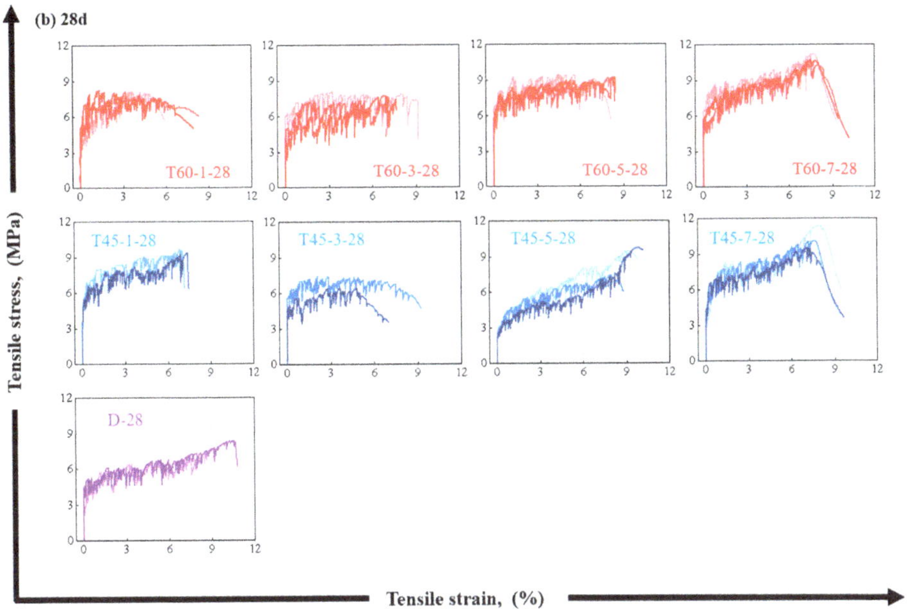

Figure 10. Stress–strain curve of uniaxial tensile test: (**a**) after aging to 7 d, (**b**) after aging to 28 d.

As shown in Figure 10, the peak tensile strength of pseudo-strain-hardening materials like HDGC is strongly associated with their stiffness and fiber bridging capability. Stiffness is reflected in the slope of the stress–strain curve during the elastic stage, while the fiber bridging capability is crucial for achieving pseudo-strain-hardening in HDGC. Notably, the fiber bridging capability alone cannot directly determine the final crack development; the fracture toughness of the matrix must also be considered [58,69].

3.2.3. Tensile Properties Parameters

Figure 11 shows the typical tensile stress–strain curves of pseudo-strain-hardening materials, which include the following tensile performance characteristic parameters: initial crack stress (σ_{ini}), initial crack strain (ε_{ini}), tensile strength (σ_{tu}), ultimate tensile strain (ε_{tu}), and strain energy density (g_{se}). The strain energy density is obtained by integrating the area under the curve from the origin to the point corresponding to the ultimate tensile strength; the specific calculation formula refers to Equation (1).

$$g_{se} = \int_{0}^{\varepsilon_{tu}} \sigma(\varepsilon_t)d\varepsilon_t,\ 0 \leq \varepsilon_t \leq \varepsilon_{tu} \quad (1)$$

Figures 12 and 13 present the effects of different curing conditions on the tensile properties of HDGC at 7-day and 28-day curing ages. Table 5 summarizes the corresponding tensile performance parameters. From Figure 12, it can be seen that curing at 45 °C has almost no impact on the 7-day initial crack strength, which remains nearly the same as the D series. In the T60 series, only a 5-day curing duration significantly improved the initial crack strength, with the lowest initial crack strength observed in T60-1-7. At the 28-day curing duration, the initial crack strength of T45-5-28, T45-7-28, T60-3-28, and T60-5-28 failed to increase. Combining this with the similar findings of the D series, it can be inferred that extending the curing duration does not necessarily enhance the initial crack strength; the early curing environment may have a greater influence on the tensile stiffness of HDGC.

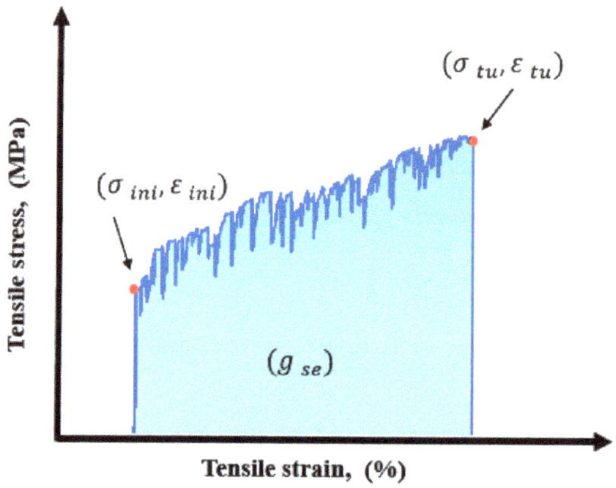

Figure 11. Characteristic parameters of direct tensile test for HDGC.

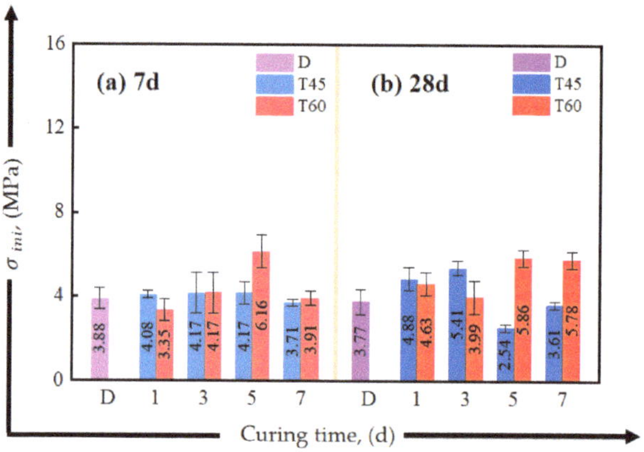

Figure 12. Effect of curing time on initial cracking strength of direct tensile test: (**a**) after aging to 7 d, (**b**) after aging to 28 d.

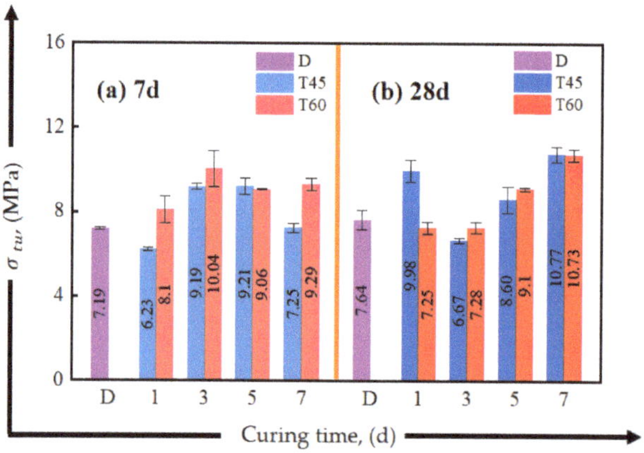

Figure 13. Effect of curing time on peak tensile strength of direct tensile test: (**a**) after aging to 7 d, (**b**) after aging to 28 d.

Table 5. Tensile properties parameters under different curing conditions.

Specimen	σ_{ini} (MPa)	ε_{ini} (%)	σ_{tu} (MPa)	ε_{tu} (%)	g_{se} ($\times 10^3$ kJ/m^3)	σ_{tu}/σ_{ini}
D-7	3.88 (0.49)	0.021 (0.006)	7.19 (0.08)	6.34 (0.47)	47.3 (0.5)	1.86
D-28	3.77 (0.60)	0.022 (0.004)	7.64 (0.44)	9.27 (0.69)	56.6 (6.3)	2.03
T45-1-7	4.08 (0.17)	0.034 (0.003)	6.23 (0.09)	8.24 (0.41)	44.2 (2.1)	1.53
T45-3-7	4.17 (0.98)	0.031 (0.007)	9.19 (0.15)	8.48 (0.98)	46.5 (1.2)	2.21
T45-5-7	4.17 (0.52)	0.021 (0.005)	9.21 (0.41)	9.74 (0.36)	74.0 (5.2)	2.21
T45-7-7	3.71 (0.17)	0.017 (0.003)	7.25 (0.21)	4.71 (0.36)	36.0 (3.1)	1.95
T60-1-7	3.35 (0.50)	0.018 (0.004)	8.09 (0.65)	8.22 (0.64)	63.8 (5.2)	2.42
T60-3-7	4.17 (0.97)	0.032 (0.007)	10.0 (0.9)	9.25 (0.57)	72.6 (5.0)	2.42
T60-5-7	6.16 (0.77)	0.043 (0.015)	9.06 (0.02)	9.38 (0.26)	85.1 (1.8)	1.47
T60-7-7	3.91 (0.34)	0.023 (0.004)	9.29 (0.30)	8.85 (0.50)	77.8 (10.1)	2.37
T45-1-28	4.87 (0.55)	0.025 (0.002)	9.99 (0.51)	7.54 (0.37)	52.1 (4.8)	2.05
T45-3-28	5.40 (0.34)	0.030 (0.004)	6.67 (0.14)	5.62 (0.92)	48.5 (11.6)	1.23
T45-5-28	2.54 (0.19)	0.022 (0.002)	8.60 (0.62)	8.45 (0.63)	58.4 (4.2)	3.39
T45-7-28	3.61 (0.19)	0.018 (0.001)	10.8 (0.4)	7.76 (0.14)	70.8 (4.8)	2.98
T60-1-28	4.63 (0.56)	0.025 (0.005)	7.25 (0.28)	5.41 (0.16)	49.7 (3.8)	1.57
T60-3-28	3.99 (0.79)	0.030 (0.002)	7.28 (0.26)	8.21 (0.70)	51.3 (4.9)	1.82
T60-5-28	5.86 (0.42)	0.029 (0.003)	9.10 (0.09)	7.44 (0.49)	66.6 (0.9)	1.55
T60-7-28	5.78 (0.40)	0.028 (0.003)	10.7 (0.3)	7.88 (0.09)	81.4 (0.92)	1.86

From Figure 13, it can be observed that the peak tensile strength of each group has increased compared to the initial cracking strength. We can specifically observe from Table 5 that the increases in the peak tensile strength relative to the initial cracking strength are 142% and 239% for T60-1-7 and T45-5-28, respectively, corresponding to σ_{tu}/σ_{ini} values of 2.42 and 3.39. These are the largest values in their respective age series. This finding is consistent with the results discussed in the stress–strain curves. Notably, T60-1-7 and T45-5-28 exhibited the greatest increase in tensile stress during the pseudo-strain-hardening phase, correspond to the groups with the lowest initial crack strength in the 7-day and 28-day series, respectively. Considering the mechanism analysis in Section 3.1.2, which suggests that oven curing conditions can induce zeolite formation in the matrix, it can be concluded that zeolite affected the matrix stiffness by altering the pore structure. Specific oven curing conditions could enhance the fiber bridging capacity, thereby improving the material's strain-hardening capability and resulting in a more significant stress increase [58,69].

The initial cracking in concrete materials occurs early and randomly [70–72], so this section will not discuss the influence of curing conditions on the material's initial cracking stress and strain. As shown in Figures 14 and 15, oven curing temperatures of 45 °C and 60 °C were beneficial for the early (7-day) ductility development of the material, especially for T45-5-7 and T60-3-7, and the peak strain increased by 53.6% and 45.9% compared to D-7, respectively. Additionally, oven curing for 5 days or more helped to enhance the tensile strain energy density of the HDGC, as shown in Figure 15. Both T60-5 and T60-7 exhibited top-tier tensile energy absorption capacity at both the 7-day and 28-day curing ages. The highest value was observed in T60-5-7 at 85.2 kJ/m^3, which is 1.8 times that of D-7.

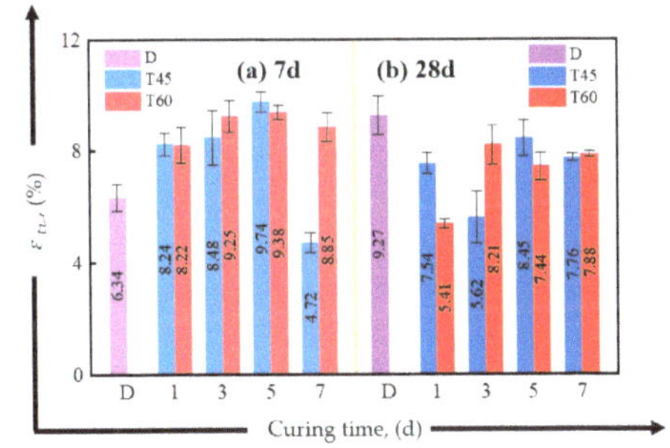

Figure 14. Effect of curing time on peak tensile strain of direct tensile test: (**a**) after aging to 7 d, (**b**) after aging to 28 d.

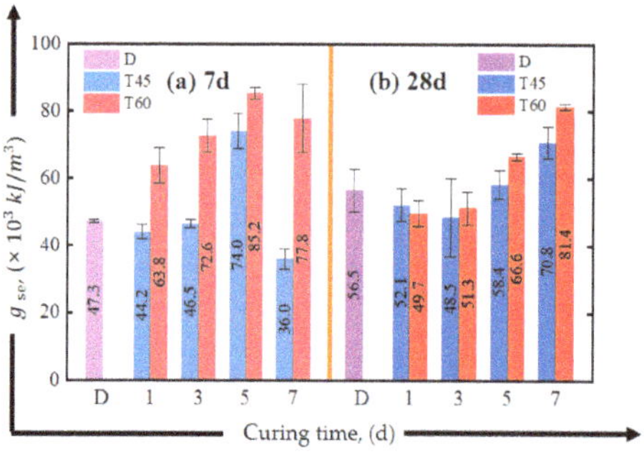

Figure 15. Effect of curing time on strain energy density of direct tensile test: (**a**) after aging to 7 d, (**b**) after aging to 28 d.

3.3. Microstructure Analysis

3.3.1. The Influence of Curing Temperature on the Development of the Matrix

Figure 16 presents the SEM micrographs and EDS analysis results for HDGC under different curing conditions. Given the significant improvements in tensile strength and energy absorption for T45 and T60 after 7 days of curing, samples were taken from the fracture surfaces of T45-7-28 and T60-7-28 tensile specimens. Taking the microstructure of D-28 shown in Figure 16 as an example, Points 1 and 3 indicate unreacted fly ash and slag, Point 2 indicates quartz sand, Point 4 indicates gel products, and the elongated objects are PE fibers. EDS analysis of the PE fibers (Point 5) is shown in Figure 14b. The Ca/Si ratio is 0.818, and the Al/Si ratio is 0.363. Based on atomic percentage data [73], the attached material on the fibers is identified as C-(N)-A-S-H gel. Notably, the strong bonding ability of C-(N)-A-S-H gel contributes to enhancing the adhesion of PE fibers within the HDGC matrix.

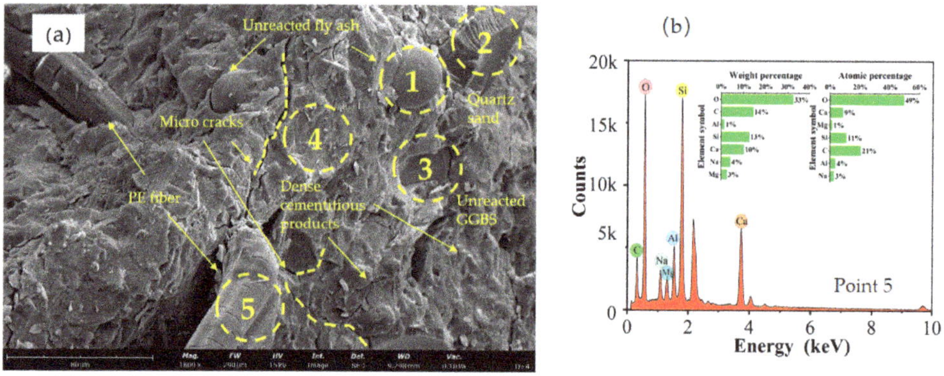

Figure 16. Microstructure analysis of Group D: (**a**) SEM image, (**b**) EDS analysis of Point 5.

Based on the SEM micrographs and EDS surface scan results in Figures 17–20, the Ca/Si ratio of the matrix is approximately 0.6, while the Al/Si ratio increases with higher curing temperatures. This is because elevated temperatures accelerate the dissolution of alumina in the precursor, promoting the precursor's reaction. Additionally, temperature changes can alter the morphology of fly ash, leading to the formation of multilayered, plate-like gel structures within the matrix, as shown in Figure 20 (T60). Overall, at the same scale, the microstructure of the materials under all three curing conditions appears to be well-developed and dense.

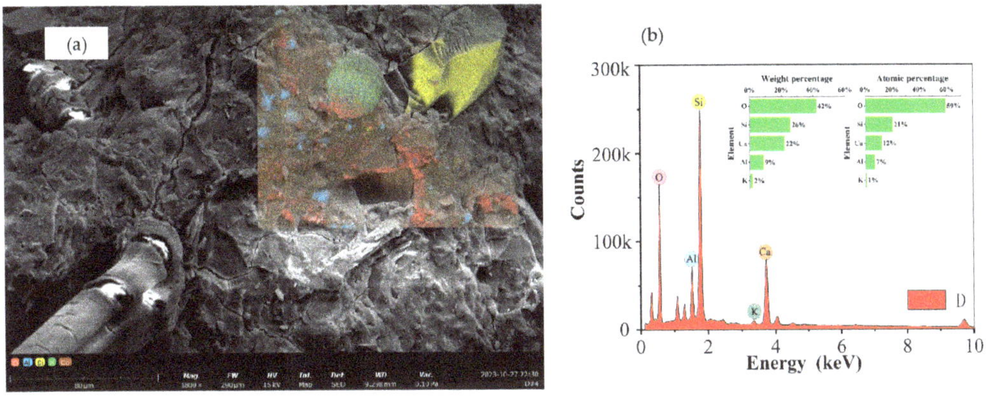

Figure 17. EDS mapping results of D-28: (**a**) surface scan schematic, (**b**) EDS results.

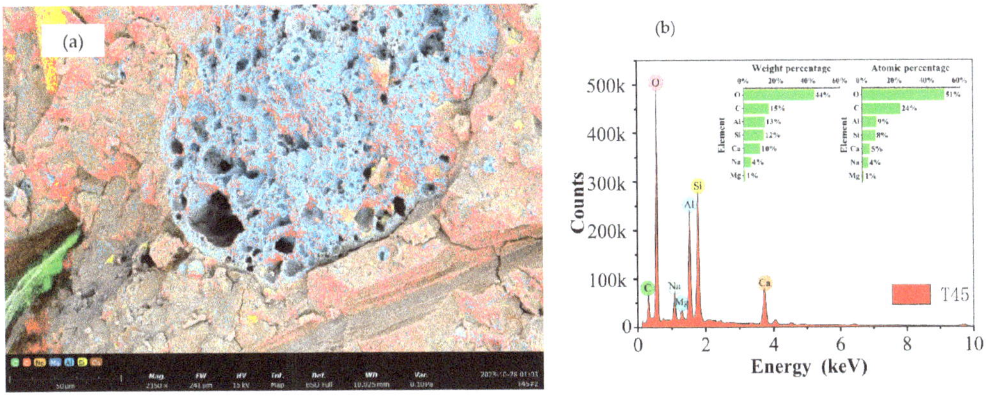

Figure 18. EDS mapping results of T45-7-28: (**a**) surface scan schematic, (**b**) EDS results.

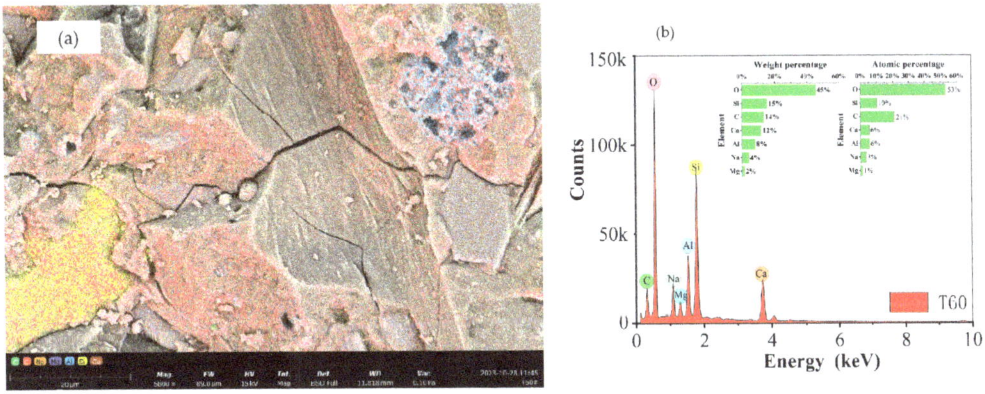

Figure 19. EDS mapping results of T60-7-28: (**a**) surface scan schematic, (**b**) EDS results.

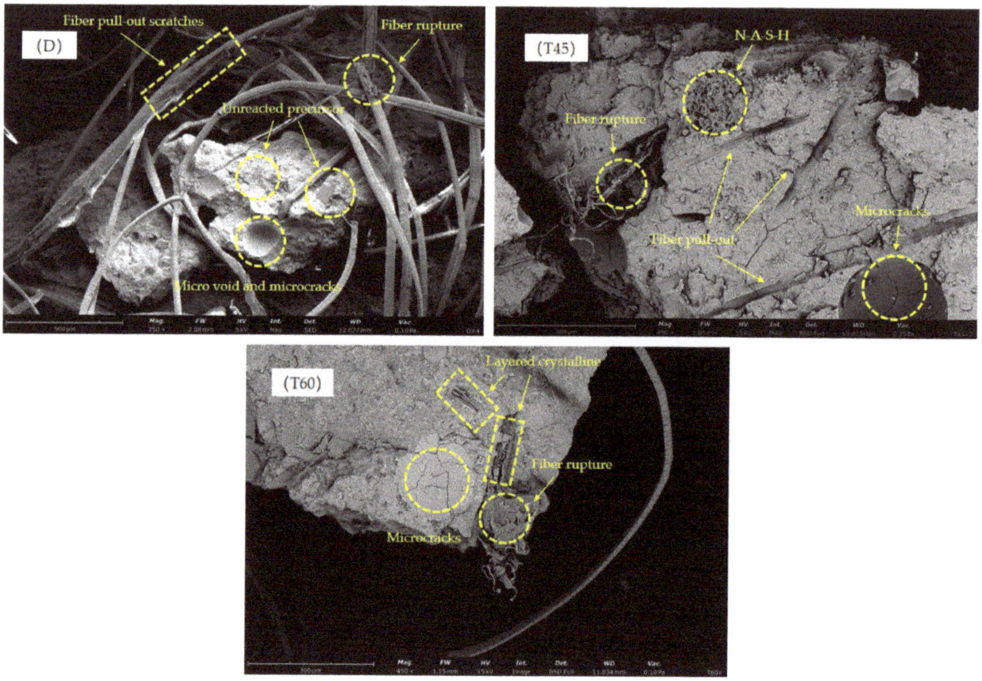

Figure 20. SEM images of the matrix and fibers: (D) D-28, (T45) T45-7-28, (T60) T60-7-28.

3.3.2. The Influence of Curing Temperature on the Bonding between Matrix and Fiber

Figure 20 shows the effect of different curing temperatures on the bonding between the HDGC matrix and fibers. The SEM images reveal some undissolved precursors and pores formed due to moisture evaporation during mixing or after the matrix has set. It is evident that these undissolved precursors and the surrounding pores have microcracks, which directly affect the density of the matrix and the bonding strength between the fibers and the matrix. As observed in Section 3.3.1, a higher Al/Si ratio due to oven curing indicates an improved precursor reaction, enhancing the matrix's density and thus the fiber bridging performance. This confirms the mechanisms discussed in Section 3.2.3, where specific oven curing conditions were found to enhance the fiber bridging ability and improve the material's pseudo-strain-hardening effects.

3.3.3. The Influence of Curing Temperature on the Fiber Failure Mode

The specific failure modes of fibers are illustrated in Figure 21. The morphology of fiber tensile failure can be categorized into two types: fiber pullout and fiber rupture.

Fiber Pullout: Under microscopic examination, a dense network of cracks is observed on the fiber surface, or there is surface delamination. The fibers are elongated and exhibit flattened or curled modes; specifically, refer to Figure 21a,b.

Fiber Rupture: This mode is characterized by fine, randomly tangled filaments, fibers fracturing into multiple strands, or local yielding reaching the extreme limit, leading to necking and rupture at the ends; specifically, refer to Figure 21c,d.

From Figure 22, it can be observed how the three curing temperatures affect the fiber failure modes. At the same observation scale, fiber pullout failure is more commonly seen in the D group with room temperature curing. This is manifested by noticeable damage on the fiber surface due to friction with the matrix, while the fiber ends remain intact. In contrast, the T45 and T60 oven-cured samples show a richer variety of fiber rupture failure modes, including fiber end necking and splitting into fine strands. Specifically, oven curing enhanced the adhesion between the fibers and the matrix, leading to a closer bond and an improved fiber bridging effect. On the other hand, this also mitigated the negative impact of low-modulus fibers on the overall modulus of HDGC, as evidenced by the increased axial compressive modulus in the T45 and T60 series, thereby increasing the rigidity of HDGC.

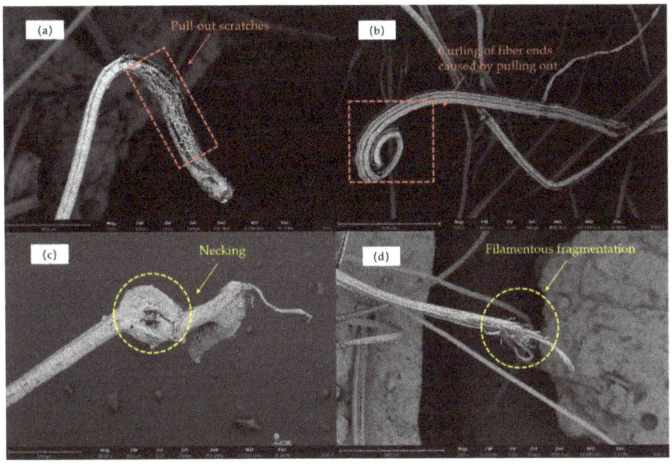

Figure 21. Fiber failure mode: (**a**) pullout scratches, (**b**) curling of fiber ends, (**c**) necking, (**d**) filamentous fragmentation.

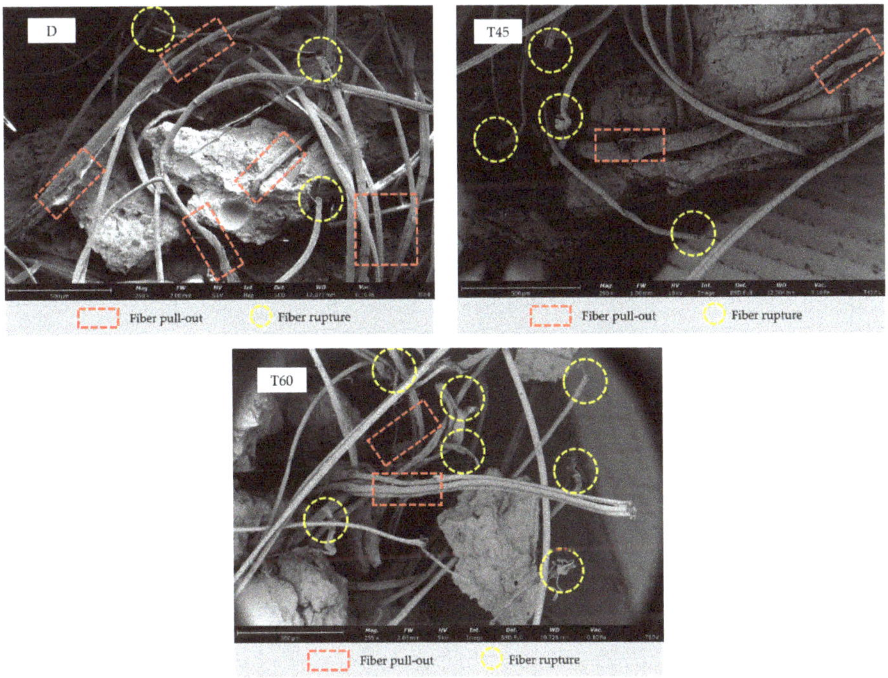

Figure 22. The effect of curing temperature on fiber failure modes: (D) D-28, (T45) T45-7-28, (T60) T60-7-28.

4. Conclusions

In this study, three curing temperatures (room temperature, 45 °C, and 60 °C), four curing durations (1 day, 3 days, 5 days, and 7 days), and final curing ages (7 days and 28 days) were investigated to study their effects on the axial compressive and tensile properties of HDGC. The mechanisms behind the impact of curing conditions on the mechanical properties of HDGC were also explored, by analyzing microstructural changes under different curing temperatures. The conclusions are as follows:

(1) Curing at 45 °C and 60 °C allows HDGC to enter a ductile failure stage after reaching compressive strength, unlike the room temperature-cured samples, which show splitting cracks and immediate failure. A 5-day heat curing at 45 °C maximizes the compressive strength of HDGC and extends its ductile failure phase. Therefore, 45 °C is recommended as the optimal oven curing temperature to improve HDGC's compressive performance.

(2) The 28-day compressive strength of HDGC cured at 60 °C is significantly lower than at 7 days. The formation of zeolite crystals from the slag and fly ash precursors at this temperature significantly affects the pore structure of the matrix, leading to a decrease in matrix strength during long-term curing and a reduction in compressive strength over time.

(3) T60-3-7 and T45-5-28 exhibited significant tensile strain hardening, yet had the lowest initial tensile cracking strength. Specific oven curing conditions can reduce the initial cracking toughness of the matrix while enhancing the fiber bridging capability, thereby increasing the degree of strain hardening in HDGC according to the criteria for pseudo-strain-hardening materials. This also helps to improve the energy dissipation capacity of the material.

(4) Higher curing temperatures promote the full reaction of the precursors, increasing the bond strength between the matrix and fibers. This enhances fiber bridging capability while reducing the negative impact of low-modulus PE fibers on the overall stiffness of the material. Oven curing improved the compressive elastic modulus of HDGC.

(5) Considering the overall effects of curing conditions on HDGC's mechanical properties, this study recommends 45 °C as the optimal curing temperature. For long-term mechanical performance, which is crucial for concrete materials, a heat curing duration of 5 to 7 days is the better choice.

Author Contributions: Conceptualization, R.L. (Ruihao Luo); methodology, R.L. (Ruihao Luo) and R.L. (Runan Liu); software, G.Q.; validation, G.Q., M.J. and Y.W.; formal analysis, R.L. (Ruihao Luo); investigation, R.L. (Ruihao Luo); resources, Y.G.; data curation, G.Q.; writing—original draft preparation, R.L. (Ruihao Luo); writing—review and editing, R.L. (Ruihao Luo) and R.L. (Runan Liu); visualization, G.Q., M.J. and Y.W.; supervision, Y.G.; project administration, Y.G.; funding acquisition, Y.G. All authors have read and agreed to the published version of the manuscript.

Funding: This research was funded by National Natural Science Foundation of China, Grant Number 12272099.

Institutional Review Board Statement: Not applicable.

Informed Consent Statement: Not applicable.

Data Availability Statement: The data presented in this study are available on request from the corresponding author. The data are not publicly available due to confidentiality issues.

Conflicts of Interest: Guang Qin, is employed by the Foshan Road and Bridge Supervision Station Corporation Limited, Minyang Jiang is employed by the Fujian Xingyuan Construction Engineering Development Corporation Limited, and Yixian Wu is employed by Fujian Chengshuo Construction Engineering Corporation Limited. The authors declare that this study received funding from the National Natural Science Foundation of China. The research was conducted autonomously, ensuring the integrity and objectivity of the findings.

References

1. Almutairi, A.L.; Tayeh, B.A.; Adesina, A.; Isleem, H.F.; Zeyad, A.M. Potential applications of geopolymer concrete in construction: A review. *Case Stud. Constr. Mater.* **2021**, *15*, e00733. [CrossRef]
2. Chen, G.; Zheng, D.-p.; Chen, Y.-w.; Lin, J.-X.; Lao, W.-j.; Guo, Y.-c.; Chen, Z.-b.; Lan, X.-w. Development of high performance geopolymer concrete with waste rubber and recycle steel fiber: A study on compressive behavior, carbon emissions and economical performance. *Constr. Build. Mater.* **2023**, *393*, 131988. [CrossRef]
3. Pan, H.; Xie, Z.; Chen, G.; Su, J.; Zhuo, K.; Chen, Z.; Lin, J.; Feng, C.; Guo, Y. Dynamic compressive behavior of high-strength engineered geopolymer composites. *J. Build. Eng.* **2023**, *80*, 108036. [CrossRef]
4. Miri, M.; Ghasemi, M.R.; Nezhad, H.B. Experimental Investigation of Corrosion Cracking in Reinforced Concrete Beams Containing Nano Wollastonite. *J. Rehabil. Civ. Eng.* **2019**, *7*, 117–138.
5. Yu, J.; Yao, J.; Lin, X.; Li, H.; Lam, J.Y.K.; Leung, C.K.Y.; Sham, I.M.L.; Shih, K. Tensile performance of sustainable Strain-Hardening Cementitious Composites with hybrid PVA and recycled PET fibers. *Cem. Concr. Res.* **2018**, *107*, 110–123. [CrossRef]
6. Şahmaran, M.; Özbay, E.; Yücel, H.; Lachemi, M.; Li, V. Effect of Fly Ash and PVA Fiber on Microstructural Damage and Residual Properties of Engineered Cementitious Composites Exposed to High Temperatures. *J. Mater. Civ. Eng.* **2011**, *23*, 1735–1745. [CrossRef]
7. Liu, J.-C.; Tan, K.H. Mechanism of PVA fibers in mitigating explosive spalling of engineered cementitious composite at elevated temperature. *Cem. Concr. Compos.* **2018**, *93*, 235–245. [CrossRef]
8. Lin, J.-X.; Song, Y.; Xie, Z.-H.; Guo, Y.-C.; Yuan, B.; Zeng, J.-J.; Wei, X. Static and dynamic mechanical behavior of engineered cementitious composites with PP and PVA fibers. *J. Build. Eng.* **2020**, *29*, 101097. [CrossRef]
9. Ranade, R.; Li, V.; Stults, M.D.; Heard, W.; Rushing, T. Composite Properties of High-Strength, High-Ductility Concrete. *ACI Mater. J.* **2013**, *110*, 413–422.
10. Zhu, B.; Pan, J.; Zhang, M.; Leung, C.K.Y. Predicting the strain-hardening behaviour of polyethylene fibre reinforced engineered cementitious composites accounting for fibre-matrix interaction. *Cem. Concr. Compos.* **2022**, *134*, 104770. [CrossRef]
11. Zhang, Z.; Qin, F.; Ma, H.; Xu, L. Tailoring an impact resistant engineered cementitious composite (ECC) by incorporation of crumb rubber. *Constr. Build. Mater.* **2020**, *262*, 120116. [CrossRef]
12. Duxson, P.; Fernández-Jiménez, A.; Provis, J.L.; Lukey, G.C.; Palomo, A.; van Deventer, J.S.J. Geopolymer technology: The current state of the art. *J. Mater. Sci.* **2007**, *42*, 2917–2933. [CrossRef]
13. Shi, C.; Jiménez, A.F.; Palomo, A. New cements for the 21st century: The pursuit of an alternative to Portland cement. *Cem. Concr. Res.* **2011**, *41*, 750–763. [CrossRef]
14. Valentini, L. Modeling Dissolution–Precipitation Kinetics of Alkali-Activated Metakaolin. *ACS Omega* **2018**, *3*, 18100–18108. [CrossRef] [PubMed]
15. Li, Z.; Xu, Y.; Wu, C.; Zhang, W.; Chen, Y.; Li, Y. Preparation and Research on Mechanical Properties of Eco-Friendly Geopolymer Grouting Cementitious Materials Based on Industrial Solid Wastes. *Materials* **2024**, *17*, 3874. [CrossRef] [PubMed]
16. Ling, Y.; Zhang, X.; Wu, Y.; Zou, W.; Wang, C.; Li, C.; Li, W. Evaluation of Bonding Behavior between Engineered Geopolymer Composites with Hybrid PE/PVA Fibers and Concrete Substrate. *Materials* **2024**, *17*, 3778. [CrossRef] [PubMed]
17. Ohno, M.; Li, V.C. A feasibility study of strain hardening fiber reinforced fly ash-based geopolymer composites. *Constr. Build. Mater.* **2014**, *57*, 163–168. [CrossRef]
18. He, P.; Wang, M.; Fu, S.; Jia, D.; Yan, S.; Yuan, J.; Xu, J.; Wang, P.; Zhou, Y. Effects of Si/Al ratio on the structure and properties of metakaolin based geopolymer. *Ceram. Int.* **2016**, *42*, 14416–14422. [CrossRef]
19. Assi, L.N.; Eddie Deaver, E.; Ziehl, P. Effect of source and particle size distribution on the mechanical and microstructural properties of fly Ash-Based geopolymer concrete. *Constr. Build. Mater.* **2018**, *167*, 372–380. [CrossRef]
20. Wan, Q.; Rao, F.; Song, S.; García, R.E.; Estrella, R.M.; Patiño, C.L.; Zhang, Y. Geopolymerization reaction, microstructure and simulation of metakaolin-based geopolymers at extended Si/Al ratios. *Cem. Concr. Compos.* **2017**, *79*, 45–52. [CrossRef]
21. Liew, Y.M.; Kamarudin, H.; Mustafa Al Bakri, A.M.; Bnhussain, M.; Luqman, M.; Khairul Nizar, I.; Ruzaidi, C.M.; Heah, C.Y. Optimization of solids-to-liquid and alkali activator ratios of calcined kaolin geopolymeric powder. *Constr. Build. Mater.* **2012**, *37*, 440–451. [CrossRef]
22. Fardjaoui, N.E.H.; Wicklein, B.; Aranda, P.; Sobrados, I.; El Berrichi, F.Z.; Ruiz-Hitzky, E. Modulation of inorganic matrices for functional nanoarchitectures fabrication: The simultaneous effect of moisture and temperature in the preparation of metakaolin based geopolymers. *Bull. Chem. Soc. Jpn.* **2018**, *91*, 1158–1167. [CrossRef]
23. Görhan, G.; Aslaner, R.; Şinik, O. The effect of curing on the properties of metakaolin and fly ash-based geopolymer paste. *Compos. Part B Eng.* **2016**, *97*, 329–335. [CrossRef]
24. Singh, G.V.P.B.; Subramaniam, K.V.L. Influence of processing temperature on the reaction product and strength gain in alkali-activated fly ash. *Cem. Concr. Compos.* **2019**, *95*, 10–18. [CrossRef]
25. *ASTM-C511*; Standard Specification for Mixing Rooms, Moist Cabinets, Moist Rooms, and Water Storage Tanks Used in the Testing of Hydraulic Cements and Concretes. ASTM Standards: West Conshohocken, PA, USA, 2021.
26. Narayanan, A.; Shanmugasundaram, P. An Experimental Investigation on Flyash-based Geopolymer Mortar under different curing regime for Thermal Analysis. *Energy Build.* **2017**, *138*, 539–545. [CrossRef]
27. Helmy, A.I.I. Intermittent curing of fly ash geopolymer mortar. *Constr. Build. Mater.* **2016**, *110*, 54–64. [CrossRef]

28. de Vargas, A.S.; Dal Molin, D.C.C.; Vilela, A.C.F.; Silva, F.J.d.; Pavão, B.; Veit, H. The effects of Na_2O/SiO_2molar ratio, curing temperature and age on compressive strength, morphology and microstructure of alkali-activated fly ash-based geopolymers. *Cem. Concr. Compos.* **2011**, *33*, 653–660. [CrossRef]
29. Mo, B.-h.; Zhu, H.; Cui, X.-m.; He, Y.; Gong, S.-y. Effect of curing temperature on geopolymerization of metakaolin-based geopolymers. *Appl. Clay Sci.* **2014**, *99*, 144–148. [CrossRef]
30. Hardjito, D. Studies of Fly Ash-Based Geopolymer Concrete. Ph.D. Thesis, Curtin University, Perth, Australia, 2005.
31. Temuujin, J.; Williams, R.P.; van Riessen, A. Effect of mechanical activation of fly ash on the properties of geopolymer cured at ambient temperature. *J. Mater. Process. Technol.* **2009**, *209*, 5276–5280. [CrossRef]
32. Palomo, A.; Grutzeck, M.W.; Blanco, M.T. Alkali-activated fly ashes: A cement for the future. *Cem. Concr. Res.* **1999**, *29*, 1323–1329. [CrossRef]
33. Chindaprasirt, P.; Chareerat, T.; Hatanaka, S.; Cao, T. High-Strength Geopolymer Using Fine High-Calcium Fly Ash. *J. Mater. Civ. Eng.* **2011**, *23*, 264–270. [CrossRef]
34. Bakria, A.M.M.A.; Kamarudin, H.; BinHussain, M.; Nizar, I.K.; Zarina, Y.; Rafiza, A.R. The Effect of Curing Temperature on Physical and Chemical Properties of Geopolymers. *Phys. Procedia* **2011**, *22*, 286–291. [CrossRef]
35. Verma, M.; Dev, N. Effect of ground granulated blast furnace slag and fly ash ratio and the curing conditions on the mechanical properties of geopolymer concrete. *Struct. Concr.* **2022**, *23*, 2015–2029. [CrossRef]
36. Kürklü, G. The effect of high temperature on the design of blast furnace slag and coarse fly ash-based geopolymer mortar. *Compos. Part B Eng.* **2016**, *92*, 9–18. [CrossRef]
37. Türker, H.T.; Balçikanli, M.; Durmuş, İ.H.; Özbay, E.; Erdemir, M. Microstructural alteration of alkali activated slag mortars depend on exposed high temperature level. *Constr. Build. Mater.* **2016**, *104*, 169–180. [CrossRef]
38. Li, X.; Wang, Z.; Jiao, Z. Influence of Curing on the Strength Development of Calcium-Containing Geopolymer Mortar. *Materials* **2013**, *6*, 5069–5076. [CrossRef]
39. Yankwa Djobo, J.N.; Elimbi, A.; Kouamo Tchakouté, H.; Kumar, S. Mechanical properties and durability of volcanic ash based geopolymer mortars. *Constr. Build. Mater.* **2016**, *124*, 606–614. [CrossRef]
40. Nasvi, M.; Pathegama, R.; Sanjayan, J. Geopolymer as well cement and variation of its mechanical behavior with curing temperature. *Greenh. Gases Sci. Technol.* **2012**, *2*, 46–58. [CrossRef]
41. Görhan, G.; Kürklü, G. The influence of the NaOH solution on the properties of the fly ash-based geopolymer mortar cured at different temperatures. *Compos. Part B Eng.* **2014**, *58*, 371–377. [CrossRef]
42. Arellano-Aguilar, R.; Burciaga-Díaz, O.; Gorokhovsky, A.; Escalante-García, J.I. Geopolymer mortars based on a low grade metakaolin: Effects of the chemical composition, temperature and aggregate:binder ratio. *Constr. Build. Mater.* **2014**, *50*, 642–648. [CrossRef]
43. Hardjito, D.; Wallah, S.E.; Sumajouw, D.M.J.; Rangan, B.V. On the development of fly ash-based geopolymer concrete. *ACI Mater. J.* **2004**, *101*, 467–472.
44. Atiş, C.D.; Görür, E.B.; Karahan, O.; Bilim, C.; İlkentapar, S.; Luga, E. Very high strength (120MPa) class F fly ash geopolymer mortar activated at different NaOH amount, heat curing temperature and heat curing duration. *Constr. Build. Mater.* **2015**, *96*, 673–678. [CrossRef]
45. Yusuf, M.O.; Megat Johari, M.A.; Ahmad, Z.A.; Maslehuddin, M. Evolution of alkaline activated ground blast furnace slag–ultrafine palm oil fuel ash based concrete. *Mater. Des.* **2014**, *55*, 387–393. [CrossRef]
46. Patankar, S.V.; Ghugal, Y.M.; Jamkar, S.S. Effect of Concentration of Sodium Hydroxide and Degree of Heat Curing on Fly Ash-Based Geopolymer Mortar. *Indian J. Mater. Sci.* **2014**, *2014*, 938789. [CrossRef]
47. Li, X.; Ma, X.; Zhang, S.; Zheng, E. Mechanical Properties and Microstructure of Class C Fly Ash-Based Geopolymer Paste and Mortar. *Materials* **2013**, *6*, 1485–1495. [CrossRef] [PubMed]
48. Guo, X.; Shi, H.; Dick, W.A. Compressive strength and microstructural characteristics of class C fly ash geopolymer. *Cem. Concr. Compos.* **2010**, *32*, 142–147. [CrossRef]
49. Rovnaník, P. Effect of curing temperature on the development of hard structure of metakaolin-based geopolymer. *Constr. Build. Mater.* **2010**, *24*, 1176–1183. [CrossRef]
50. Bakharev, T.; Sanjayan, J.G.; Cheng, Y.-B. Alkali activation of Australian slag cements. *Cem. Concr. Res.* **1999**, *29*, 113–120. [CrossRef]
51. Li, Z.; Liu, S. Influence of Slag as Additive on Compressive Strength of Fly Ash-Based Geopolymer. *J. Mater. Civ. Eng.* **2007**, *19*, 470–474. [CrossRef]
52. Oh, J.E.; Monteiro, P.J.M.; Jun, S.S.; Choi, S.; Clark, S.M. The evolution of strength and crystalline phases for alkali-activated ground blast furnace slag and fly ash-based geopolymers. *Cem. Concr. Res.* **2010**, *40*, 189–196. [CrossRef]
53. Islam, A.; Alengaram, U.J.; Jumaat, M.Z.; Bashar, I.I. The development of compressive strength of ground granulated blast furnace slag-palm oil fuel ash-fly ash based geopolymer mortar. *Mater. Design (1980–2015)* **2014**, *56*, 833–841. [CrossRef]
54. Lv, Y.; Wang, C.; Han, W.; Li, X.; Peng, H. Study of the Mechanical Properties and Microstructure of Alkali-Activated Fly Ash–Slag Composite Cementitious Materials. *Polymers* **2023**, *15*, 1903. [CrossRef]
55. Ismail, I.; Bernal, S.A.; Provis, J.L.; San Nicolas, R.; Hamdan, S.; van Deventer, J.S.J. Modification of phase evolution in alkali-activated blast furnace slag by the incorporation of fly ash. *Cem. Concr. Compos.* **2014**, *45*, 125–135. [CrossRef]

56. Puertas, F.; Martínez-Ramírez, S.; Alonso, S.; Vázquez, T. Alkali-activated fly ash/slag cements: Strength behaviour and hydration products. *Cem. Concr. Res.* **2000**, *30*, 1625–1632. [CrossRef]
57. Nath, S.K.; Maitra, S.; Mukherjee, S.; Kumar, S. Microstructural and morphological evolution of fly ash based geopolymers. *Constr. Build. Mater.* **2016**, *111*, 758–765. [CrossRef]
58. Su, J.-y.; Chen, G.; Pan, H.-s.; Lin, J.-X.; Zhang, J.; Zhuo, K.-x.; Chen, Z.-b.; Guo, Y.-c. Rubber modified high strength-high ductility concrete: Effect of rubber replacement ratio and fiber length. *Constr. Build. Mater.* **2023**, *404*, 133243. [CrossRef]
59. *GB/T-50146-2014;* Technical Code for Application of Fly Ash Concrete. China Architecture & Building Press: Beijing, China, 2014.
60. *GB/T-18046-2017;* Ground Granulated Blast Furnace Slag Used for Cement, Mortar and Concrete. General Administration of Quality Supervision, Inspection and Quarantine, The People's Republic of China: Beijing, China, 2017.
61. Tosun-Felekoğlu, K.; Felekoğlu, B.; Ranade, R.; Lee, B.Y.; Li, V.C. The role of flaw size and fiber distribution on tensile ductility of PVA-ECC. *Compos. Part B Eng.* **2014**, *56*, 536–545. [CrossRef]
62. *ASTM-C496;* Standard Test Method for Splitting Tensile Strength of Cylindrical Concrete Specimens. ASTM Standards: West Conshohocken, PA, USA, 2011.
63. *JC/T2461-2018;* Standard Test Method for the Mechanical Properties of Ductile Fiber Reinforcedcementitious Composites. Ministry of Industry and Information Technology of the People's Republic of China: Beijing, China, 2018.
64. Nasir, M.; Johari, M.A.M.; Maslehuddin, M.; Yusuf, M.O.; Al-Harthi, M.A. Influence of heat curing period and temperature on the strength of silico-manganese fume-blast furnace slag-based alkali-activated mortar. *Constr. Build. Mater.* **2020**, *251*, 118961. [CrossRef]
65. Khale, D.; Chaudhary, R. Mechanism of geopolymerization and factors influencing its development: A review. *J. Mater. Sci.* **2007**, *42*, 729–746. [CrossRef]
66. Heah, C.Y.; Kamarudin, H.; Bakri, A.M.M.A.; Binhussain, M.; Luqman, M.; Nizar, I.K.; Ruzaidi, C.M.; Liew, Y.M. Effect of Curing Profile on Kaolin-based Geopolymers. *Phys. Procedia* **2011**, *22*, 305–311. [CrossRef]
67. Zhang, B.; Yuan, P.; Guo, H.; Deng, L.; Li, Y.; Li, L.; Wang, Q.; Liu, D. Effect of curing conditions on the microstructure and mechanical performance of geopolymers derived from nanosized tubular halloysite. *Constr. Build. Mater.* **2021**, *268*, 121186. [CrossRef]
68. Palomo, A.; Blanco-Varela, M.T.; Granizo, M.L.; Puertas, F.; Vazquez, T.; Grutzeck, M.W. Chemical stability of cementitious materials based on metakaolin. *Cem. Concr. Res.* **1999**, *29*, 997–1004. [CrossRef]
69. Kanda, T.; Li, V.C. Multiple Cracking Sequence and Saturation in Fiber Reinforced Cementitious Composites. *Jpn. Concr. Inst.* **1998**, *9*, 19–33.
70. Lin, J.-X.; Chen, G.; Pan, H.-s.; Wang, Y.-c.; Guo, Y.-c.; Jiang, Z.-x. Analysis of stress-strain behavior in engineered geopolymer composites reinforced with hybrid PE-PP fibers: A focus on cracking characteristics. *Compos. Struct.* **2023**, *323*, 117437. [CrossRef]
71. Lin, J.-X.; Luo, R.-H.; Su, J.-Y.; Guo, Y.-C.; Chen, W.-S. Coarse synthetic fibers (PP and POM) as a replacement to steel fibers in UHPC: Tensile behavior, environmental and economic assessment. *Constr. Build. Mater.* **2024**, *412*, 134654. [CrossRef]
72. Wille, K.; El-Tawil, S.; Naaman, A.E. Properties of strain hardening ultra high performance fiber reinforced concrete (UHP-FRC) under direct tensile loading. *Cem. Concr. Compos.* **2014**, *48*, 53–66. [CrossRef]
73. Samarakoon, M.H.; Ranjith, P.G.; Duan, W.H.; De Silva, V.R.S. Properties of one-part fly ash/slag-based binders activated by thermally-treated waste glass/NaOH blends: A comparative study. *Cem. Concr. Compos.* **2020**, *112*, 103679. [CrossRef]

materials

MDPI

Article

Evaluation of Axial Compressive and Tensile Properties of PE/PVA Hybrid Fiber Reinforced Strain-Hardening Geopolymer Composites

Jingen Guo [1,2,3], Ji Shi [1,2], Liuhuo Wang [1,2], Chengyong Huang [1,2], Xiongwu Tao [1,2], Chaosen Li [4] and Zhanbiao Chen [4,*]

1 Huizhou Power Supply Bureau, Guangdong Power Grid Co., Ltd., Huizhou 516001, China; guo19830526@163.com (J.G.); 13829984272@139.com (J.S.); firehuo_0403@126.com (L.W.); 13502201000@139.com (C.H.); 13500175094@139.com (X.T.)
2 China Southern Power Grid Co., Ltd., Guangzhou 510663, China
3 Guangdong Power Grid Co., Ltd., Grid Planning Research Center, Guangzhou 510050, China
4 School of Civil and Transportation Engineering, Guangdong University of Technology, Guangzhou 510006, China; yyyee09007@126.com
* Correspondence: zbchen@gdut.edu.cn

Abstract: The strain-hardening geopolymer composite (SHGC) is a new type of fiber concrete with excellent ductility and environmental friendliness. However, the high cost of fibers greatly limits its widespread application. This paper proposes the use of untreated low-cost polyvinyl alcohol (PVA) fibers and polyethylene (PE) fibers to develop a low-cost, high-performance SHGC. Axial compression and axial tension tests were conducted on the SHGC with different PE fiber volume fractions (1%, 1.5%, and 2%) and different PVA fiber replacement ratios (0%, 25%, 50%, 75%, and 100%) to investigate the hybrid effects of fibers with different surface properties and to reveal the mechanism of fiber hybridization on the mechanical behavior of SHGCs. The results show that increasing the PE fiber volume fraction improves the compressive and tensile ductility of the SHGC while increasing the PVA fiber replacement rate impacts the strength indicators positively due to the good interface effect formed between its hydrophilic surface and the matrix. When the PVA fiber replacement ratio is 100%, the compressive strength (93.4 MPa) of the SHGC is the highest, with a 21.1% increase compared to the control group. However, the tensile strength shows a trend of first increasing and then decreasing with the increase in the PVA fiber replacement ratio, reaching the highest at a 25% replacement ratio, with a 12.5% increase compared to the control group. Furthermore, a comprehensive analysis of the economic and environmental performance of the SHGC indicates that a 25% PVA fiber replacement ratio results in the best overall economic benefits and relatively low actual costs, although the effect of fiber hybridization on carbon emission indicators is not significant. This paper provides new ideas and a theoretical basis for designing low-cost SHGCs.

Keywords: SHGC; non-oil-coated PVA fiber; hybrid; compressive; tensile; cost analysis

1. Introduction

Concrete is widely used due to its advantages, such as its ease of pouring and convenient sourcing. However, the cement used as a binding material emits a significant amount of carbon dioxide during its production process [1–3], causing serious environmental harm. Therefore, finding alternative materials is particularly urgent. Geopolymer is a novel binding material obtained by the alkaline activation of industrial solid wastes rich in silicon and aluminum (such as fly ash (FA), ground granulated blast furnace slag (GGBS), and red mud (RM) [4–6]). This material features high compressive strength, good impermeability, and excellent chemical resistance [7–9]. Moreover, its production process reduces carbon emissions by 80% compared to cement [10], making it a promising alternative. However, concrete produced with geopolymers remains brittle and has limitations in practical applications [11]. Current research indicates that adding materials such as rubber aggregates [12], nano-additives [13,14], and fibers [15,16] appears to be an effective method for mitigating the brittleness of concrete.

In recent years, scholars have developed a new type of fiber-reinforced concrete, called strain-hardening cementitious composite (SHCC), based on micromechanics principles [17–19]. Under tensile loading, this new concrete can achieve ultimate tensile strains of over 3%, with crack widths limited to less than 100 μm [20–22]. The application of this new concrete effectively addresses the brittleness issue commonly found in ordinary concrete. Leveraging the micro-design criteria of the SHCC, combining low-carbon geopolymers and high-performance fibers appears to be a promising approach to mitigate brittleness in geopolymers and reduce carbon emissions in the construction industry. Shaikh [23] first proposed the use of fly ash-based geopolymers as a complete substitute for cement to develop high-ductility strain-hardening geopolymer composites (SHGCs) in 2013, which showed even better deformability than SHCCs. Subsequently, numerous scholars have researched and optimized the performance of SHGCs [6,24–27]. Ohno and Li [28] utilized oil-coated polyvinyl alcohol (PVA) fibers and a fly ash-based geopolymer matrix to prepare a SHGC with a compressive strength reaching 27.6 MPa and tensile strain capacity of 4.3%. Lao et al. [29] employed ultra-high molecular weight polyethylene (UHMWPE) fibers, a slag-fly ash geopolymer matrix, and seawater sea sand to develop a seawater sea sand SHGC, achieving a compressive strength and ultimate tensile strain of over 140 MPa and 8%, respectively.

The trade-off for improving the mechanical properties of SHGCs is their high cost and high carbon footprint. The currently developed SHGC typically requires high-performance synthetic fibers such as oil-coated PVA fibers and PE fibers [20,30,31]. These high-performance fibers are expensive and have a significant carbon footprint [32], which limits the widespread adoption and application of SHGC materials. For example, oil-coating treatment on PVA fibers can reduce the chemical bonding between PVA fibers and the matrix, thereby enhancing the tensile deformability of SHCCs [33]. However, this process significantly increases the price of PVA fibers, making their proportion in the total material cost of a SHGC as high as 70% [32]. Therefore, reducing the fiber costs is an effective way to improve the economic feasibility of the SHCC/SHGC. Zhang et al. [34] used non-oil-coated PVA fibers to prepare a SHCC in an attempt to lower the material costs. Tensile tests comparing oil-coated PVA and non-oil-coated PVA on SHCCs showed that using non-oil-coated PVA reduced the tensile strength and deformability of the SHCC but still exhibited some strain-hardening capability. Furthermore, considering the hybrid use of low-cost fibers (for example, regenerated fibers such as polyethylene terephthalate (PET), polyamide (PA), and polypropylene (PP) [35]) with high-performance fibers can effectively reduce SHCC/SHGC costs [36–40]. Yu et al. [38] studied a typical SHCC (M45 in Li [41]) and found that replacing the PVA fibers with different proportions of recycled PET fibers negatively affected the tensile properties of the SHCC to varying degrees but improved its economic and environmental sustainability. Lin et al. [31] investigated the tensile mechanical behavior of a SHGC by replacing the PE fibers with different proportions of PP fibers. They observed a decrease in tensile strength with the use of PP fibers, but the ultimate tensile strain showed optimal performance at a 50% replacement ratio (9.71%), effectively enhancing the economic benefits of SHGCs.

However, current research on preparing SHGCs with hybrid fibers mostly leans towards using fibers with similar interface properties, which still has certain limitations. There is considerable room for improvement in the design of low-cost SHGCs. This study designs a SHGC using untreated low-cost PVA fibers mixed with high-performance PE fibers. Axial compressive and tensile tests were conducted on a SHGC with varying PE fiber volume fractions (1.0%, 1.5%, and 2.0%) and PVA fiber replacement ratios (0%, 25%, 50%, 75%, and 100%) to investigate the hybrid effects and reinforcement mechanisms of fibers with different surface characteristics (hydrophilic PVA fibers and hydrophobic PE fibers) on the mechanical behavior of a SHGC. Additionally, a comprehensive analysis of the economic and environmental benefits of the SHGC under different fiber parameters was performed. This study aims to provide reliable experimental data and theoretical foundations for designing cost-effective, high-performance SHGCs.

2. Experimental Program

2.1. Materials

SHGC's basic mix proportion referenced prior research by Lin et al. [31] and is primarily composed of precursors, quartz sand, alkaline activators, retarders, and synthetic fibers. The precursors consist of ground granulated blast furnace slag (GGBS) and fly ash (FA), classified according to the GB/T 18046 [42] and ASTM C618 [43] standards, with the GGBS categorized as an S105 grade and FA as Class F. Figure 1a shows the particle size distribution of slag, fly ash, and quartz sand, which exhibits a characteristic of "overall wide distribution with local narrow distribution". This ensures good fluidity of the fresh matrix and good density of the hardened SHGC. Table 1 shows the chemical composition of the GGBS and FA, with the GGBS providing a significant amount of calcium, thereby ensuring the strength of the SHGC. Alkali-activator comprises sodium hydroxide solution (concentration: 10 mol/L) and sodium silicate solution (modulus: 2.25, the mass ratio of SiO_2, Na_2O, and H_2O is 29.99:13.75:56.26), facilitating geopolymerization. $BaCl_2$ serves as a retarder, delaying the setting time of fresh SHGC paste to meet the workability requirements. The synthetic fibers used in this study include PE fibers and non-oil-coated PVA fibers, characterized by their physical parameters listed in Table 2. These data are sourced from the raw material manufacturer. Figure 1b–f shows the microscopic images of the raw materials, such as the precursors, quartz sand, and synthetic fibers. The equipment used for the scanning electron microscope (SEM) analysis was the Phenom Pharos G2, produced by Phenom-World B.V. (Eindhoven, the Netherlands). FA particles are spherical; thus, their incorporation can produce a "ball bearing effect", which positively contributes to the flowability of the SHGC. Figure 2 shows the XRD patterns of the GGBS and FA. The patterns indicate that the main crystalline phase of the GGBS is gehlenite, while the primary crystalline phases of FA are mulite and quartz. Additionally, both the GGBS and FA exhibit broad, amorphous peaks in the XRD patterns, which suggests that they possess good reactivity.

Figure 1. (**a**) Particle size distribution curves of GGBS, FA, and quartz sand; (**b–f**) microscopic images of raw materials. (**b**) GGBS; (**c**) FA; (**d**) quartz sand; (**e**) PE fibers; (**f**) PVA fibers.

Table 1. Chemical composition of GGBS and FA (unit: wt%).

Materials	Al_2O_3	SiO_2	CaO	Fe_2O_3	K_2O	SO_3	MgO	TiO_2	Other
FA	31.14	53.96	4.01	4.16	2.03	0	1.01	1.13	2.53
GGBS	17.70	34.50	34.00	1.03	0	1.64	6.01	0	5.12

Table 2. Physical properties of PE and PVA fibers.

Fiber Type	Density (g/cm^3)	Strength (MPa)	Elastic Modulus (GPa)	Length (mm)	Diameter (μm)	Elongation at Break (%)
PE	0.97	2500	120	12	20	3.7
PVA	1.29	1830	40	12	15	6.9

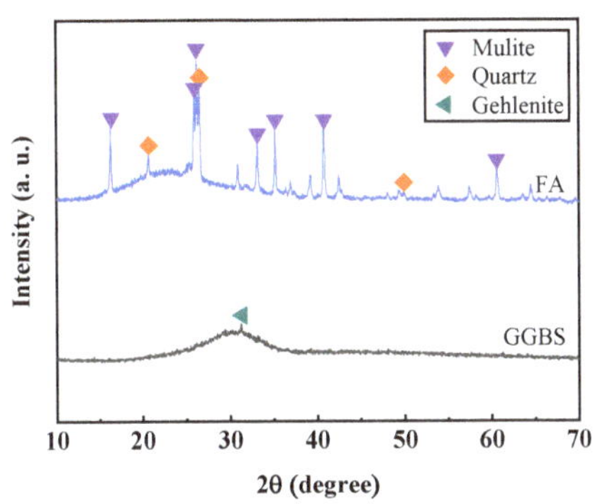

Figure 2. XRD patterns of GGBS and FA.

2.2. Mix Proportions and Specimens

This study employed cylindrical and dumbbell-shaped specimens for the axial compressive and tensile tests, respectively, with the specific dimensions shown in Figure 3. The study variables included PE fiber volume fractions (1.0%, 1.5%, and 2.0%) and PVA fiber replacement ratios (0%, 25%, 50%, 75%, and 100%). PVA fibers were used to replace PE fibers by volume, with a fixed fiber volume fraction of 2%. The specific mix proportions are detailed in Table 3.

Table 3. Mix proportions of SHGC (unit: kg/m^3, with fibers represented by volume fraction).

Mix IDs	GGBS	FA	Quartz Sand	Alkali-Activator	Water	$BaCl_2$	PE Fiber (%)	PVA Fiber (%)
U1.0-P0.0							1.0	0.0
U1.5-P0.0							1.5	0.0
U2.0-P0.0							2.0	0.0
U1.5-P0.5	849.7	364.1	242.8	242.8	72.8	12.1	1.5	0.5
U1.0-P1.0							1.0	1.0
U0.5-P1.5							0.5	1.5
U0.0-P2.0							0.0	2.0.

Note: The specimen grouping is represented in the Ux-Py format, where U stands for ultra-high molecular weight polyethylene (UHMWPE) fibers, P stands for PVA fibers, and x and y represent the volume fractions of the two fibers, respectively. For example, U1.0-P1.0 denotes a specimen containing 1.0% PE fiber and 1.0% PVA fiber.

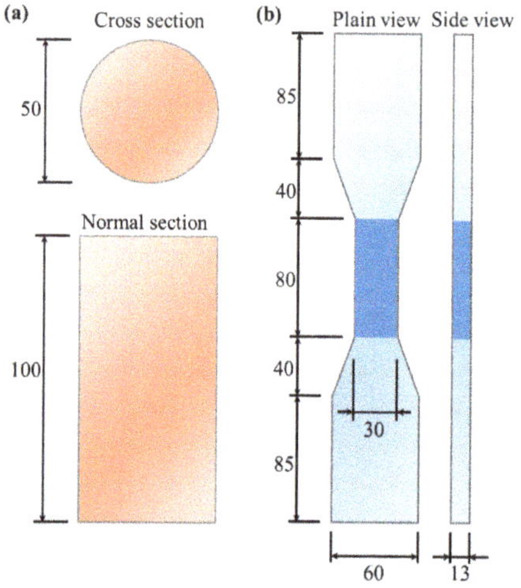

Figure 3. Experimental specimen dimensions (unit: millimeter, mm). (**a**) Cylinder specimen; (**b**) dumbbell-shaped specimen.

The specimen preparation process is as follows: initially, GGBS, FA, quartz sand, and $BaCl_2$ are added to a planetary mixer and stirred at a speed of 75 rpm for 2 min. Then, alkaline activators are added and stirred at 75 rpm for 1 min (notably, the alkaline activator, composed of sodium hydroxide solution and sodium silicate solution, needs to be mixed and cooled for 24 h before use). After thorough mixing, additional water is added and stirred at 75 rpm for 1 min. Finally, the PE and PVA fibers are sequentially added and stirred for 3 min to obtain the SHGC paste. The fresh paste is poured into prepared molds, covered with plastic film, and cured at room temperature for 24 h before being transferred to a water tank for further curing for 28 days. Figure 4 illustrates the specimen preparation process.

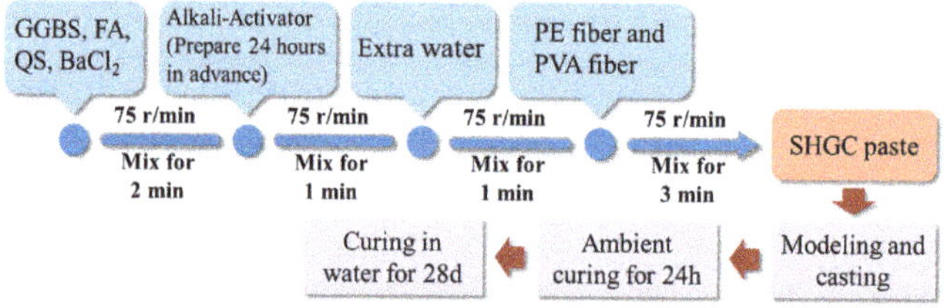

Figure 4. Process of SHGC specimen preparation.

2.3. Test Methods

According to ASTM C1437 [44], the flowability of the fresh SHGC pastes was tested. Using a steel ruler, the diameters in two perpendicular directions of the spread paste were measured to determine its flowability.

For the axial compressive test, following ASTM C109/C109M [45], a displacement-controlled loading method was employed at a rate of 0.2 mm/min. The compressive tests were conducted using the electro-hydraulic servo pressure testing machine, model C088-01, produced by MATEST (Bergamo, Italy). The load data were collected through the built-in sensors of the machine. Before loading, strain gauges were attached longitudinally to the cylindrical specimens to measure the longitudinal strains and determine the elastic modulus. Additionally, two linear variable differential transformers (LVDTs) were set up to measure the axial displacements of the specimens throughout the loading process, with a gauge length of 50 mm. During loading, data for the displacement, strain, and load were synchronously collected using a TDS540 data acquisition system at a frequency of 1 Hz.

The number of test samples was three. The specific experimental setup is illustrated in Figure 5a.

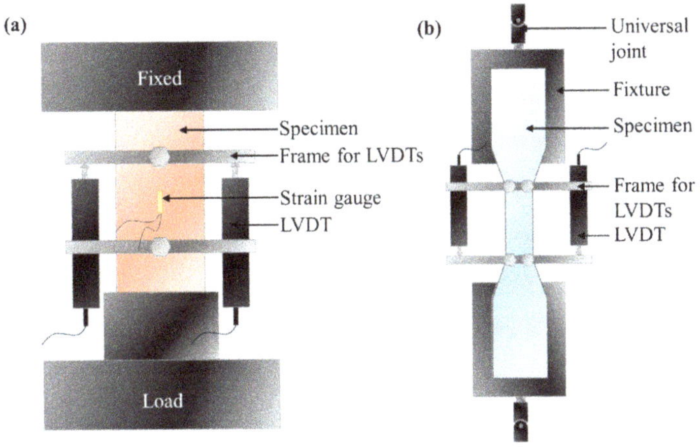

Figure 5. Experimental setup: (**a**) axial compressive test; (**b**) axial tensile test.

The axial tensile test, following JSCE (2008) [46], was conducted using a controlled loading method at a rate of 0.5 mm/min. The tensile tests were conducted using the YNS-Y3000 electro-hydraulic servo universal testing machine produced in China. The load data were collected through the built-in sensors of the machine. The specimen was secured at both ends with universal joints connected to the loading end of the testing machine to prevent eccentric tension during the test. Linear variable differential transformers (LVDTs) were placed on both sides of the specimen to measure the longitudinal displacements during tension, with a gauge length of 80 mm. Additionally, during the tensile test, the displacement data were collected using a TDS540 data acquisition system, while the load data were obtained directly from the universal testing machine, both at a frequency of 1 Hz. The number of test samples was three. The specific experimental setup is depicted in Figure 5b.

To analyze the microstructure and composition of the samples, we used a scanning electron microscope (SEM). During the testing, we employed the backscattered electrons (BSE) mode to enhance the resolution of the sample's elemental composition and phase distribution. We prepared two cubic samples, each with an edge length of approximately 7 mm, from the fracture section of the tensile specimens. One sample was used to observe the fiber distribution and failure modes, with an acceleration voltage of 5 kV applied to avoid damage to the fibers by the electron beam. The other sample's observation surface was polished to a smooth finish using silicon carbide sandpaper up to a 7000-grit level. Elemental distribution of the fiber–matrix interfacial transition zone (ITZ) was obtained using the EDS line scan mode at an acceleration voltage of 15 kV. For EDS testing, C, Ca, and Si were selected to differentiate between the fibers, matrix, and ITZ.

3. Results and Discussion

3.1. Flowability

Figure 6 illustrates the flowability of fresh SHGC. As the volume fraction of the PE fibers increases from 1% to 2%, the flowability of the fresh SHGC paste decreases from 177.5 mm to 157.5 mm, marking an 11.3% reduction. This reduction is attributed to the tendency of flexible PE fibers to bend and interfere with the internal components of the fresh paste, thereby reducing the flowability [47]. Increasing the content of PE fibers exacerbates this negative effect. Furthermore, Figure 6 shows that increasing the replacement ratios of PVA fibers significantly reduces the flowability of the fresh SHGC paste. For instance, U0.0-

P2.0 exhibits a 7.9% decrease in flowability compared to U2.0-P0.0. Unlike hydrophobic PE fibers, the hydroxyl groups on the surface of PVA fibers may form physical–chemical bonds with the polar geopolymer matrix [48], potentially increasing the interaction between fibers and the matrix, thereby reducing the paste's flowability. Optimal flowability is crucial for minimizing internal defects in SHGC. Therefore, selecting appropriate fiber volume fractions and PVA replacement ratios is essential.

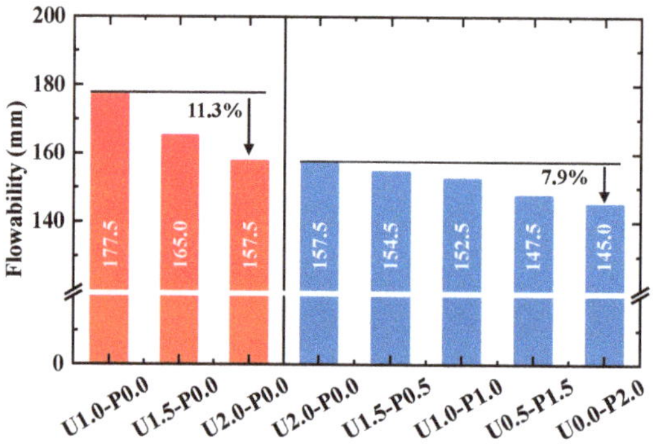

Figure 6. Flowability of fresh SHGC.

3.2. *Compressive Behavior*

3.2.1. Failure Mode of Compressive Specimens

Figure 7 depicts the compressive failure modes of the SHGC. We used red lines in the figure to depict and highlight the cracks. Under axial compressive loading, microcracks initiate and gradually propagate within the concrete, connecting and closing with an increasing load. As the load increases, these microcracks accumulate and form macroscopic cracks when the load reaches its peak. At this point, strain energy is suddenly released, causing the SHGC specimen to rapidly lose its load-bearing capacity. In Figure 7, it is evident that the primary cracks in the SHGC are predominantly oriented longitudinally, exhibiting a mode of failure characterized by longitudinal splitting, indicating significant brittleness. However, the U2.0-P0.0 specimen shows the smallest crack inclination angle, suggesting an oblique shear failure mode. This indicates that the inclusion of high-strength PE fibers at a volume fraction of 2% helps mitigate the compressive brittleness of the SHGC. Although the hydrophilicity of PVA fibers allows them to form a good interfacial bond with the matrix (as evidenced by the microscopic analysis in Section 3.3.1), their relatively low tensile strength makes them prone to rupture when macro-cracks develop in the compressive specimens. This, in turn, causes the SHGC to exhibit brittle failure characteristics.

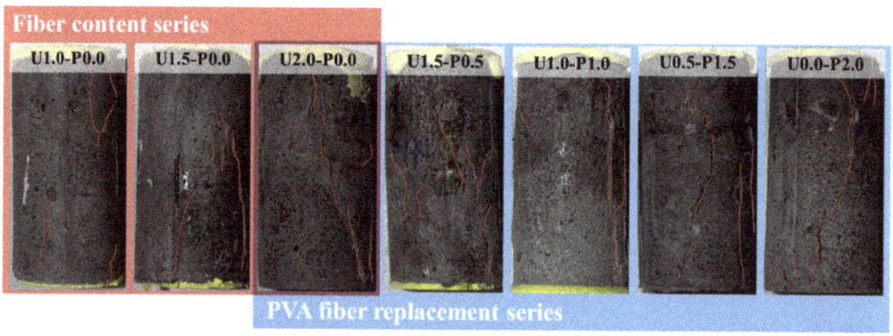

Figure 7. Compressive failure modes of SHGC.

3.2.2. Compressive Stress–Strain Curves

Figure 8 depicts the compressive stress–strain curves of the SHGC. In Figure 8, it can be observed that the stress–strain curves, except for U2.0-P0.0, exhibit a sudden drop in stress after reaching the peak, indicating significant brittle failure characteristics consistent with the longitudinal splitting failure mode. Figure 8a shows that as the volume fraction of PE fibers increases, the descending branch of the curve gradually becomes more gradual, albeit with a decrease in peak strength. Compared to the brittle behavior of U1.0-P0.0, which loses its load-bearing capacity immediately after peaking, U2.0-P0.0 shows a slow decline in stress after reaching the peak, indicating improved ductility. A sufficient fiber content effectively suppresses the development of internal defects in the specimen, thereby enhancing compressive ductility. Figure 8b indicates that as the replacement ratio of the PVA fibers increases, the descending branch of the curve becomes steeper, exhibiting brittle behavior similar to that observed with a low PE fiber content, albeit with an increase in peak strength. The limited strength of PVA fibers and their strong interface effect with the matrix make them more prone to fracture at the crack surfaces rather than pull-out failure (discussed in Section 3.3.1), leading to post-peak brittleness in the SHGC.

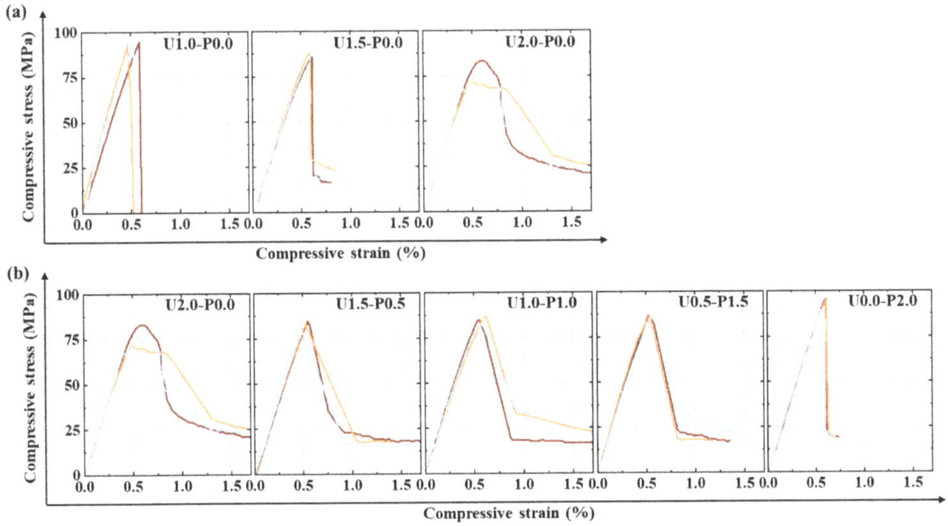

Figure 8. Compressive stress–strain curves of SHGC: (**a**) fiber content series; (**b**) PVA fiber replacement series.

3.2.3. Compressive Properties

From the compressive stress–strain curves in Figure 8, various performance parameters of SHGC can be obtained, including the peak strength, peak strain, and elastic modulus. The elastic modulus is calculated according to ASTM C469/C469M [49] (Equation (1)). The specific results are shown in Table 4. The values in parentheses represent the standard deviation.

$$E = \frac{\sigma_2 - \sigma_1}{\varepsilon_2 - 0.000050} \quad (1)$$

where, σ_1 is the stress corresponding to a longitudinal strain of 0.000050; $\sigma_2 = 0.4\sigma_{CM}$, σ_{CM} is the peak stress; ε_2 is the strain corresponding to σ_2.

Table 4. Compressive properties.

Mix IDs	Compressive Strength (MPa)	Peak Strain (%)	Elastic Modulus (GPa)
U1.0-P0.0	93.9 (0.49)	0.532 (0.05)	19.42 (1.90)
U1.5-P0.0	87.0 (0.83)	0.606 (0.01)	15.97 (0.32)
U2.0-P0.0	77.1 (4.69)	0.565 (0.05)	16.99 (0.10)
U1.5-P0.5	83.2 (1.43)	0.565 (0.02)	16.96 (0.16)
U1.0-P1.0	84.7 (1.80)	0.578 (0.03)	16.46 (0.42)
U0.5-P1.5	87.0 (0.41)	0.552 (0.03)	17.49 (0.52)
U0.0-P2.0	93.4 (2.30)	0.598 (0.01)	16.29 (0.35)

Figure 9a shows the compressive strength of the SHGC under different PE fiber contents and PVA fiber replacement ratios. With an increasing PE fiber content, the compressive strength of the SHGC decreases approximately linearly. Compared to U1.0-P0.0, U2.0-P0.0 exhibits a decrease in compressive strength by 17.9%. This is attributed to a significant reduction in flowability when the PE fiber content reaches 2%, potentially increasing internal defects in the SHGC [50]. Additionally, the weaker synergistic interaction between the hydrophobic PE fibers and the matrix before cracking, along with excessive PE fibers, may contribute to defects themselves, thereby leading to a decrease in compressive strength. In contrast, replacing PE fibers with PVA fibers effectively enhances the compressive strength. As the PVA fiber replacement ratio increases, the compressive strength gradually improves. Compared to U2.0-P0.0, U0.0-P2.0 shows an increase in compressive strength by 21.1%. This enhancement may be due to the strong interfacial bonding between PVA fibers and the matrix, which enhances their synergistic effect with the matrix and effectively restrains crack propagation, thereby improving the compressive strength of the SHGC [51].

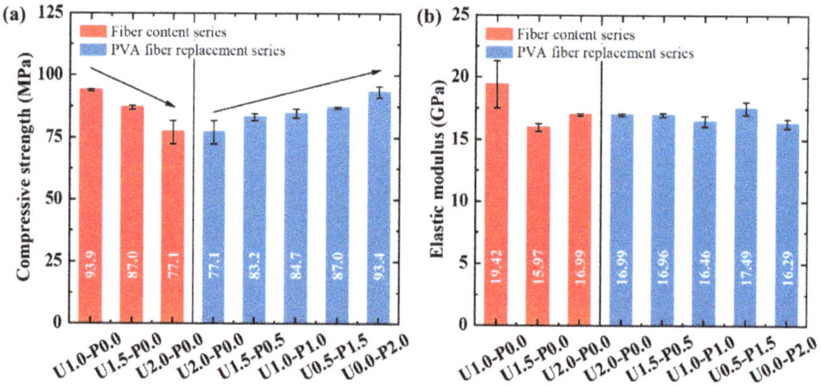

Figure 9. Compressive performance of SHGC under different PE fiber contents and PVA fiber replacement ratios. (**a**) Compressive strength; (**b**) elastic modulus.

Figure 9b presents the elastic modulus of the SHGC under different PE fiber contents and PVA fiber replacement ratios. Increasing the PE fiber content adversely affects the elastic modulus, while the replacement of PE fibers with PVA fibers shows minimal impact on the elastic modulus. Sakulich and Li [52] and Lei et al. [53] found, through nanoindentation tests, that the lateral elastic modulus of PE and PVA fibers is much lower than that of the matrix. Therefore, the inherently lower elastic modulus of synthetic fibers is the primary reason for the reduction in the compressive elastic modulus of the SHGC.

3.3. *Tensile Behavior*

3.3.1. Failure Mode of Tensile Specimens

Figure 10 illustrates the axial tensile failure modes of typical SHGC specimens. Under tensile loading, the SHGC initially cracks at the weakest section, followed by stress redistribution, where the tension is borne by the fibers. As the load increases, the SHGC continues to develop cracks, ultimately leading to a failure mode characterized by multiple crack propagation. The bridging capability of fibers directly influences the tensile failure mode [54]. With an increasing PE fiber content, the number of cracks gradually increases while the residual crack width decreases. Partial replacement of the PE fibers with PVA fibers results in a reduction in the number of cracks in the SHGC. When the PVA fiber replacement ratio exceeds 75%, specimens U0.5-P1.5 and U0.0-P2.0 exhibit a sharp decrease in crack number, indicating a pronounced brittle tensile failure mode.

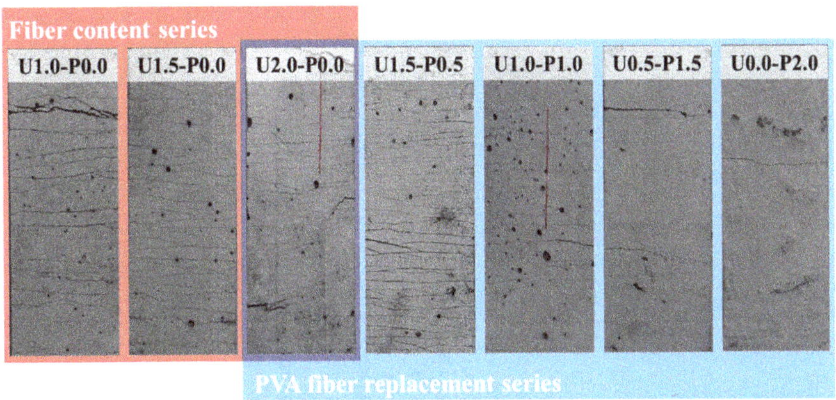

Figure 10. Typical Axial Tensile Failure Modes of SHGC Specimens.

Using a microscope (TD-2KH, produced by Sanqtid (Shenzhen, China), residual crack widths of the axial tensile failure specimens were measured, and the number of cracks and crack density (ratio of the crack number to measured length, mm^{-1}) were statistically analyzed, as shown in Figure 11. It is noteworthy that specimens U0.5-P1.5 and U0.0-P2.0, which exhibited brittle failure, are not included in the statistics. Figure 11a presents the crack density of the SHGC. With an increasing PE fiber content, the crack density gradually increases. Compared to U1.0-P0.0, U2.0-P0.0 shows a 189% increase in crack density, indicating that an increasing fiber content effectively enhances the fiber bridging capability, which is consistent with the findings of Wang et al. [55]. However, replacing 50% of the PE fibers with PVA fibers results in a 65.4% reduction in crack density. Figure 11b displays the residual crack width of the SHGC. Increasing the PE fiber content effectively suppresses crack propagation, with U2.0-P0.0 having a residual crack width of only 43.3% compared to U1.0-P0.0. Replacement with PVA fibers has a minimal effect on the residual crack width.

Figure 12 shows the microstructural features of the SHGC tensile specimen fracture surfaces. Figure 12a displays the fracture surface morphology of specimens U1.5-P0.5, U2.0-P0.0, and U0.0-P2.0. It can be observed that the fiber distribution on the fracture surface of U0.0-P2.0 is sparser, indicating a tendency for PVA fibers to undergo fracture failure at the crack surface, whereas PE fibers tend to experience pull-out failure due to their higher strength. Figure 12b presents the micrographs of the fiber distribution at the fracture surface of specimens U1.5-P0.5, U2.0-P0.0, and U0.0-P2.0. It is evident that the fracture surface of U2.0-P0.0 predominantly exhibits pull-out failure of the PE fibers, whereas a significant proportion of the PVA fibers undergo fracture failure at the fracture surface of U0.0-P2.0. This contributes to the reduced saturation cracking performance of the SHGC with an increasing PVA fiber content. Figure 12c shows an enlarged view of the fiber ends at the fracture surface of specimen U1.5-P0.5, which was mixed with both PE and PVA fibers, revealing the pull-out and fracture failures of PE and PVA fibers, respectively. Figure 12d illustrates the microstructure of the fibers and matrix, showing a loose interface transition zone (ITZ) between the PE fibers and the matrix, whereas the PVA fibers, which possess hydrophilic characteristics, exhibit tight bonding with the matrix. Linear scans of the fiber–matrix ITZ using an energy-dispersive spectrometer (EDS), distinguishing elements C, Al, and Si, are shown in Figure 12e. In Figure 12e, it can be observed that the ITZ width between the PVA fibers and the matrix is smaller compared to that between the PE fibers and the matrix, indicating better interface properties between the PVA fibers and the matrix, enabling an improved stress transfer and synergistic effects with the matrix.

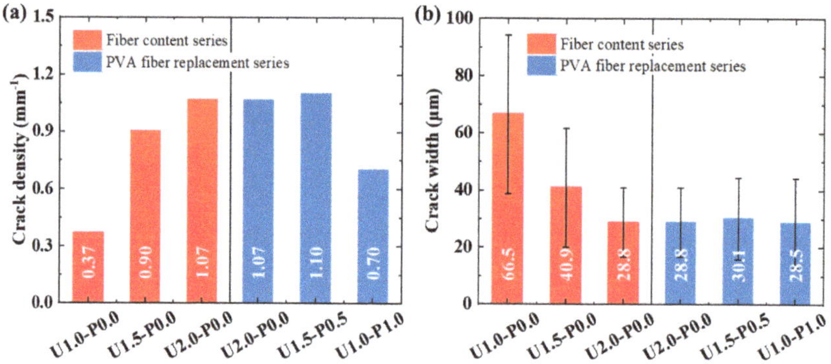

Figure 11. Crack characteristic parameters of typical SHGC specimens. (**a**) Crack density; (**b**) residual crack width.

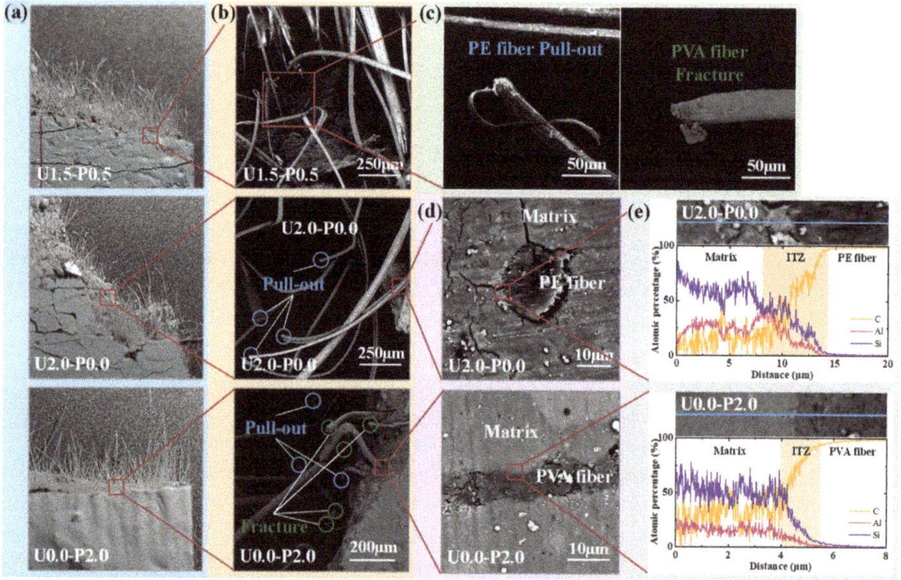

Figure 12. SHGC tensile specimen fracture surface microanalysis. (**a**) Fracture surface morphology of the specimen; (**b**) SEM images of the specimen fracture surface; (**c**) pull-out failure characteristics of

PE fibers and fracture failure characteristics of PVA fibers; (**d**) microstructure of fibers, matrix, and ITZ (interfacial transition zone); (**e**) element distribution at fibers, matrix, and ITZ obtained by linear scanning using EDS.

3.3.2. Tensile Stress–Strain Curves

Figure 13 shows the stress–strain curves of the SHGC. In Figure 13, it can be seen that, except for U0.5-P1.5 and U0.0-P2.0, which experienced brittle failure, the stress–strain curves of the SHGC exhibit strain-hardening after cracking, accompanied by "sawtooth" fluctuations due to stress redistribution. Figure 13a indicates that with an increase in the PE fiber volume fraction, both tensile strength and ultimate tensile strain significantly improve. Figure 13b demonstrates that with a 25% substitution of PVA fibers, the tensile strength increases, but the tensile strain decreases noticeably. Further increasing the substitution rate of PVA fibers to 50% enhances the initial cracking strength but results in a significant decrease in tensile strength and ultimate tensile strain. The tensile strain energy density is represented by the area under the stress–strain curve and is calculated using Equation (2):

$$g_{se} = \int_{0}^{\varepsilon_u} \sigma(\varepsilon)d\varepsilon \quad (2)$$

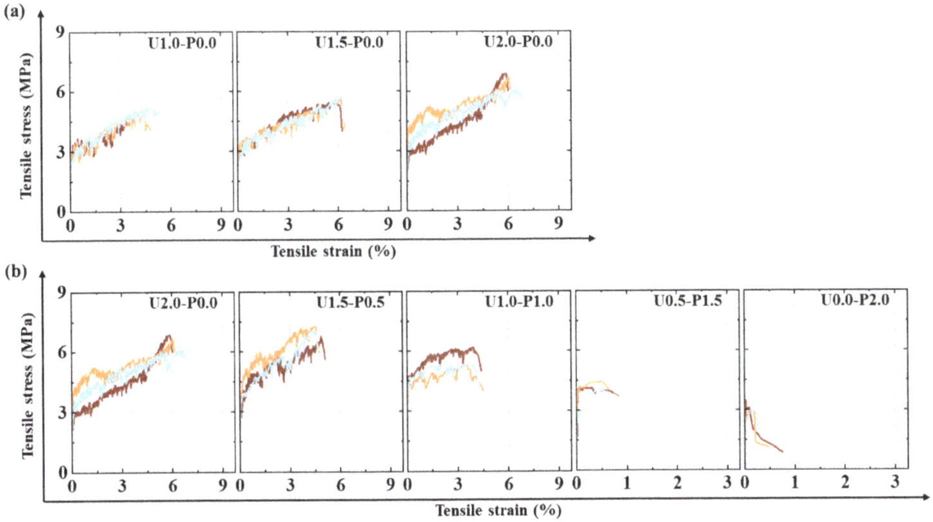

Figure 13. SHGC tensile stress–strain curves: (**a**) Fiber content series; (**b**) PVA fiber replacement series.

3.3.3. Tensile Properties

From the tensile stress–strain curves in Figure 13, the tensile performance parameters of the SHGC can be obtained, including the initial cracking strength (σ_c), initial cracking strain (ε_c), tensile strength (σ_u), ultimate tensile strain (ε_u), and tensile strain energy density (g_{se}). The specific results are presented in Table 5.

Figure 14 presents the tensile performance parameters of the SHGC under different PE fiber contents and PVA fiber replacement rates. Figure 14a displays the initial cracking strength of the SHGC under various fiber parameters. As the PE fiber content increases, there is a rising trend in the initial cracking strength, with U2.0-P0.0 showing a 19.2% increase compared to U1.0-P0.0. Partial or complete replacement of the PE fibers with PVA fibers effectively enhances the initial cracking strength, exhibiting a trend of increase followed by a decrease. Due to the high tensile modulus of PE fibers reaching 120 GPa—much higher than the matrix—increasing their content can enhance the initial cracking strength. The hydrophilicity of PVA fibers allows for better synergistic effects with the matrix before cracking occurs in the SHGC, effectively increasing the initial cracking strength. The initial cracking strength of U1.0-P1.0 reaches 4.3 MPa. However, excessive replacement of the PVA fibers may introduce defects due to a loss of flowability, thereby negatively affecting the initial cracking strength, with a decrease observed after a replacement rate of 50%. Additionally, Figure 14a indicates that when the overall fiber volume content is 2% (including the different PVA fiber replacement rates), the initial cracking strength of the SHGC exhibits significant dispersion. The initial cracking strength in tension is primarily controlled by the matrix. The increased internal porosity and its uneven distribution at high fiber volume fractions lead to a greater dispersion in the matrix's initial cracking strength.

Table 5. Tensile properties.

Mix IDs	σ_c (MPa)	ε_c (%)	σ_u (MPa)	ε_u (%)	g_{se} (kJ/m^3)
U1.0-P0.0	2.50 (0.27)	0.024 (0.017)	4.92 (0.30)	4.74 (0.90)	167.0 (34.2)
U1.5-P0.0	2.47 (0.08)	0.027 (0.007)	5.74 (0.20)	6.16 (0.09)	262.2 (7.3)
U2.0-P0.0	2.98 (0.62)	0.069 (0.009)	6.26 (0.88)	6.07 (0.11)	299.4 (36.5)
U1.5-P0.5	3.46 (0.55)	0.057 (0.026)	7.04 (0.26)	4.70 (0.21)	252.7 (19.5)
U1.0-P1.0	4.30 (0.20)	0.050 (0.034)	5.64 (0.47)	3.80 (0.13)	194.8 (26.2)
U0.5-P1.5	3.74 (0.74)	0.032 (0.017)	4.19 (0.21)	0.80 (0.13)	17.1 (4.6)
U0.0-P2.0	3.10 (0.33)	0.025 (0.016)	3.10 (0.33)	0.52 (0.23)	4.3 (1.6)

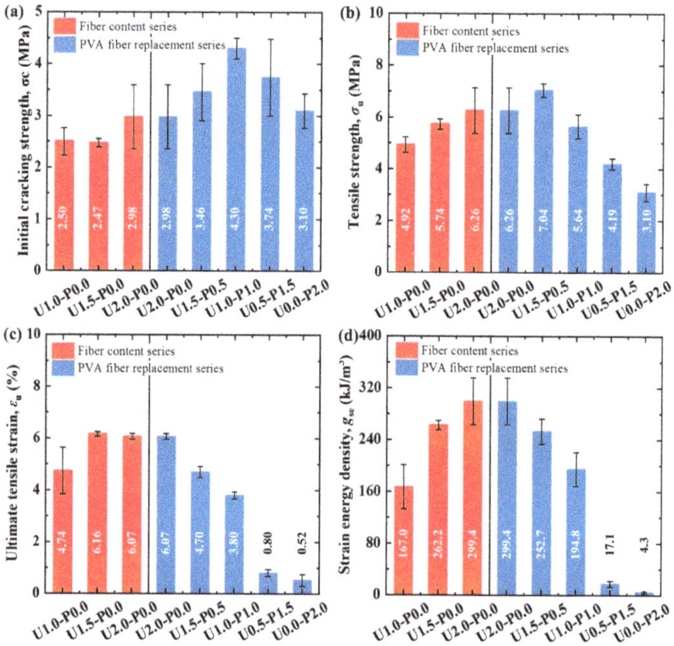

Figure 14. Tensile performance indicators of SHGC. (**a**) Initial cracking strength; (**b**) tensile strength; (**c**) ultimate tensile strain; (**d**) strain energy density.

Figure 14b shows the tensile strength of the SHGC under different fiber parameters. The variation trend of tensile strength is similar to that of the initial cracking strength, except that the tensile strength begins to decrease after a PVA fiber replacement ratio of 25%. The lower strength of PVA fibers and their propensity to fracture rather than pull out at the SHGC cracks (as noted in Section 3.3.1) contribute to the degradation of tensile strength. Unlike the initial cracking strength, the tensile strength at the ultimate state is primarily controlled by the fiber's bridging performance. The high viscosity of the matrix ensures good fiber dispersion uniformity, allowing the tensile strength of the SHGC to exhibit good consistency.

Figures 14c and 13d, respectively, illustrate the ultimate tensile strain and tensile strain energy density of the SHGC under different fiber parameters. The SHGC with 2% PE fibers exhibited the highest tensile deformation capability and strain energy density. PVA fibers tend to fracture prematurely during tension, reducing both the deformation and energy absorption capabilities, as evidenced by the consistent patterns in the residual crack number and width changes.

3.4. Economic and Environmental Analyses

Reducing the content of PE fibers and partially replacing them with cost-effective PVA fibers can enhance the economic and environmental sustainability of SHGCs. Furthermore, analysis of the experimental results indicates that both methods positively impact certain mechanical properties. Therefore, considering mechanical performance better reflects the influence of the fiber parameter variations on the economic and environmental performance of the SHGC.

Table 6 presents the carbon emissions and material costs used for the calculations, with the carbon emission factor for barium chloride currently unverifiable and therefore not considered. The material costs were based on the local purchase prices and converted to USD using the exchange rate from July 2024. The carbon emission index (CI) and comprehensive cost index (CCI) [56] were employed as environmental performance assessment metrics, calculated using Equations (3) and (4), respectively. Table 7 and Figure 15 summarize the comprehensive carbon emissions and costs of the SHGC per cubic meter under different fiber parameters, where both the carbon emissions and cost indices have been normalized.

$$CI_{\mathrm{i}} = \frac{M_{\mathrm{i}}}{CE} \tag{3}$$

$$CCI_{\mathrm{i}} = \frac{M_{\mathrm{i}}}{Cost} \tag{4}$$

Table 6. Carbon emission factors and material costs of raw materials.

Materials	GGBS	FA	NaOH	Na_2SiO_3	$BaCl_2$	QS	PE Fibers	PVA Fibers	Water
Cost (USD/t)	48.7	26.4	2783.4	417.5	3340.1	55.7	55,667.6	6262.6	0.42
Carbon emission factor (kg CO_2/kg)	0.019 [56]	0.009 [56]	1.915 [56]	0.804 [57]	-	0.010 [56]	2.000 [58]	1.710 [59]	0.8 [59]

Table 7. Comprehensive carbon emissions and cost indices (normalized) of SHGC under different fiber parameters.

Mix IDs	*Cost*	*CE*	CI_c	CI_t	CI_ε	CI_g	CCI_c	CCI_t	CCI_ε	CCI_g
U1.0-P0.0	1.00	1.00	1.00	1.00	1.00	1.00	1.00	1.00	1.00	1.00
U1.5-P0.0	1.30	1.02	0.90	1.14	1.27	1.53	0.90	1.00	1.21	0.90
U2.0-P0.0	1.60	1.05	0.78	1.21	1.22	1.71	0.80	0.80	1.12	0.80
U1.5-P0.5	1.34	1.05	0.84	1.36	0.94	1.44	1.07	0.74	1.13	1.07
U1.0-P1.0	1.09	1.06	0.85	1.08	0.76	1.10	1.05	0.74	1.07	1.05
U0.5-P1.5	0.84	1.06	0.87	0.80	0.16	0.10	1.02	0.20	0.12	1.02
U0.0-P2.0	0.58	1.06	0.94	0.59	0.10	0.02	1.08	0.19	0.04	1.08

Note: CI_c, CI_t, CI_ε, and CI_g, respectively, denote comprehensive carbon emission indices considering compressive strength, tensile strength, ultimate tensile strain, and tensile strain energy density. CCI_c, CCI_t, CCI_ε, and CCI_g, respectively, denote comprehensive cost indices considering compressive strength, tensile strength, ultimate tensile strain, and tensile strain energy density.

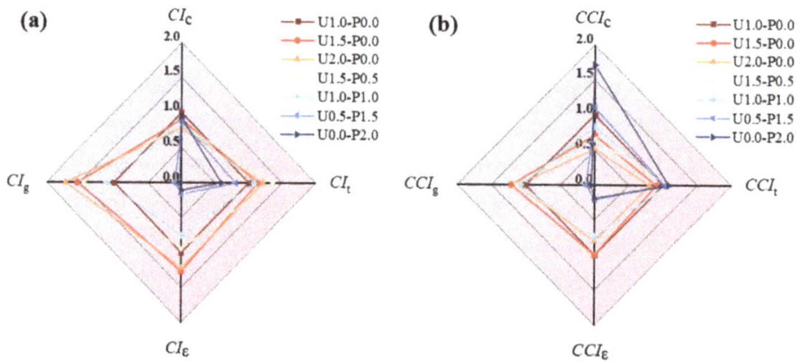

Figure 15. Radar chart analyses of comprehensive carbon emissions and cost indices of SHGC under different fiber parameters. (**a**) Comprehensive carbon emission indices; (**b**) comprehensive cost indices.

In this context, M_i represents the mechanical performance indicators, including compressive strength (MPa), tensile strength (MPa), ultimate tensile strain (%), and tensile strain energy density (kJ/m^3).

From Figure 15a, it is evident that increasing the PE fiber content and partially replacing the PE fibers with PVA fibers results in higher comprehensive cost indices. Specifically, U1.5-P0.0 with 1.5% and U2.0-P0.0 with 2% PE fibers, as well as U1.5-P0.5 with 25% PVA fiber replacement, exhibit significantly higher cost-effectiveness in terms of the enveloping area on the radar chart. Each of these configurations shows different mechanical performance advantages, reflecting varying strengths in the comprehensive cost index. Figure 15b illustrates that U1.0-P0.0, using 1% PE fibers, excels in comprehensive environmental indices, with the largest enveloping area on the radar chart, indicating optimal comprehensive environmental benefits. However, due to PVA's inherently higher carbon emission factor, the environmental benefits of partially or completely replacing the PE fibers with PVA fibers in the SHGC are not prominently enhanced. This effect becomes more pronounced when the PVA fiber replacement rate exceeds 75%, as indicated by a significant reduction in the radar chart's enveloping area. Comprehensive economic and environmental indices are not the sole criteria for decision-making. When designing SHGC compositions, it is crucial to consider the actual requirements and balance the mechanical performance, economic benefits, environmental impacts, and other performance indicators in a multidimensional manner [54].

4. Conclusions

This paper investigates the axial compressive and tensile tests of strain-hardening geopolymer composites (SHGCs) with varying fiber parameters (the fiber volume fraction and PVA fiber replacement ratio). It explores the influence mechanisms of fiber parameters on static mechanical behaviors and comprehensively analyzes the economic and environmental benefits of the SHGC. The main conclusions obtained are as follows:

Increasing the volume fraction of high-performance PE fibers can effectively inhibit crack formation, thereby enhancing the post-peak ductility of the SHGC specimens under axial compression loads. However, increasing the volume fraction of PVA fibers results in brittle failure behavior under compression for the SHGC. This is because the relatively low strength of PVA fibers causes them to fracture at the cracked sections, and they are unable to effectively dissipate the strain energy during compressive failure.

The hydrophobic PE fibers have weak synergy with the matrix; an excessive amount of PE fibers can become defects themselves, leading to a degradation in compressive strength. The compressive strength of the SHGC decreased by 17.9% when the PE fiber content increased from 1% to 2%. In contrast, increasing the volume fraction of PVA fibers can effectively improve compressive strength due to their hydrophilic nature, which results in better interfacial bonding with the matrix, enhancing their synergistic action. When the PVA fiber replacement rate is 100%, the compressive strength is 21.1% higher than when the replacement rate is 0%.

Increasing the PE fiber content comprehensively improves the tensile properties of the SHGC, including the initial cracking strength, tensile strength, ultimate tensile strain, and tensile strain energy. However, increasing the PVA fiber replacement rate only improves the tensile strength of the SHGC (with a maximum tensile strength of 7.04 MPa at a 25% PVA fiber replacement rate) while negatively impacting the tensile deformation capacity and energy dissipation capacity. When the PVA fiber replacement rate exceeds 50%, the multiple cracking ability of the SHGC significantly declines, resulting in brittle tensile failure.

A comprehensive analysis of the production costs, carbon emissions, and mechanical performance indicators of the SHGC indicates that a PVA fiber replacement ratio of 25% achieves great comprehensive economic benefits with lower actual costs. The small difference in carbon emission factors between PVA and PE fibers means that the method of PVA fiber replacement has a minimal impact on carbon emission indices.

Author Contributions: Conceptualization, Z.C.; methodology, J.G. and C.L.; software, J.S. and L.W.; validation, J.G. and C.H.; formal analysis, X.T.; investigation, J.S.; resources, J.G.; data curation, C.L.; writing—original draft preparation, J.G.; writing—review and editing, Z.C.; visualization, J.S.; supervision, L.W.; project administration, C.H.; funding acquisition, Z.C. All authors have read and agreed to the published version of the manuscript.

Funding: This research was funded by the Huizhou Power Supply Bureau, Guangdong Power Grid Co., Ltd., China Southern Power Grid Co., Ltd., grant number 0313002023030201XG00061.

Data Availability Statement: The data presented in this study are available on request from the corresponding author. The data are not publicly available due to confidentiality issues.

Conflicts of Interest: Authors Jingen Guo, Ji Shi, Liuhuo Wang, Chengyong Huang and Xiongwu Tao were employed by the companies Huizhou Power Supply Bureau, Guangdong Power Grid Co., Ltd., and China Southern Power Grid Co., Ltd. Author Jingen Guo was also employed by Guangdong Power Grid Co., Ltd., Grid Planning Research Centers. The remaining authors declare that the research was conducted in the absence of any commercial or financial relationships that could be construed as a potential conflict of interest.

References

1. Ostovari, H.; Müller, L.; Skocek, J.; Bardow, A. From Unavoidable CO_2 Source to CO_2 Sink? A Cement Industry Based on CO_2 Mineralization. *Environ. Sci. Technol.* **2021**, *55*, 5212–5223. [CrossRef] [PubMed]
2. Rootzén, J.; Johnsson, F. Managing the costs of CO_2 abatement in the cement industry. *Clim. Policy* **2016**, *17*, 781–800. [CrossRef]
3. Winnefeld, F.; Leemann, A.; German, A.; Lothenbach, B. CO_2 storage in cement and concrete by mineral carbonation. *Curr. Opin. Green Sustain. Chem.* **2022**, *38*, 100672. [CrossRef]
4. Rees, C.A.; Provis, J.L.; Lukey, G.C.; van Deventer, J.S. The mechanism of geopolymer gel formation investigated through seeded nucleation. *Colloids Surfaces A Physicochem. Eng. Asp.* **2008**, *318*, 97–105. [CrossRef]
5. Zakira, U.; Zheng, K.; Xie, N.; Birgisson, B. Development of high-strength geopolymers from red mud and blast furnace slag. *J. Clean. Prod.* **2023**, *383*, 135439. [CrossRef]
6. Lin, J.-X.; Liu, R.-A.; Liu, L.-Y.; Zhuo, K.-Y.; Chen, Z.-B.; Guo, Y.-C. High-strength and high-toughness alkali-activated composite materials: Optimizing mechanical properties through synergistic utilization of steel slag, ground granulated blast furnace slag, and fly ash. *Constr. Build. Mater.* **2024**, *422*, 135811. [CrossRef]
7. Gharzouni, A.; Vidal, L.; Essaidi, N.; Joussein, E.; Rossignol, S. Recycling of geopolymer waste: Influence on geopolymer formation and mechanical properties. *Mater. Des.* **2016**, *94*, 221–229. [CrossRef]
8. Lingyu, T.; Dongpo, H.; Jianing, Z.; Hongguang, W. Durability of geopolymers and geopolymer concretes: A review. *Rev. Adv. Mater. Sci.* **2021**, *60*, 1–14. [CrossRef]
9. Hu, W.; Nie, Q.; Huang, B.; Shu, X. Investigation of the strength development of cast-in-place geopolymer piles with heating systems. *J. Clean. Prod.* **2019**, *215*, 1481–1489. [CrossRef]
10. Kim, Y.; Chae, T. Effect of addition of as-received IGCC slag in making geopolymer. *J. Ceram. Process. Res.* **2018**, *19*, 378–382.
11. Ohno, M.; Li, V.C. An integrated design method of Engineered Geopolymer Composite. *Cem. Concr. Compos.* **2018**, *88*, 73–85. [CrossRef]

12. Chen, G.; Zhuo, K.-X.; Luo, R.-H.; Lai, H.-M.; Cai, Y.-J.; Xie, B.-X.; Lin, J.-X. Fracture behavior of environmentally friendly high-strength concrete using recycled rubber powder and steel fibers: Experiment and modeling. *Case Stud. Constr. Mater.* **2024**, *21*, e03501. [CrossRef]
13. Zhang, P.; Sha, D.; Li, Q.; Zhao, S.; Ling, Y. Effect of Nano Silica Particles on Impact Resistance and Durability of Concrete Containing Coal Fly Ash. *Nanomaterials* **2021**, *11*, 1296. [CrossRef]
14. Golewski, G.L. Enhancement fracture behavior of sustainable cementitious composites using synergy between fly ash (FA) and nanosilica (NS) in the assessment based on digital image processing procedure. *Theor. Appl. Fract. Mech.* **2024**, *131*, 104442. [CrossRef]
15. Xiao, S.-H.; Liao, S.-J.; Zhong, G.-Q.; Guo, Y.-C.; Lin, J.-X.; Xie, Z.-H.; Song, Y. Dynamic properties of PVA short fiber reinforced low-calcium fly ash—Slag geopolymer under an SHPB impact load. *J. Build. Eng.* **2021**, *44*, 103220. [CrossRef]
16. Brugaletta, F.; Becher, A.F.; Rostagno, D.L.; Kim, J.; Medina, J.I.F.; Ziejewska, C.; Marczyk, J.; Korniejenko, K. The Different Properties of Geopolymer Composites Reinforced with Flax Fibers and Carbon Fibers. *Materials* **2024**, *17*, 2633. [CrossRef] [PubMed]
17. Kanda, T.; Li, V. *A New Micromechanics Design Theory for Pseudo Strain Hardening*; University of Michigan: Ann Arbor, MI, USA, 2003.
18. Li, V. From Micromechanics to Structural Engineering—The Design of Cementitious Composites for Civil Engineering Applications. *Doboku Gakkai Ronbunshu* **1993**, *471*, 1–12. [CrossRef]
19. Li, V.C. On Engineered Cementitious Composites—Tailored Composites Through Micromechanical Modeling. *J. Adv. Concr. Technol.* **2003**, *1*, 215–230. [CrossRef]
20. Su, J.-Y.; Chen, G.; Pan, H.-S.; Lin, J.-X.; Zhang, J.; Zhuo, K.-X.; Chen, Z.-B.; Guo, Y.-C. Rubber modified high strength-high ductility concrete: Effect of rubber replacement ratio and fiber length. *Constr. Build. Mater.* **2023**, *404*, 133243. [CrossRef]
21. de Oliveira, A.M.; Silva, F.d.A.; Fairbairn, E.d.M.R.; Filho, R.D.T. Coupled temperature and moisture effects on the tensile behavior of strain hardening cementitious composites (SHCC) reinforced with PVA fibers. *Mater. Struct.* **2018**, *51*, 65. [CrossRef]
22. Li, J.; Weng, J.; Yang, E.-H. Stochastic model of tensile behavior of strain-hardening cementitious composites (SHCCs). *Cem. Concr. Res.* **2019**, *124*, 105856. [CrossRef]
23. Shaikh, F.U.A. Deflection hardening behaviour of short fibre reinforced fly ash based geopolymer composites. *Mater. Des.* **2013**, *50*, 674–682. [CrossRef]
24. Zhang, S.; Li, V.C.; Ye, G. Micromechanics-guided development of a slag/fly ash-based strain-hardening geopolymer composite. *Cem. Concr. Compos.* **2020**, *109*, 103510. [CrossRef]
25. Trindade, A.C.C.; Curosu, I.; Liebscher, M.; Mechtcherine, V.; Silva, F.d.A. On the mechanical performance of K- and Na-based strain-hardening geopolymer composites (SHGC) reinforced with PVA fibers. *Constr. Build. Mater.* **2020**, *248*, 118558. [CrossRef]
26. Wang, Y.; Wang, Y.; Zhang, M. Effect of sand content on engineering properties of fly ash-slag based strain hardening geopolymer composites. *J. Build. Eng.* **2021**, *34*, 101951. [CrossRef]
27. Nematollahi, B.; Qiu, J.; Yang, E.-H.; Sanjayan, J. Micromechanics constitutive modelling and optimization of strain hardening geopolymer composite. *Ceram. Int.* **2017**, *43*, 5999–6007. [CrossRef]
28. Ohno, M.; Li, V.C. A feasibility study of strain hardening fiber reinforced fly ash-based geopolymer composites. *Constr. Build. Mater.* **2014**, *57*, 163–168. [CrossRef]
29. Lao, J.-C.; Huang, B.-T.; Xu, L.-Y.; Khan, M.; Fang, Y.; Dai, J.-G. Seawater sea-sand Engineered Geopolymer Composites (EGC) with high strength and high ductility. *Cem. Concr. Compos.* **2023**, *138*, 104998. [CrossRef]
30. Trindade, A.C.C.; Heravi, A.A.; Curosu, I.; Liebscher, M.; de Andrade Silva, F.; Mechtcherine, V. Tensile behavior of strain-hardening geopolymer composites (SHGC) under impact loading. *Cem. Concr. Compos.* **2020**, *113*, 103703. [CrossRef]
31. Lin, J.-X.; Chen, G.; Pan, H.-S.; Wang, Y.-C.; Guo, Y.-C.; Jiang, Z.-X. Analysis of stress-strain behavior in engineered geopolymer composites reinforced with hybrid PE-PP fibers: A focus on cracking characteristics. *Compos. Struct.* **2023**, *323*, 117437. [CrossRef]
32. Zhong, H.; Zhang, M. Effect of recycled tyre polymer fibre on engineering properties of sustainable strain hardening geopolymer composites. *Cem. Concr. Compos.* **2021**, *122*, 104167. [CrossRef]
33. Li, V.; Wu, C.; Wang, S.X.; Ogawa, A.; Saito, T. Interface tailoring for strain-hardening polyvinyl alcohol-engineered cementitious composite (PVA-ECC). *ACI Mater. J.* **2002**, *99*, 463–472.
34. Zhang, Z.; Zhang, Q. Matrix tailoring of Engineered Cementitious Composites (ECC) with non-oil-coated, low tensile strength PVA fiber. *Constr. Build. Mater.* **2018**, *161*, 420–431. [CrossRef]
35. Coppola, L.; Bellezze, T.; Belli, A.; Bianco, A.; Blasi, E.; Cappello, M.; Caputo, D.; Chougan, M.; Coffetti, D.; Coppola, B.; et al. New Materials and Technologies for Durability and Conservation of Building Heritage. *Materials* **2023**, *16*, 1190. [CrossRef] [PubMed]
36. Lin, J.-X.; Su, J.-Y.; Pan, H.-S.; Peng, Y.-Q.; Guo, Y.-C.; Chen, W.-S.; Sun, X.-L.; Yuan, B.-X.; Liu, G.-T.; Lan, X.-W. Dynamic compression behavior of ultra-high performance concrete with hybrid polyoxymethylene fiber and steel fiber. *J. Mater. Res. Technol.* **2022**, *20*, 4473–4486. [CrossRef]
37. Lin, J.-X.; Song, Y.; Xie, Z.-H.; Guo, Y.-C.; Yuan, B.; Zeng, J.-J.; Wei, X. Static and dynamic mechanical behavior of engineered cementitious composites with PP and PVA fibers. *J. Build. Eng.* **2020**, *29*, 101097. [CrossRef]
38. Yu, J.; Yao, J.; Lin, X.; Li, H.; Lam, J.Y.K.; Leung, C.K.Y.; Sham, I.M.L.; Shih, K. Tensile performance of sustainable Strain-Hardening Cementitious Composites with hybrid PVA and recycled PET fibers. *Cem. Concr. Res.* **2018**, *107*, 110–123. [CrossRef]

39. Peng, Y.-Q.; Zheng, D.-P.; Pan, H.-S.; Yang, J.-L.; Lin, J.-X.; Lai, H.-M.; Wu, P.-Z.; Zhu, H.-Y. Strain hardening geopolymer composites with hybrid POM and UHMWPE fibers: Analysis of static mechanical properties, economic benefits, and environmental impact. *J. Build. Eng.* **2023**, *76*, 107315. [CrossRef]
40. Lin, J.-X.; Luo, R.-H.; Su, J.-Y.; Guo, Y.-C.; Chen, W.-S. Coarse synthetic fibers (PP and POM) as a replacement to steel fibers in UHPC: Tensile behavior, environmental and economic assessment. *Constr. Build. Mater.* **2024**, *412*, 134654. [CrossRef]
41. Li, V. *Engineered Cementitious Composites (ECC)—Material, Structural, and Durability Performance*; University of Michigan: Ann Arbor, MI, USA, 2011.
42. *GB/T-18046*; Ground Granulated Blast Furnace Slag Used for Cement, Mortar and Concrete. China Architecture and Building Press: Beijing, China, 2017. (In Chinese)
43. *ASTM-C618*; Coal Fly Ash and Raw or Calcined Natural Pozzolan for Use in Concrete. Annual Book of ASTM Standards: West Conshohocken, PA, USA, 2012.
44. *ASTM-C1437*; Flow of Hydraulic Cement Mortar. Annual Book of ASTM Standards: West Conshohocken, PA, USA, 2013.
45. *ASTM-C109/C109M*; Compressive Strength of Hydraulic Cement Mortars (Using 2-in. or [50-mm] Cube Specimens). Annual Book of ASTM Standards: West Conshohocken, PA, USA, 2020.
46. Yokota, H.; Rokugo, K.; Sakata, N. *Recommendations for Design and Construction of High Performance Fiber Reinforced Cement Composite with Multiple Fine Cracks*; Japan Society of Civil Engineers: Tokyo, Japan, 2008; pp. 1–16.
47. Wang, W.; Shen, A.; Lyu, Z.; He, Z.; Nguyen, K.T.Q. Fresh and rheological characteristics of fiber reinforced concrete—A review. *Constr. Build. Mater.* **2021**, *296*, 123734. [CrossRef]
48. Kong, H.-J.; Bike, S.G.; Li, V.C. Constitutive rheological control to develop a self-consolidating engineered cementitious composite reinforced with hydrophilic poly(vinyl alcohol) fibers. *Cem. Concr. Compos.* **2003**, *25*, 333–341. [CrossRef]
49. *ASTM C469/C469M-22*; Standard Test Method for Static Modulus of Elasticity and Poisson's Ratio of Concrete in Compression. ASTM International: West Conshohocken, PA, USA, 2014.
50. Tran, N.P.; Gunasekara, C.; Law, D.W.; Houshyar, S.; Setunge, S. Microstructural characterisation of cementitious composite incorporating polymeric fibre: A comprehensive review. *Constr. Build. Mater.* **2022**, *335*, 127497. [CrossRef]
51. Alrefaei, Y.; Dai, J.-G. Tensile behavior and microstructure of hybrid fiber ambient cured one-part engineered geopolymer composites. *Constr. Build. Mater.* **2018**, *184*, 419–431. [CrossRef]
52. Sakulich, A.R.; Li, V.C. Nanoscale characterization of engineered cementitious composites (ECC). *Cem. Concr. Res.* **2011**, *41*, 169–175. [CrossRef]
53. Lei, D.-Y.; Guo, L.-P.; Chen, B.; Curosu, I.; Mechtcherine, V. The connection between microscopic and macroscopic properties of ultra-high strength and ultra-high ductility cementitious composites (UHS-UHDCC). *Compos. Part B Eng.* **2019**, *164*, 144–157. [CrossRef]
54. Lee, S.K.; Oh, T.; Banthia, N.; Yoo, D.-Y. Optimization of fiber aspect ratio for 90 MPa strain-hardening geopolymer composites (SHGC) with a tensile strain capacity over 7.5%. *Cem. Concr. Compos.* 2023; 139, 105055. [CrossRef]
55. Wang, Y.; Zhong, H.; Zhang, M. Experimental study on static and dynamic properties of fly ash-slag based strain hardening geopolymer composites. *Cem. Concr. Compos.* **2022**, *129*, 104481. [CrossRef]
56. Chen, G.; Zheng, D.-P.; Chen, Y.-W.; Lin, J.-X.; Lao, W.-J.; Guo, Y.-C.; Chen, Z.-B.; Lan, X.-W. Development of high performance geopolymer concrete with waste rubber and recycle steel fiber: A study on compressive behavior, carbon emissions and economical performance. *Constr. Build. Mater.* **2023**, *393*, 131988. [CrossRef]
57. Xu, L.-Y.; Qian, L.-P.; Huang, B.-T.; Dai, J.-G. Development of artificial one-part geopolymer lightweight aggregates by crushing technique. *J. Clean. Prod.* **2021**, *315*, 128200. [CrossRef]
58. Shoji, D.; He, Z.; Zhang, D.; Li, V.C. The greening of engineered cementitious composites (ECC): A review. *Constr. Build. Mater.* **2022**, *327*, 126701. [CrossRef]
59. Wang, H.; Xu, J.; Song, Y.; Hou, M.; Xu, Y. Development of low-cost engineered cementitious composites using Yellow River silt and unoiled PVA fiber. *Constr. Build. Mater.* **2024**, *425*, 136063. [CrossRef]

materials

MDPI

Article

Effects of Rice Husk Ash Particle Size and Luxan Value Influence on Mortar Properties and Proposal of Hydration Ratio Measurement Method

Junho Kim [1,*], Hikaru Fumino [2] and Manabu Kanematsu [1]

1 Department of Architecture, Faculty of Science and Technology, Tokyo University of Science, Noda City 278-8510, Japan; manabu@rs.tus.ac.jp
2 National Institute of Technology (KOSEN), Oyama College, Oyama City 323-0806, Japan; fumino@oyama-ct.ac.jp
* Correspondence: jh-kim@rs.tus.ac.jp

Abstract

A fundamental study has been conducted on the effective utilization of rice husk ash (RHA) in concrete. RHA is an agricultural byproduct characterized by silicon dioxide as its main component, with a content of 90% or more and a porous structure that absorbs water during mixing, thereby reducing fluidity. The quality of RHA varies depending on the calcination environment; however, the effect is not consistent. In this study, the pore structure was modified, and fluidity was improved by adjusting the particle size of the RHA. From a quality control perspective, this study aims to classify grades using Luxan values. While the characterization of RHA is based on Luxan values, the methodology for measuring its hydration response has not been reviewed. The test methods used in this study are as follows. To test the raw materials, density, specific surface area, XRF, SEM, and isothermal adsorption–desorption curves were measured, and fluidity was measured in fresh mortar. In a hardened mortar, compressive strength and drying shrinkage length change rate were measured. In addition, XRD and TG were measured for specimens after the compressive strength test. The selective dissolution method was used to measure the hydration rate. By adjusting the particle size of RHA to 45 μm, fluidity was enhanced. The relationship between the Luxan value and the basic properties of the mortar indicates that higher Luxan values correspond to greater compressive strength and drying shrinkage. We believe that the method used in this experiment can be used to quantify RHA.

Keywords: rice husk ash; pozzolanic material; particle size; hydration ratio

1. Introduction

Rice husks are an agricultural byproduct with an estimated annual discharge of 120 million tons globally [1]. The husks contain approximately 15–20% SiO_2 by mass [2]. RHA (rice husk ash) has about 90% silica, so it is a pozzolanic reaction material such as fly ash and silica fume, which are conventionally used concrete admixtures. According to previous studies, RHA contains a large amount of silica and can induce a pozzolan reaction, such as fly ash and silica fume. In addition, in fire resistance tests, the fire resistance performance was higher than that of ordinary concrete [3–9]. As such, RHA is actively being reviewed for use in concrete to reduce the environmental load and recycle resources. However, processing methods for rice husks have not yet been established, and existing treatment methods are particularly challenging due to the substantial amount of amorphous silica they contain. Recent research has identified a method for producing RHA with a high content of amorphous silica by controlling the burning temperature and calcination process [10,11]. The RHA obtained via this method exhibits a SiO_2 content nearly equivalent to that of silica fume and is currently being investigated as a potential new pozzolanic material.

In addition, when RHA is substituted for cement in concrete, a positive effect is shown, and the effect according to the substitution rate is also considered. Regarding the use of RHA, the method of substituting for cement is generally performed, and different properties are shown depending on the substitution method. However, it is reported that the substitution amount of RHA is generally 10% to 20% [12–21]. The properties of RHA concrete have been reported to increase long-term compressive strength and decrease autogenous shrinkage and durability [22–25]. Breugel V. K. and Ye G. Tuan V. N. substituted 10% and 20% RHA and compared it with silica fume. When the particle size of RHA was smaller, the long-term compressive strength at 91 days was 13% higher, and the autogenous shrinkage was reduced by 90% [22]. Additionally, Van V. et al. compared silica fume with a 10% replacement of RHA and reported that the rate of change in self-shrinkage length was about half that of silica fume [24]. In addition, experiments to confirm the effect of adjusting the particle size of RHA have been reported, and silica fume exhibits comparable mechanical properties according to the adjustment of the particle size [26–29]. Adjusting the particle size showed an increase in compressive strength of nearly two times [26]. Xu W. et al. controlled the particle size to 7 μm, which showed a compressive strength twice as high as that of basic RHA [27]. In the study by Venkatanarayanan H. K. and Rangaraju P. R., the size of RHA was adjusted to less than 10 μm, and it showed a compressive strength 10% higher than that of RHA without particle size adjustment [29].

There are two main factors that affect the production quality of RHA as an admixture in concrete. First, temperature and time are essential factors in the calcination process. Andres Salas et al. reported the effect of the change in the calcination method. RHA, to which chemical and calcination treatment was applied, had more amorphous properties and showed positive results in increasing the compressive strength of concrete [30]. Lingling Hu et al. [31] performed hydrochloric acid treatment as a pretreatment and experimented with changing calcination temperatures. In this study, after pretreatment with hydrochloric acid, calcination at 600 °C was used and showed higher performance compared to other admixtures. Hossein Beidaghy Dizaji et al. [32] studied the effect of changes in the RHA pretreatment method on physical properties, and when the pretreatment method was applied, the specific surface area was higher. In addition, pretreatment in the calcination process showed a higher SiO_2 production rate. Ayesha Siddik et al. [33] reported that producing RHA with more than 80% amorphous silica was possible when incinerated at 700 °C. In addition, it has been reported that the smaller the RHA particles, the more active the pozzolan reaction. RHA particles of 20 mm or less at 30% or less are reportedly optimal. Sung-Hoon Kang et al. [34] reported that calcination at 650 °C for 2 h is effective in producing RHA with an amorphous silica content of 90% or more. In addition, black RHA and white RHA were used in calcination, and white RHA positively affected compressive strength. White and black RHA come from different firing temperatures, where firing at 650 °C produces white RHA. Deepa G Nair et al. [35] studied amorphous silica according to calcination temperature. Amorphous silica with high RHA calcined at 500 °C and 700 °C was contained and showed good pozzolanic reaction. In addition, it was reported that crystals were formed at 900 °C. Jose James and M. Subba Rao [36] tested the reactivity of RHA according to the calcination temperature, and as a result, RHA calcined at around 500 °C had the highest reactivity. Alias Mohd Yusof et al. [37] studied the conditions for producing RHA. The calcination process and washing of RH were used as variables;

the calcination in a plug flow reactor produced fewer impurities than calcination in open burning, and the washing process with water resulted in fewer impurities. Ru-Shan Bie et al. [11] investigated the effect of changes in calcination temperature and calcination time on compressive strength when RHA is produced. It has been reported that the lower the firing temperature and the longer the firing temperature, the more positive the compressive strength. M.F.M. Zain et al. [38] studied the calcination method of RHA and reported that high-quality RHA could be obtained by supplying sufficient air.

Conversely, no straightforward approach exists to quantify the activity of RHA as a pozzolanic material, aside from measurement methods employing X-ray diffraction (XRD) or X-ray fluorescence spectrometry (XRF). Consequently, the current situation presents a challenging context for determining RHA quality. Luxán presented a method for measuring the reaction of pozzolanic materials [39] to gauge material performance [11,40].

RHA is reported to have pores as small as 10 μm [11], which can adversely affect compressive strength and fluidity when it is used as a replacement in concrete [41]. While the use of fly ash as a pozzolanic material is widespread, with research focusing on its particle size and fluidity [42], similar research on RHA in this context has not been conducted. Additionally, C-S-H content reportedly increases when RHA is partially replaced with cement [43]. Although many studies have focused on RHA, most of the literature is concerned with its preprocessing [31,32,44–46]. No methods have been proposed to improve fluidity other than using superplasticizers (SP) or increasing the water content in the formulation. Moreover, studies on the reaction ratio and measurement method for RHA remain unclear [47].

A synthesis of RHA studies reveals three key areas requiring further investigation: the absence of a reliable method for measuring the pozzolanic reaction of RHA, the need to address fluidity issues associated with RHA, and the necessity of developing a robust approach for quantifying RHA quality. Accordingly, we have focused our efforts on developing solutions to these problems.

The objective of this study is to investigate the potential for increasing the use of RHA in concrete. To address the poor fluidity of RHA, which has been identified as a disadvantage under certain conditions, RHA particle size is adjusted to determine the effects of this adjustment. Additionally, this study aims to clarify the relationship between RHA quality and Luxan values. Furthermore, a method for measuring the hydration rate of RHA is presented.

2. Materials and Methods

2.1. Specimen Overview

In this study, three types of RHA with Luxan values of 1.1, 2.8, and 4.2 were used. The grades were chosen as 2.8 for universally produced products, 1.1 for low-quality products, and 4.2 for high-quality products. Two types of particle sizes—less than 45 μm and fine pulverization—were employed. The test specimens were fabricated using a steel mold with dimensions of 40 × 40 × 160 mm. Six levels were prepared: one mix with varying levels of RHA and another mix with only ordinary Portland cement (OPC). The mixture compositions for each level are presented in Table 1. The RHA replacement rate for cement was set at 15%. Figure 1 shows the three types of RHA used, black, gray, and white, with Luxan values of 1.1, 2.8, and 4.2, referred to as B, G, and W, respectively. The properties of the materials used in the study are listed in Table 2. Mixing, curing methods, and compressive strength measurements were performed in accordance with JIS R 5201 [48]. Cement and RHA were thoroughly mixed, and water was added. After 24 h, the specimens were demolded and cured in water. Compressive strength tests were conducted at various ages. The dry shrinkage test was performed one week after demolding, with measurements taken in a controlled climate of 20 °C and 60% humidity.

Table 1. Mixture design of mortar specimen.

Sample	RHA		W/C (%) [1]	Unit Weight (g) [2]			
	Luxan Value	Main Particle Size (μm)		W	S	OPC	RHA
OPC	-	-	50	225	1350	450.0	-
B45	1.1	~45				382.5	67.5
B		-					
G45	2.8	~45					
G		-					
W45	4.2	~45					

[1] W/C: water to cement ratio. [2] The unit weight used to make the 3 test pieces.

(**a**) Luxan value: 1.1 (BRHA) (**c**) Luxan value: 2.8 (GRHA)

(**b**) Luxan value: 1.1 (BRHA45) (**d**) Luxan value: 2.8 (GRHA45) (**e**) Luxan value: 4.2 (WRHA45)

Figure 1. RHA according to the change in calcination conditions.

Table 2. Material characteristics of mortar specimen.

Materials	Symbol	Characteristics
Ordinary Portland Cement	OPC	Density: 3.15 g/cm^3
Sand	S	Density: 1.29–1.31 g/cm^3
Black RHA	B	Luxan Value: 1.1, Rapid Calcination at 600 °C, Particle Size: ~45 μm, Raw Material, Calcination: 180 min
Grey RHA	G	Luxan Value: 2.8, Rapid Calcination at 550 °C, Particle Size: ~45 μm, Raw Material, Calcination: 180 min
White RHA	W	Luxan Value: 4.2, Cooling After Calcination at 500 °C, Particle Size: ~45 μm, Raw Material, Calcination: 180 min

Table 3 shows the basic properties of OPC. Table 4 lists the details of the paste specimens used in the hydration rate experiments. Three types of main particles were used, and the Luxan value was tested for two particle sizes. The RHA replacement rate was set at 15%, and the W/C was set to 50%.

Table 3. Characteristics of OPC.

	Setting Time		Compressive Strength (MPa)			Chemical Composition				
	Initial Setting	Final Setting	3 d	7 d	28 d	MgO	SO_3	Ignition Loss	Total Alkali	Insoluble Residue
OPC	2 h 30 min	4 h	30.4	48.3	67.0	0.92	1.98	2.39	0.57	0.021

Table 4. Mixture design of paste specimen.

Sample	RHA		W/C (%) [1]	Unit Weight (g) [2]		
	Luxan Value	Main Particle Size (μm)		W	OPC	RHA
OPC	-	-	50	225	450.0	-
B45	1.1	~45			382.5	67.5
B100		~100				
B		-				
G45	2.8	~45				
G100		~100				
G		-				

[1] W/C: water to cement ratio. [2] The unit weight used to make the 3 test pieces.

2.2. Particle Size Control

A ball mill was used to control the particle size of the RHA. Two ceramic balls with diameters of 10 mm and 3 mm were used, and both the mill and the balls were made of ceramic. To achieve a particle size of 45 microns, the mill was set to a spinning speed of 600 rpm for 60 min. The particle size distribution curve obtained after particle size adjustment is shown in Figure 2. The experiment was performed in accordance with JIS R 5201 [48]. The particle size was measured using a laser diffraction/scattering particle size analyzer (LA-950V2, Kyoto, Japan), with the main particle size of the raw material being 500 μm and the adjusted material being 45 μm. Figure 3 shows the scanning electron microscopy (SEM, JEOL/JCM-6000, JEOL Ltd., Tokyo, Japan) image of the raw material and the resulting image after particle size control. It is evident that the voids present in the RHA particles are eliminated after particle size control. As shown in Table 5, the results of the X-ray fluorescence (XRF, Rigaku/ZSX) analysis indicate that particle size control had no effect on mineral fracture.

Table 5. Measurement results of XRF of RHA (mass %).

Specimen	B	B45	G	G45
MgO	0.1	0.1	0.2	0.2
Al_2O_3	0.0	0.1	0.0	0.1
SiO_2	92.0	92.7	90.4	91.6
P_2O_5	0.4	0.4	0.6	0.5
SO_3	0.2	0.1	0.3	0.3
K_2O	5.8	5.4	6.4	5.5
CaO	1.0	0.8	1.4	1.2

Table 5. *Cont.*

Specimen	B	B45	G	G45
Cr_2O_3	0.0	0.0	0.0	0.0
MnO	0.3	0.2	0.3	0.3
Fe_2O_3	0.1	0.1	0.1	0.2
NiO	0.0	0.0	0.0	0.0
ZnO	0.0	0.0	0.0	0.0
SnO_2	0.0	0.0	0.0	0.0
L.O.I.	0.9	2.3	1.4	2.5

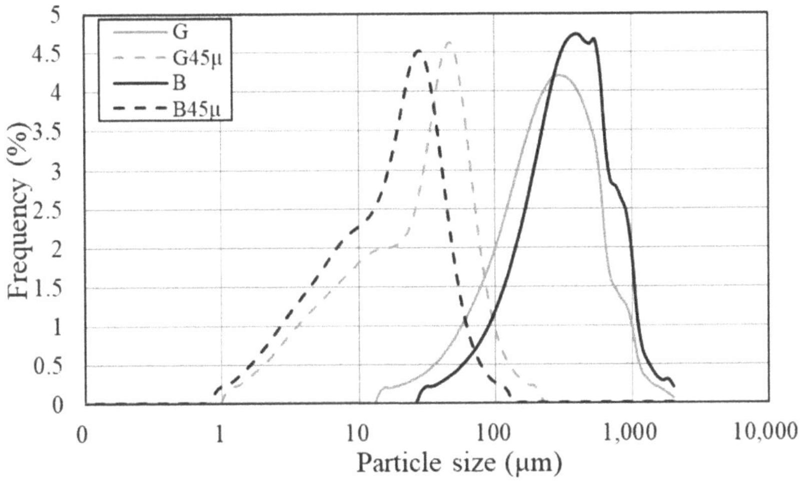

Figure 2. Particle size distribution.

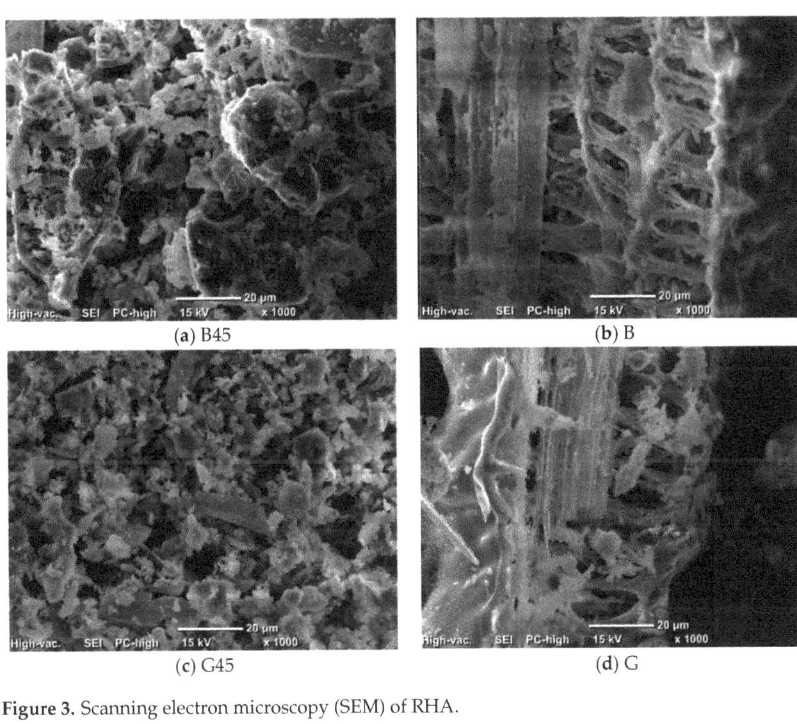

Figure 3. Scanning electron microscopy (SEM) of RHA.

2.3. Test of Compressive Strength

A compressive strength test was performed in accordance with JIS R 5201 [48]. The compressive strength was measured at 7, 28, and 91 d, and the average of three tests at each age was taken as the final compressive strength, calculated using Equation (1). The loading speed was set at a rate of 50 ± 10 N per second to obtain the maximum load. The specimen used for the compressive strength test was a 40 × 40 × 160 mm sample, which was divided into three pieces of 40 × 40 × 40 mm for each specimen.

$$C = W/1600 \tag{1}$$

where C: Compressive strength (N/mm^2), and W: Maximum load (N).

2.4. Test of Drying Shrinkage

Drying shrinkage was determined by curing the specimens under water for 7 d after pouring the mortar, with the initial point of drying starting after 7 d. The drying shrinkage was measured for 91 d using an embedded gauge. This experiment was also performed in accordance with JIS R 5201 [48].

2.5. Test for Density

Density was tested following the JIS R 5201 standard [48]. Measurements were conducted using a Le Chatelier flask and kerosene. The sample used for density measurement had an unadjusted RHA of 45 microns.

2.6. Water Vapor Isothermal Adsorption Test

A water vapor isothermal adsorption test was conducted to evaluate the changes in pore structure after particle size control. An isothermal adsorption curve was obtained using a water vapor adsorption method with an equilibration time of 60 s. Measurements were taken at increments of 5% relative humidity (RH) in the range of 5–98%, with the maximum value of 98% RH being repeated. The specific surface area resulting from water vapor adsorption was calculated using the Brunauer–Emmett–Teller theory based on data derived from the desorption process within the RH range of 10–35% [49]. In this method, the area occupied by water was determined to be 0.125 nm.

2.7. Thermogravimetric Differential Thermal Analysis (TG-DTA)

To determine the quantity of calcium hydroxide generated through the hydration of cement in the pozzolanic reaction with RHA, Joule differential thermal analysis was conducted. The temperature range was set between 30 and 1000 °C, with a heating rate of 20 °C per minute. A TG/DTA 7200, manufactured by Hitachi (Tokyo, Japan), was used for the analysis, with samples adjusted to under 300 μm. α-Alumina was used as the reference sample, and the sample weight was set to 20 mg. Additionally, 120 mL/min of nitrogen gas was utilized for TG.

CH and CC were calculated according to Equations (2) and (3) below.

Calcium hydroxide generally dehydrates (decomposes into CaO and H_2O) at around 400–500 °C. The weight change is denoted as $WL_{Ca(OH)2}$, and the molecular weights of calcium hydroxide, $m_{Ca(OH)2} = 74$, and that of water, $m_{(H2O)} = 18$, are used to determine the amount of calcium hydroxide.

$$Ca(OH)_{2,measured} = WL_{Ca(OH)_2} \times \frac{m_{Ca(OH)_2}}{m_{(H_2O)}} = WL_{Ca(OH)_2} \times \frac{74}{18} \tag{2}$$

Calcium carbonate is generally decarburized (decomposed into CaO and CO_2) at about 600–700 °C. The weight change is determined as WL_{CaCO3}, and calcium hydroxide is determined using the molecular weight of calcium carbonate, m_{CaCO3} = 100 and CO_2, m_{CO2} = 44.

$$CaCO_{3,measured} = WL_{CaCO_3} \times \frac{m_{CaCO_3}}{m_{CO_2}} = WL_{CaCO_3} \times \frac{100}{44} \tag{3}$$

2.8. X-Ray Diffraction (XRD) Analysis

For XRD measurements, the samples collected from each curing age were pulverized to a particle size of less than 300 μm. The measurement conditions included a voltage of 30 kV, a tube current of 10 mA, a step width of 0.01°, and a diffraction angle range of 2θ = 5° to 75°. Additionally, 10% α-alumina was mixed with the samples for analysis using the Rietveld method. TOPAS V5 (manufactured by Bruker AXS, Billerica, MA, USA) was used for Rietveld analysis, and the data were interpreted based on the XRD measurements. The amount of amorphous content was determined by comparing the amount of calcium hydroxide in TG-DTA with the amount of reference material (α-alumina).

2.9. Methodology for the Quantification of RHA

The selective dissolution method was employed to quantify the hydration rate of RHA, following a methodology previously used for quantifying fly ash [14]. The powdered sample was treated with a solution containing 2N hydrochloric acid and 5% sodium carbonate to dissolve cement and cement hydrates. The mass of each sample was measured after drying.

2.10. Determination of Unreacted RHA from Hardened Cement Paste

Table 6 presents the results of the hydration-rate experiments. Reaction rate tests were conducted using the selective dissolution method proposed by Osawa et al., which effectively recovered unreacted fly ash [47]. Since RHA primarily consists of SiO_2 and exhibits properties similar to fly ash, the quantity of residual RHA was determined using a comparable method. A solution of 2N hydrochloric acid and 5% aqueous sodium carbonate was added to the powder sample to dissolve the cement and cement hydrates. The mass of each sample was then measured after drying. The insoluble residual rates for each sample were 93.07% and 85.44%, respectively. Reaction rates were corrected using the calculated insoluble residual rates. The specimens were cured in water, and measurements were taken at each curing age.

Table 6. Measurement results of insoluble residual rate of RHA.

Specimen	Weight Before Dissolution (g)	Weight After Dissolution (g)	Weight After/Before Dissolution (%)
BRHA	1.01	0.94	93.07
GRHA	1.03	0.88	85.44

2.11. Calculation of Luxan

The Luxan value, proposed by Luxan, is a simple method for measuring the pozzolanic reaction and is known to be effective in measuring the amount of amorphous silica [39,40]. Generally, a higher Luxan value indicates a higher active value.

After determining the following two measurements, the result is derived by substituting it in Equation (4).

(1) The electrical conductivity is measured before adding 40 mL of calcium hydroxide (value a; mS/cm).

(2) RHA sample (1.0 g) is added to the calcium hydroxide mixture, and the electrical conductivity is measured after 2 min (value b; mS/cm).

$$\text{Luxan} = \text{value a} - \text{value b} \quad (4)$$

For the preparation of the saturated calcium hydroxide solution used in this study, 0.64 g of calcium hydroxide reagent was added to 500 mL of pure water and mixed in an environment of 40 °C.

3. Results

This section is divided into three subsections. First, basic materials are reviewed to verify the impact of particle size adjustments. Second, the fundamental characteristics of the mortar, including flow, compressive strength, and drying shrinkage, are examined. Finally, investigations involving the assessment of methodologies used for quantifying the RHA hydration rate are discussed. The results of the fundamental characteristics of the mortar experiments represent the average of three specimens.

3.1. Basic Materials

3.1.1. Density

Figure 4 presents the data obtained from the density measurements of various powder materials. This study hypothesized that the pores in RHA would be destroyed during particle size adjustment. However, the density measurement results indicated that the density of RHA at 45 microns exhibited only a slight increase, suggesting that an increase in weight per unit volume contributed to the observed effect.

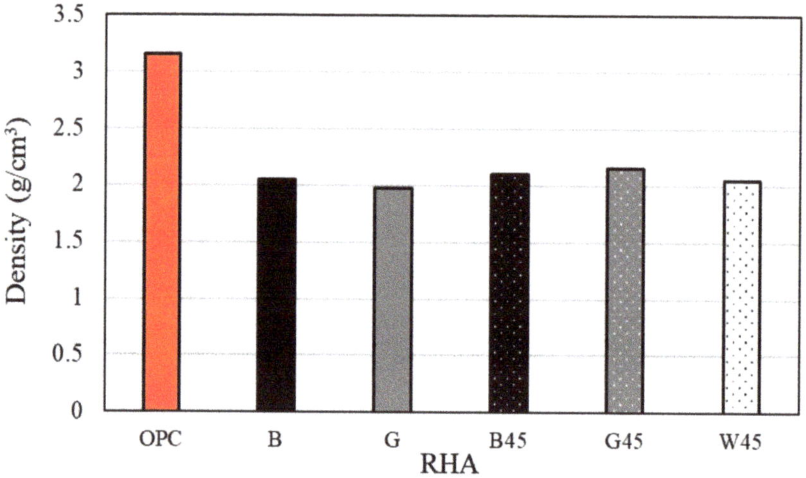

Figure 4. Density according to RHA.

Previous studies have indicated that an increase in the substitution rate of RHA is associated with a reduction in density, which is attributable to the presence of pores in RHA [4,16,24,50]. Additionally, the density of secondary pores (0.1–100 μm) increases with the substitution rate of RHA, and the density of these secondary pores is sensitive to changes in the proportion of RHA [12]. Further research has also shown that the pore structure of RHA reduces the density of materials [22].

3.1.2. Water Vapor Isothermal Adsorption Test Results

As shown in Figure 5, the isothermal adsorption–desorption curve yielded the following results: specimens with elevated Luxan values exhibited a notable increase in weight due to water vapor adsorption, indicating the presence of a substantial quantity of mesopores. Additionally, the RHA sample with a 45 μm particle size adjustment demonstrated comparable outcomes to the raw RHA in terms of color. In terms of weight change, the 45 μm particle size-adjusted specimen had a lower weight compared to the raw RHA material, suggesting that the microstructure of the RHA was affected. This outcome is attributed to the influence of the porous RHA structure, which operated in a manner similar to that previously described [12,22].

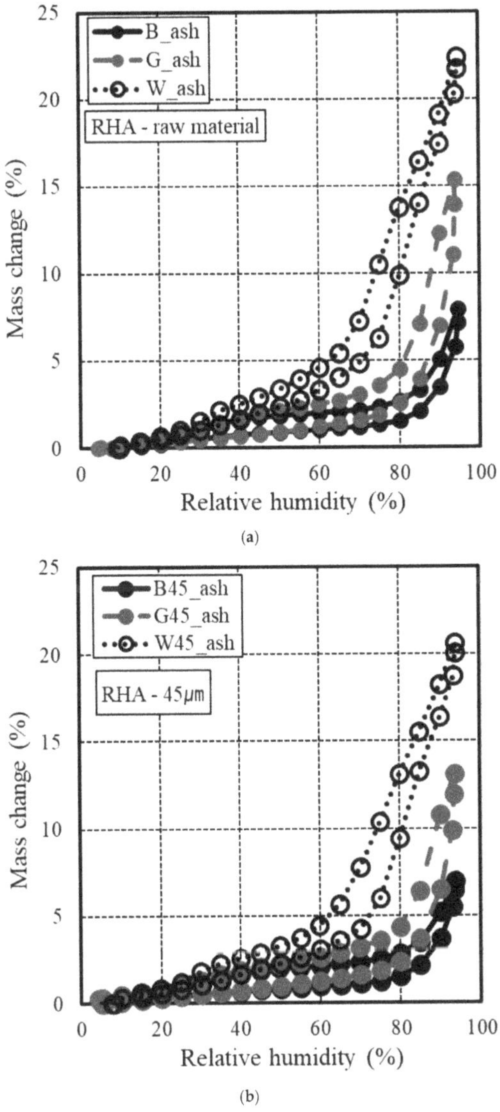

Figure 5. Isothermal adsorption–desorption curve for RHA. (**a**) raw material, (**b**) 45 μm.

Figure 6 shows the results for specific surface areas. The specific surface area, calculated using the BET method from the isothermal adsorption/desorption curves in the RH range of 10~35% during desorption, revealed that the two raw material samples, B and G, exhibited nearly identical results, while W had a larger specific surface area compared to the others. Additionally, the particle size-adjusted samples generally had a higher specific surface area than the raw material samples, with B45 and G45 showing results similar to those of W, and W45 exhibiting a more pronounced increase.

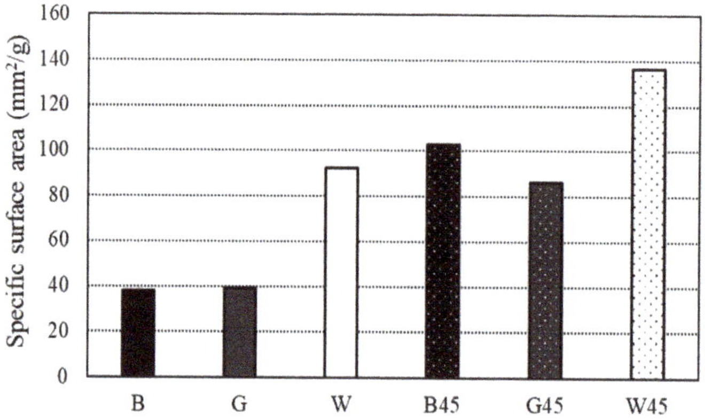

Figure 6. Specific surface area by BET.

3.2. Fundamental Characteristics of the Mortar

3.2.1. Flow Test

Figures 7 and 8 present the results and a photograph of the flow experiment, respectively. As shown in Figure 8, flow measurements could not be conducted on the test specimen with RHA before particle size adjustment due to moisture absorption. The specimens containing 45 μm RHA exhibited reduced absorption of mixing water, resulting in improved flow. B, with a low Luxan value, demonstrated a high flow value, which inversely correlated with the maximum weight change observed in the isothermal adsorption curve (Figure 5). This phenomenon is attributed to the reduction in RHA micropores identified during grain-size adjustment [11], which positively influenced fluidity. Additionally, other studies have reported that the pore structure of RHA absorbs water, a finding confirmed in this study [22,24,47].

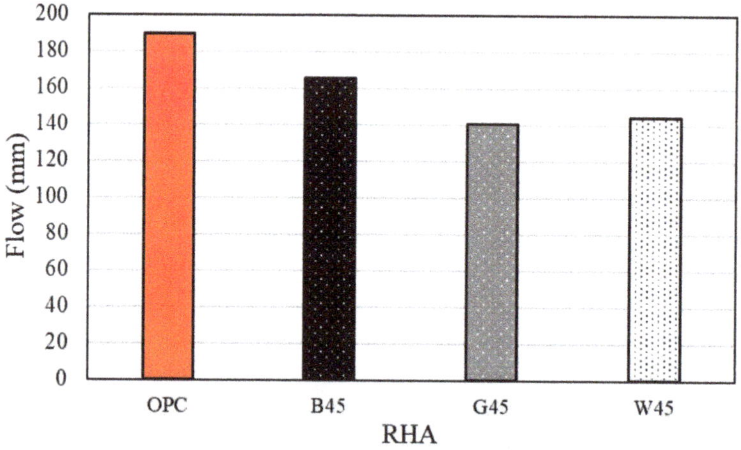

Figure 7. Flow according to RHA replacement.

Figure 8. Flow test photographs according to RHA replacement.

Previous literature has used RHA with particle sizes up to 75 μm, which often led to increased formulation complexities. However, the present study used RHA with particle sizes of 46 μm or less and demonstrated improved fluidity [31]. Additionally, as shown in Figure 3, the particle size control ruptured the RHA pores.

3.2.2. Compressive Strength TEST

Figure 9 illustrates the variation in compressive strength with respect to RHA substitution. Overall, specimens containing RHA with adjusted particle sizes exhibited higher compressive strength than those with raw RHA. By adjusting the RHA particles, high compressive strength was achieved at all ages. In addition, the samples before particle control showed no difference between B and G. However, after particle control, B showed higher compressive strength.

This trend is attributed to the increased specific surface area resulting from particle size adjustment. Comparative analysis of RHA specimens B45, G45, and W45 revealed that B45 consistently demonstrated the highest compressive strength at all curing ages, outperforming the other two specimens. Additionally, RHA with adjusted particle sizes showed a tendency for enhanced compressive strength, likely due to improved bonding with aggregates [26,51]. The compressive strengths of specimens G45 and W45 were similar. Slope analysis of the graph in Figure 9 showed that the RHA specimens with adjusted particle sizes exhibited a significantly greater increase in compressive strength from 7 to 28 d compared to the OPC. However, the increase in compressive strength from 28 to 91 d was not significantly different from that observed for the OPC. These findings suggest that the pozzolanic reaction was most active between 7 and 28 d. The rate of increase in compressive strength was a function of curing age. Reducing the RHA particle size led to a higher rate of increase in compressive strength from 7 to 28 d. The data indicate that finer RHA particles resulted in a greater increase in compressive strength during this period. Furthermore, the relationship between the Luxan value and compressive strength suggests that the rate of strength increase correlates with the Luxan value, indicating that the Luxan value is proportional to the pozzolanic reaction of amorphous silica. The increase in compressive strength due to RHA is attributed to both the filler effect and pozzolanic reaction [52]. At 28 d of curing, the compressive strength of specimens substituted with raw RHA decreased, consistent with findings from other studies. This decrease is believed to result from an increased number of voids associated with RHA use [4,10,16].

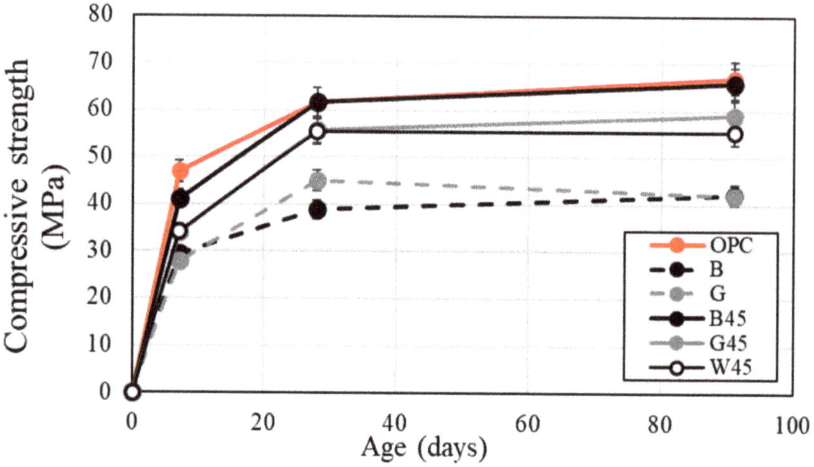

Figure 9. Compressive strength according to RHA replacement.

Additionally, it has been reported that the rate of increase in compressive strength at long-term curing ages is higher than at initial curing ages, which is consistent with findings from previous studies [14,19,28,53]. It was also noted that the number of voids and water absorption decreased inversely with compressive strength [53].

Breugel et al. [22] reported that the curing reaction occurs within the pores of RHA, suggesting that the reaction is a complex process involving both pozzolanic and curing reactions within the RHA voids. Zerbino et al. [26] and Giaccio et al. [51] described the characteristics of RHA with and without particle size adjustment, noting that strong bonding to the aggregate and cement matrix was achieved when the RHA was substituted in appropriate amounts. Therefore, it is concluded that adjusting the particle size appropriately can yield a similar performance to OPC using only cement.

3.2.3. Dry Shrinkage Test

Figure 10a displays the dry shrinkage results for each specimen based on the average data obtained from embedded gauges on three specimens. Except for G45, the RHA specimens exhibited dry shrinkage results that were either similar to or lower than those of the OPC. The RHA test specimens with particle control showed lower drying shrinkage than the test specimens without particle control. In addition, B showed lower drying shrinkage overall than G. A comparison of the dry shrinkage strain between raw RHA and 45 μm RHA specimens revealed that the latter had a greater tendency towards larger dry shrinkage. B45 showed compressive strength comparable to OPC but with reduced drying shrinkage relative to OPC. Figure 10b illustrates the drying shrinkage strain at 80 d. The results for RHA specimens B, G, B45, and G45 indicated that a lower Luxan value corresponded to a smaller dry shrinkage strain. This observation aligns with the trends reported in the literature [12,14].

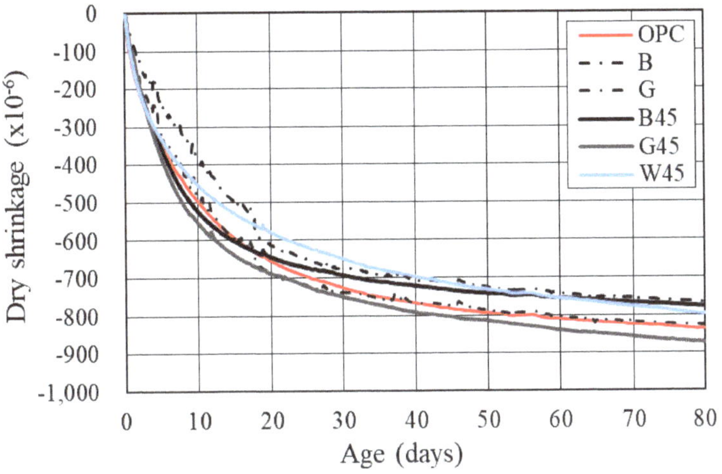

(**a**)

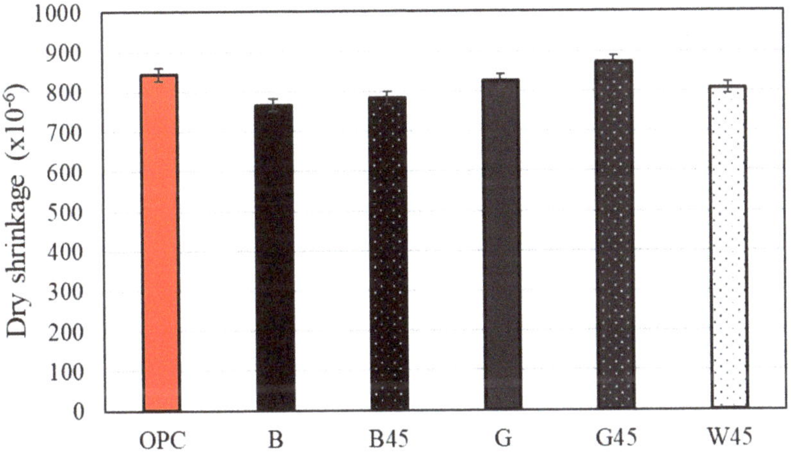

(**b**)

Figure 10. Dry shrinkage according to RHA replacement. (**a**) dry shrinkage; (**b**) drying shrinkage strain.

In a previous study, RHA substituted for calcined dolomite resulted in decreased dry shrinkage as RHA content increased [14]. The dry shrinkage results in this study were similar to those of the plain specimens. Similarly, Wang et al. [12] reported that pore volume, which decreased with curing age, tended to increase with a higher RHA substitution ratio. This suggests that the hydration products generated by the pozzolanic reaction may fill the pores, explaining the observed increase in compressive strength with curing age compared to the plain specimens.

In the autogenous shrinkage experiment by Breugel et al. [22], it was reported that RHA with a particle size of 5.6 μm or larger reduces autogenous shrinkage because the curing reaction occurs within the pores of the RHA. Therefore, proper particle size control is expected to positively impact drying shrinkage.

Additionally, studies by Viet-Thien-An Van et al. [24] and Gemma Rodriguez de Sensale et al. [54] reported that RHA substitution contributed to a reduction in autogenous shrinkage. Similarly, Hu et al. [31] found that the creep rate decreased with an increased substitution rate of RHA. Although the experimental conditions varied, the results suggest that reducing capillary voids had a beneficial effect, with RHA substitution showing improvements when the particle size was greater than 10 μm.

3.2.4. Thermogravimetric/Differential Thermal Analysis (TG-DTA)

Figure 11 displays the calculation results for the amount of $Ca(OH)_2$ produced, as determined by TG-DTA. The quantity of $Ca(OH)_2$ was calculated based on the mass loss observed between 350 and 450 °C on the thermogravimetric (TG) curve. The results indicated that RHA substitutions led to specimens with lower amounts of calcium hydroxide compared to those with OPC. This reduction is attributed to the consumption of calcium hydroxide during the pozzolanic reaction. Furthermore, a comparison of raw RHA and 45 μm RHA specimens showed that, except for B45, the 45 μm RHA specimens had lower concentrations of calcium hydroxide. Despite this, the B45 specimen exhibited the highest compressive strength among the RHA-substituted specimens, suggesting the need for further investigation into the relationship between the RHA pozzolanic reaction and compressive strength. The data indicated that calcium hydroxide consumption through the pozzolanic reaction was consistent with trends observed in previous studies [8,21,47,54,55].

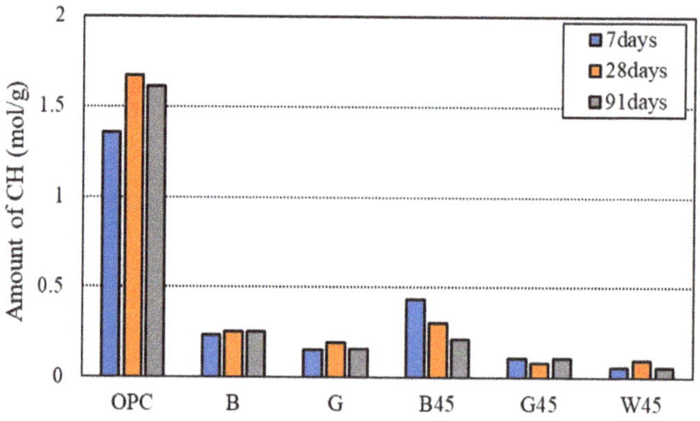

Figure 11. Amount of calcium hydroxide according to RHA replacement.

Additionally, it has been reported that a reduction in calcium hydroxide and a decrease in pores in the interfacial zone are associated with improved physical properties [21]. In an experiment by Rêgos et al. [23], varying the amount of amorphous silica with a 20% RHA substitution showed that a higher amount of amorphous silica increased the pozzolanic reaction due to greater consumption of calcium hydroxide. These findings correlate with the Luxan value used in this study, reflecting similar trends.

Tuan et al. [8] reported that water absorbed by the porous structure of RHA can enhance the hydration reaction of cement, leading to improved strength development. Additionally, both calcium hydroxide and RHA are known to undergo a pozzolanic reaction. This effect is believed to contribute to the simultaneous decrease in calcium hydroxide and increase in compressive strength observed in this study.

In cases where RHA had a high Luxan value, more calcium hydroxide was consumed, and higher compressive strength was achieved at later ages. This suggests that the Luxan value is proportionally related to the amount of amorphous silica. Conversely, in the mechanism of strength development by RHA, it was observed that the water absorbed by RHA mesopores, rather than the consumption of calcium hydroxide, lowered the W/B ratio of the mixture. This reduction in W/B ratio is believed to positively influence the pozzolanic response over long curing periods, as reported by Viet-Thien-An Van et al. [24].

3.2.5. X-Ray Diffraction (XRD)

The objective of this study was to determine whether the particle size of RHA influenced the minerals produced during the hydration reaction and to assess whether the pozzolanic reaction led to the consumption of CH.

Figure 12 presents the XRD measurement results for specimens with RHA replacement. Quartz was consistently detected at all levels and ages, as the samples were obtained from mortar test specimens. Additionally, the portlandite peak near 18° on the horizontal axis confirmed the presence of portlandite in the B45 specimens at each age. Although a minimal peak was observed at 91 d in other specimens, these results were largely consistent with the calcium hydroxide quantities derived from the TG-DTA results.

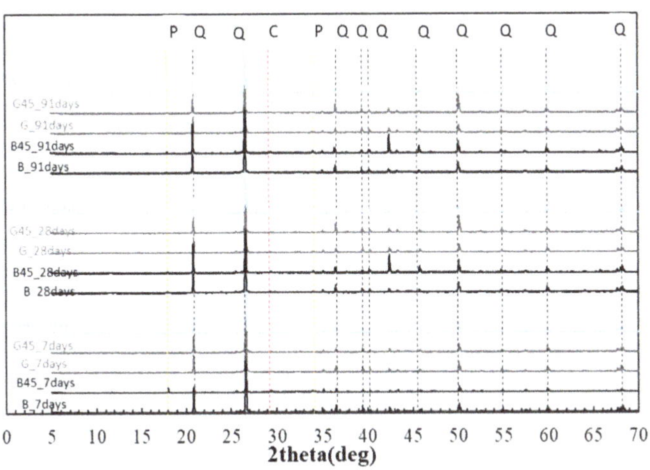

Figure 12. XRD of RHA (P: portlandite, C: calcite, and Q: quartz.

Figure 13 displays the results of the Rietveld method for quantification. The quantity of quartz remained consistent, with the exception of the results for the 7 d curing age.

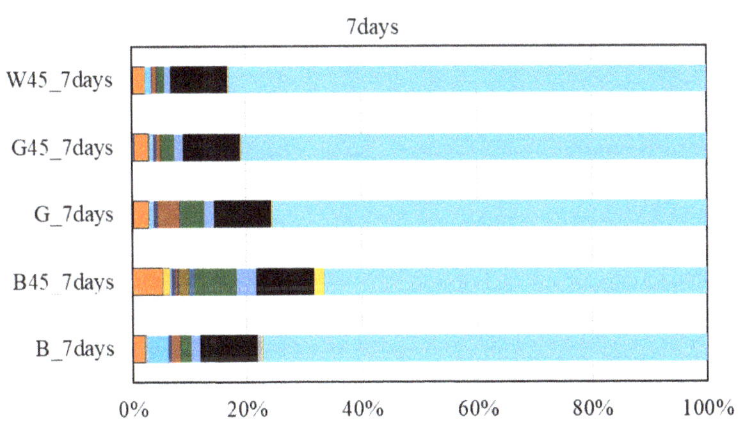

Figure 13. *Cont.*

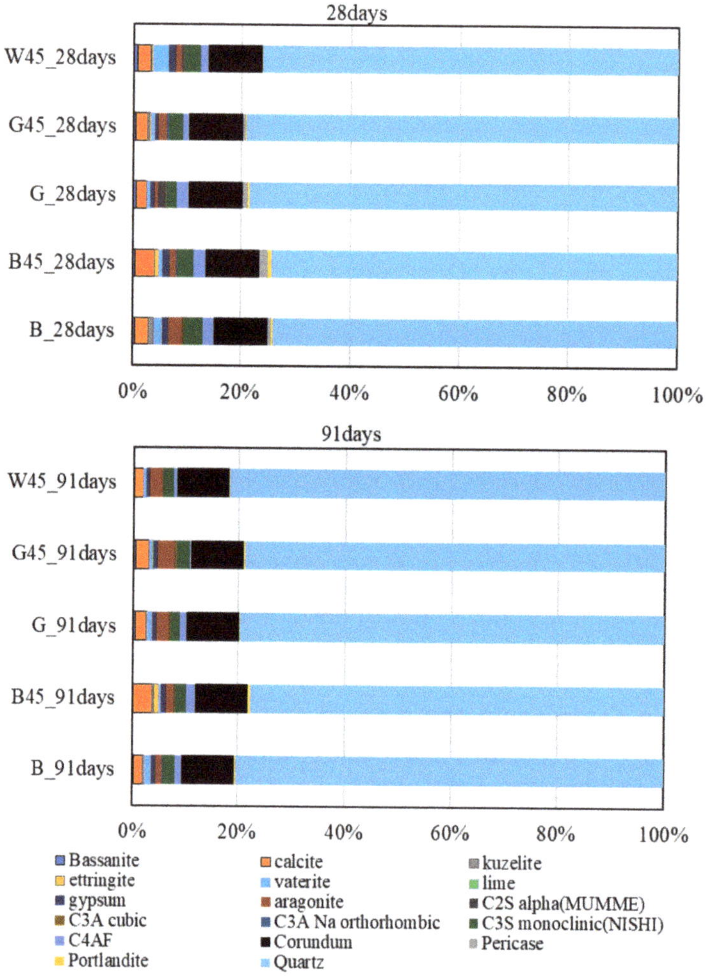

Figure 13. Rietveld method for RHA.

3.2.6. Results of the RHA Reaction Rate Test

Figure 14 presents the RHA reaction rate test results using the selective dissolution method. For all test specimens, excluding the raw material and G45, the amount of unreacted RHA tended to decrease as the curing age increased. This indicates that the hydration reaction of RHA progressed from 7 to 28 d. Comparing the raw material and G, the difference in reaction rate at 28 d was only about 4%, with values at 7 d being similar to those of uncrushed G. For G45, approximately 85% had reacted by 7 d, indicating improved reactivity of RHA due to particle size adjustment.

Regarding the Luxan value, G exhibited a higher reaction rate than B in the 45 μm specimen. However, B reacted more than G in the raw material and 100 μm samples. Above 100 μm, the difference in reaction rate was approximately 10%, suggesting that the reaction rate could vary with curing age. Therefore, further investigations are warranted.

In terms of particle size, the difference in reaction rate between raw and crushed RHA was significant. Particle size adjustment notably affected the reaction rate even at early curing ages. Smaller particle sizes of RHA, which increase the specific surface area, promoted the hydration reaction. Thus, adjusting the particle size improved the reactivity of RHA.

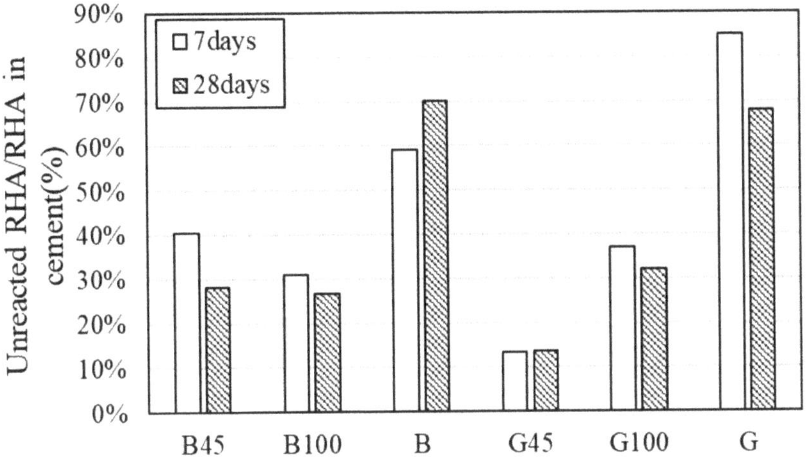

Figure 14. RHA reaction rate test results using the selective dissolution method.

3.2.7. Calcium Hydroxide Determination

Figure 15 presents the quantitative results for calcium hydroxide and calcium carbonate calculated from the TG-DTA measurement data. For B, the calcium hydroxide content tended to increase with material age. Since B had a low Luxan value, the production of calcium hydroxide through hydration reactions surpassed its consumption through the pozzolanic reaction. In contrast, no increase in calcium hydroxide was observed in G, suggesting that the pozzolanic reaction in G was more advanced than in BRHA. However, the differences in the quantitative results for the hydrates were minimal. Future investigations should focus on the quantitative amounts and reaction rates of RHA at extended curing periods when the pozzolanic reaction becomes more active.

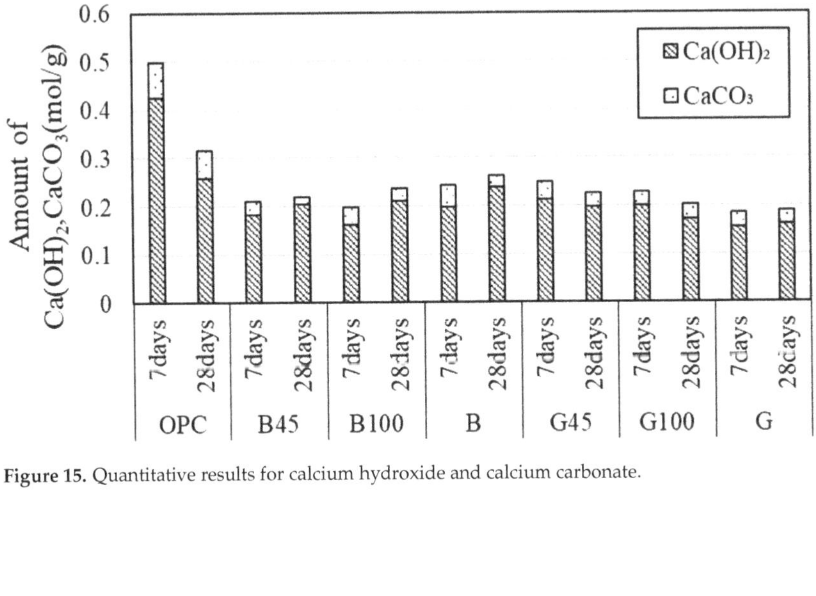

Figure 15. Quantitative results for calcium hydroxide and calcium carbonate.

4. Discussion

4.1. Effect of RHA Particle Size on Mortar

Figure 16 illustrates the relationship between specific surface area and particle size calculated using BET theory from the water vapor adsorption test data. The specific surface area increased when RHA was powdered. Adjusting the particle size of RHA disrupted its pore structure and improved flowability. Additionally, Figure 17 demonstrates a positive correlation between compressive strength and particle size. Reducing the particle size increased the specific surface area of RHA, which, in turn, contributed to enhanced compressive strength. Another study by Xu et al. [41] found that increasing the specific surface area of RHA improved its mechanical properties and degree of hydration.

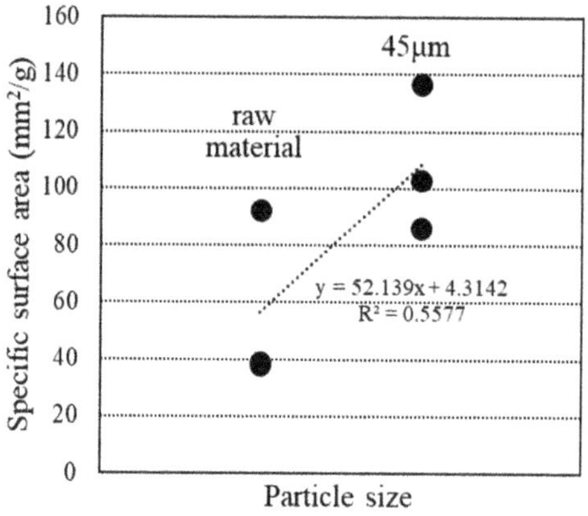

Figure 16. Relationship between particle size and specific surface area.

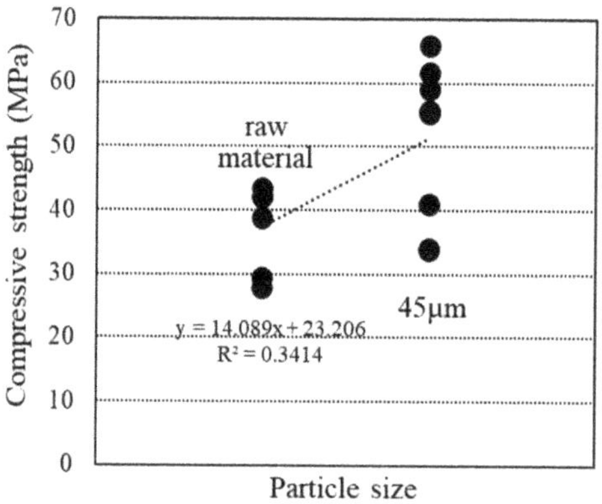

Figure 17. Relationship between particle size and compressive strength.

4.2. Effect of Luxan Value on Mortar

Figure 18 shows the relationship between the amount of calcium hydroxide produced and the Luxan value, as calculated from the TG curve obtained from the TG-DTA data. The amount of calcium hydroxide produced decreased as the Luxan value increased, which was attributed to the pozzolanic reaction between the amorphous silica in the RHA.

Figure 19 illustrates the relationship between the Luxan value and the rate of increase in compressive strength at 28 d compared to 7 d. A higher Luxan value was associated with a greater rate of increase, indicating that an increase in the Luxan value accelerated the pozzolanic reaction and resulted in a significant rise in compressive strength.

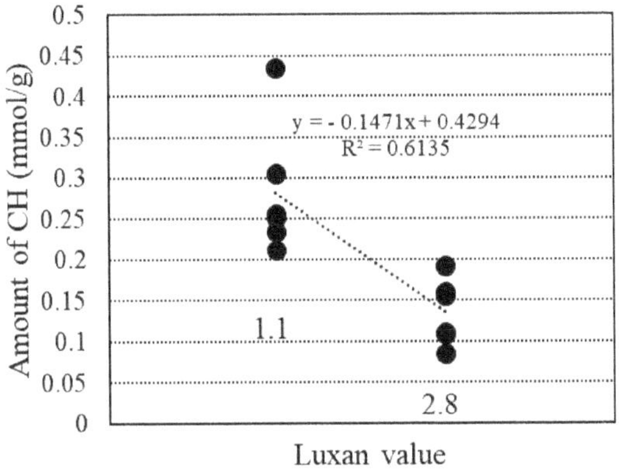

Figure 18. Relationship between CH amount and Luxan value.

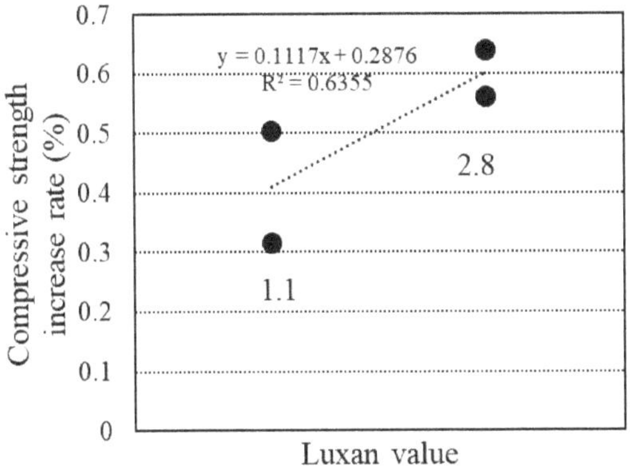

Figure 19. Relationship between compressive strength and Luxan value.

Figure 20 reveals a slight correlation between dry shrinkage at 80 d and the Luxan value. According to a study by He et al. [20], a higher RHA ratio generally reduced dry shrinkage strain; however, an upper limit effect was observed, with no significant difference in dry shrinkage strain above a certain RHA amount. Additionally, lower Luxan values had a minimal effect on reducing dry shrinkage at the same RHA ratio, although this effect was smaller compared to the influence of the RHA ratio used in this study [8].

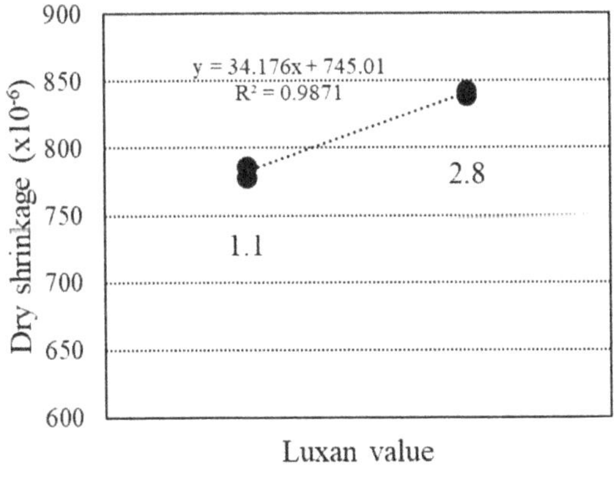

Figure 20. Relationship between dry shrinkage and Luxan value.

With the exception of the relationship between Luxan value and dry shrinkage, the correlations were relatively weak. This may be due to the limited amount of data, indicating the need for additional data collection.

5. Conclusions

This study aimed to identify an effective method for processing RHA. To achieve this, it was necessary to determine the Luxan value for powdered RHA and adjust the particle size to address limitations associated with RHA, such as reduced compressive strength and increased particle size at early curing ages. Additionally, we demonstrated the feasibility of quantifying the hydration rate. The following conclusions were obtained:

(1) Compressive Strength: Adjusting the particle size leads to an increase in compressive strength proportional to the reduction in particle size. In addition, it showed a tendency to increase in proportion to the specific surface area.

(2) Effect of particle size control: Compressive strength increased 36% for B and 28% for G with particle size control. Dry shrinkage decreased 3.8% with particle size control and 4.5% without particle size control.

(3) Flow: Reducing the grain size of RHA improves the flow value, and the flow rate exceeds that of specimens without particle size adjustments. Adjusting the particle size effectively resolves fluidity issues associated with RHA mortar.

(4) High Luxan values: RHA with a high Luxan value, indicating elevated pozzolanic reactivity, is more effective in increasing compressive strength.

(5) Low Luxan value: RHA with a low Luxan value decreases the degree of drying shrinkage.

(6) The reaction rate varies significantly between raw RHA and powdered RHA, indicating that the powdering process substantially affects the reactivity of RHA, even during the initial curing period. We propose that adjusting the particle size of RHA can enhance its reactivity.

In future research, we aim to investigate whether the pH of recycled aggregate can be used to activate the pozzolanic reaction of RHA and contribute to the development of environmentally friendly concrete.

Author Contributions: Conceptualization, J.K.; methodology, H.F.; validation, J.K. and H.F.; formal analysis, J.K. and H.F.; investigation, J.K. and H.F.; resources, M.K.; data curation, J.K. and H.F.; writing—original draft preparation, J.K. and H.F.; writing—review and editing, M.K.; project administration, J.K.; funding acquisition, M.K. All authors have read and agreed to the published version of the manuscript.

Funding: This study received no external funding.

Institutional Review Board Statement: Not applicable.

Informed Consent Statement: Not applicable.

Data Availability Statement: The original contributions presented in the study are included in the article, further inquiries can be directed to the corresponding author.

Acknowledgments: We would like to thank Hokuriku Techno Co., Ltd. for providing the calcined RHA for our experiments with RHA.

Conflicts of Interest: The authors declare no conflicts of interest.

References

1. Feng, Q. *Study on the Fundamental Properties of Rice Husk Ash and Its Application*; Hachinohe Institute of Technology Bulletin; Hachinohe Institute of Technology: Aomori, Japan, 2004; Volume 24.
2. Kalderis, D.; Bethanis, S.; Paraskeva, P.; Diamadopoulos, E. Production of activated carbon from bagasse and rice husk by a single-stage chemical activation method at low retention times. *Bioresour. Technol.* **2008**, *99*, 6809–6816. [CrossRef]
3. Mahdi, N.S.; Babu, V.D.; Hossiney, N.; Mohd, M.A.B.A. Strength and durability properties of geopolymer paver blocks made with fly ash and brick kiln rice husk ash. *Case Stud. Constr. Mater.* **2022**, *16*, e00800. [CrossRef]
4. Ketov, A.; Rudakova, L.; Vaisman, I.; Ketov, I.; Haritonovs, V.; Sahmenko, G. Recycling of rice husks ash for the preparation of resistant, lightweight and environment-friendly fired bricks. *Constr. Build. Mater.* **2021**, *302*, 124385. [CrossRef]
5. Zhang, Z.; Yang, F.; Liu, J.C.; Wang, S. Eco-friendly high strength, high ductility engineered cementitious composites (ECC) with substitution of fly ash by rice husk ash. *Cem. Concr. Res.* **2020**, *137*, 106200. [CrossRef]
6. Henry, C.S.; Lynam, J.G. Embodied energy of rice husk ash for sustainable cement production. *Case Stud. Chem. Environ. Eng.* **2020**, *2*, 100004. [CrossRef]
7. Nuaklong, P.; Jongvivatsakul, P.; Pothisiri, T.; Sata, V.; Chindaprasirt, P. Influence of rice husk ash on mechanical properties and fire resistance of recycled aggregate high-calcium fly ash geopolymer concrete. *J. Clean. Prod.* **2020**, *252*, 119797. [CrossRef]
8. Tuan, V.N.; Ye, G.; Breugel, V.K.; Copuroglu, O. Hydration and microstructure of ultra high performance concrete incorporating rice husk ash. *Cem. Concr. Res.* **2011**, *41*, 1104–1111. [CrossRef]
9. Sensale, G.R. Strength development of concrete with rice-husk ash. *Cem. Concr. Compos.* **2005**, *28*, 158–160. [CrossRef]
10. Liu, D.; Zhang, W.; Lin, H.; Li, Y.; Lu, H.; Wang, Y. A green technology for the preparation of high capacitance rice husk-based activated carbon. *J. Clean. Prod.* **2016**, *112*, 1190–1198. [CrossRef]
11. Bie, R.; Song, X.; Liu, Q.; Ji, X.; Chen, P. Studies on effects of burning conditions and rice husk ash (RHA) blending amount on the mechanical behavior of cement. *Cem. Concr. Compos.* **2015**, *55*, 162–168. [CrossRef]
12. Wang, J.; Fu, J.; Song, W.; Zhang, Y. Effect of rice husk ash (RHA) dosage on pore structural and mechanical properties of cemented paste backfill. *J. Mater. Res. Technol.* **2022**, *17*, 840–851. [CrossRef]
13. Iftikhar, B.; Alih, C.S.; Vafaei, M.; Elkotb, M.A.; Shutaywi, M.; Javed, M.F.; Deebani, W.; Khan, M.I.; Aslam, F. Predictive modeling of compressive strength of sustainable rice husk ash concrete:Ensemble learner optimization and comparison. *J. Clean. Prod.* **2022**, *348*, 131285. [CrossRef]
14. Shaaban, M. Properties of concrete with binary binder system of calcined dolomite powder and rice husk ash. *Heliyon* **2021**, *7*, e06311. [CrossRef] [PubMed]
15. Rattanachu, P.; Toolkasikorn, P.; Tangchirapat, W.; Chindaprasirt, P.; Jaturapitakkul, C. Performance of recycled aggregate concrete with rice husk ash as cement binder. *Cem. Concr. Compos.* **2020**, *108*, 103533. [CrossRef]
16. Garrett, D.T.; Cardenas, E.H.; Lynam, G.J. Sugarcane bagasse and rice husk ash pozzolans: Cement strength and corrosion effects when using saltwater. *Curr. Res. Green Sustain. Chem.* **2020**, *1–2*, 7–13. [CrossRef]
17. Kishore, R.; Bhikshma, V.; Prakashc, J.P. Study on Strength Characteristics of High Strength Rice Husk Ash Concrete. *Procedia Eng.* **2011**, *14*, 2666–2672. [CrossRef]
18. Fapohunda, C.; Akinbile, B.; Shittu, A. Structure and properties of mortar and concrete with rice husk ash as partial replacement of ordinary Portland cement—A review. *Int. J. Sustain. Built Environ.* **2017**, *6*, 675–692. [CrossRef]
19. Siddique, R.; Singh, K.; Singh, M.; Corinaldesi, V.; Rajor, A. Properties of bacterial rice husk ash concrete. *Constr. Build. Mater.* **2016**, *121*, 112–119. [CrossRef]
20. He, Z.; Li, L.; Du, S. Creep analysis of concrete containing rice husk ash. *Cem. Concr. Compos.* **2017**, *80*, 190–199. [CrossRef]
21. Zhang, H.M.; Lastra, R.; Malhotra, M.V. Rice-husk ash paste and concrete: Some aspects of hydration and the microstructure of the interfacial zone between the aggregate and paste. *Cem. Concr. Res.* **1996**, *26*, 963–977. [CrossRef]
22. Breugel, V.K.; Ye, G.; Tuan, V.N. Ultra high performance concrete made with rice husk ash for reduced autogenous shrinkage. In Proceedings of the 4th International FIB Congress 2014, Mumbai, India, 10–14 February 2014; pp. 10–14.
23. Rêgo, S.H.J.; Nepomuceno, A.A.; Figueiredo, P.E.; Hasparyk, P.N. Microstructure of cement pastes with residual rice husk ash of low amorphous silica content. *Constr. Build. Mater.* **2015**, *80*, 56–68. [CrossRef]
24. Van, V.; Rößler, C.; Bui, D.; Ludwig, H. Rice husk ash as both pozzolanic admixture and internal curing agent in ultra-high performance concrete. *Cem. Concr. Compos.* **2014**, *53*, 270–278. [CrossRef]
25. Saraswathy, V.; Song, H.W. Corrosion performance of rice husk ash blended concrete. *Constr. Build. Mater.* **2006**, *21*, 1779–1784. [CrossRef]
26. Zerbino, R.; Giaccio, G.; Isaia, G.C. Concrete incorporating rice-husk ash without processing. *Constr. Build. Mater.* **2011**, *25*, 371–378. [CrossRef]

27. Xu, W.; Lo, Y.T.; Ouyang, D.; Memon, S.A.; Xing, F.; Wang, W.; Yuan, X. Effect of rice husk ash fineness on porosity and hydration reaction of blended cement paste. *Constr. Build. Mater.* **2015**, *89*, 90–101. [CrossRef]
28. Alex, J.; Dhanalakshmi, J.; Ambedkar, B. Experimental investigation on rice husk ash as cement replacement on concrete production. *Constr. Build. Mater.* **2016**, *127*, 353–362. [CrossRef]
29. Venkatanarayanan, H.K.; Rangaraju, P.R. Effect of grinding of low-carbon rice husk ash on the microstructure and performance properties of blended cement concrete. *Cem. Concr. Compos.* **2015**, *55*, 348–363. [CrossRef]
30. Salas, A.; Delvasto, S.; de Gutierrez, R.M.; Lange, D. Comparison of two processes for treating rice husk ash for use in high performance concrete. *Cem. Concr. Res.* **2009**, *39*, 773–778. [CrossRef]
31. Hu, L.; He, Z.; Zhang, S. Sustainable use of rice husk ash in cement-based materials: Environmental evaluation and performance improvement. *J. Clean. Prod.* **2020**, *264*, 121744. [CrossRef]
32. Dizaji, B.H.; Zeng, T.; Holzig, H.; Bauer, J.; Kloß, G.; Enke, D. Ash transformation mechanism during combustion of rice husk and rice straw. *Fuel* **2022**, *307*, 121768. [CrossRef]
33. Siddika, A.; AlMamun, A.; Alyousef, R.; Mohammadhosseini, H. State-of-the-art-review on rice husk ash: A supplementary cementitious material in concrete. *J. King Saud Univ. Eng. Sci.* **2021**, *33*, 294–307. [CrossRef]
34. Kang, S.H.; Hong, S.G.; Moon, J. The use of rice husk ash as reactive filler in ultra-high performance concrete. *Cem. Concr. Res.* **2019**, *115*, 389–400. [CrossRef]
35. Nair, D.G.; Fraaij, A.; Klaassen, A.A.K.; Kentgens, A.P.M. A structural investigation relating to the pozzolanic activity of rice husk ashes. *Cem. Concr. Res.* **2007**, *38*, 861–869. [CrossRef]
36. James, J.; Rao, M.S. Reactivity of Rice Husk Ash. *Cem. Concr. Res.* **1986**, *16*, 296–302. [CrossRef]
37. Yusof, A.M.; Nizam, N.A.; Rashid, N.A.A. Hydrothermal conversion of rice husk ash to faujasite-types and NaA-type of zeolites. *J. Porous Mater.* **2010**, *17*, 39–47. [CrossRef]
38. Zain, M.F.M.; Islam, M.N.; Mahmud, F.; Jamil, M. Production of rice husk ash for use in concrete as a supplementary cementitious material. *Constr. Build. Mater.* **2011**, *25*, 798–805. [CrossRef]
39. Luxán, M.P.; Madruga, F.; Saavedra, J. Rapid evaluation of pozzolanic activity of natural products by conductivity measurement. *Cem. Concr. Res.* **1989**, *19*, 63–68. [CrossRef]
40. Sugita, S.; Shoya, M. A fundamental study on the effective use for rice husk ash as a pozzolanic material. *Jpn. J. JSCE* **1995**, *29*, 43–53. [CrossRef] [PubMed]
41. Xu, W.; Lo, T.Y.; Wang, W.; Ouyang, D.; Wang, P.; Xing, F. Pozzolanic reactivity of silica fume and ground rice husk ash as reactive silica in a cementitious system: A comparative study. *Materials* **2016**, *9*, 146. [CrossRef] [PubMed]
42. Doi, K.; Shirahama, N.; Yamashita, M. Characteristics of fly ash adjusted in particle size by classification. *Cem. Sci. Concr. Technol.* **2017**, *71*, 626–632. [CrossRef]
43. Ishiguro, S. Utilization of rice husk ash as an admixture in high-strength concrete. *Bull. Fac. Bioresour. Mie Univ.* **1999**, *22*, 63–69.
44. Li, Z.; Lu, P.; Song, C.; Che, Y.; Wang, X.; Lyu, Z.; Liu, Y. Developing Ni/SAPO-11 deoxygenation catalysts using yttrium modified SAPO-11 prepared by rice husk ash. *Fuel* **2024**, *366*, 131385. [CrossRef]
45. Zou, J.; Ma, X.; Zhao, M.; Guan, Q.; Chen, D.; Chao, Y. Ternary geopolymer synthesized using self-made rice husk ash-based water glass for removing Cu2+ and Cr3+ from wastewater. *Mater. Lett.* **2024**, *367*, 136556. [CrossRef]
46. Kim, S.J.; Park, S.J.; Jo, S.H.; Yoon, S.J.; Moon, J.H.; Ra, H.W.; Yoon, S.M.; Lee, J.G.; Choi, H.G.; Kim, J.K.; et al. Low-NOx emission and amorphous siliceous ash production from rice husk combustion in circulating fluidized bed system. *Fuel* **2024**, *374*, 132441. [CrossRef]
47. Farid, A.S.; Zaheer, M.M. Production of new generation and sustainable concrete using rice husk ash (RHA): A review. *Mater. Today Proc.* 2023, *in press*. [CrossRef]
48. *JISR5201*; Physical Testing Methods for Cement. Japanese Standards Association: Tokyo, Japan, 2015. Available online: https://www.jisc.go.jp/app/jis/general/GnrJISNumberNameSearchList?toGnrJISStandardDetailList (accessed on 16 December 2024).
49. Washio, K. Surface properties analyses by gas adsorption and mercury intrusion methods. *Imaging Soc. Jpn.* **2007**, *46*, 482–488. [CrossRef]
50. Mahdi, N.S.; Babu, R.V.D.; Shivakumar, M.; Abdullah, B.A.M.M. Mitigation of environmental problems using brick kiln rice husk ash in geopolymer composites for sustainable development. *Curr. Res. Green Sustain. Chem.* **2021**, *4*, 100193. [CrossRef]
51. Giaccio, G.; de Sensale, G.R.; Zerbino, R. Failure mechanism of normal and high-strength concrete with rice-husk ash. *Cem. Concr. Compos.* **2007**, *29*, 566–574. [CrossRef]
52. Otsuka, T.; Mori, S.; Ishikawa, M.; Sakai, E. Relation between mineral compositions of fly ash and its pozzolanic reaction. *Cem. Sci. Concr. Technol.* **2009**, *63*, 16–21. [CrossRef]
53. Msinjili, N.S.; Schmidt, W.; Rogge, A.; Kuhne, H.C. Performance of rice husk ash blended cementitious systems with added superplasticizers. *Cem. Concr. Compos.* **2017**, *83*, 202–208. [CrossRef]

54. De Sensale, G.R.; Ribeiro, A.B.; Gonçalves, A. Effects of RHA on autogenous shrinkage of Portland cement pastes. *Cem. Concr. Compos.* **2008**, *30*, 892–897. [CrossRef]
55. Raheem, A.A.; Anifowose, M.A. Effect of ash fineness and content on consistency and setting time of RHA blended cement. *Mater. Today Proc.* **2023**, *86*, 18–23. [CrossRef]

materials

MDPI

Article

Strategies for Minimizing Environmental Impact in Construction: A Case Study of a Cementitious 3D Printed Lost Formwork for a Staircase †

Sophie Viktoria Albrecht *, Stefan Hellerbrand, Florian Weininger and Charlotte Thiel

Department of Construction Materials, OTH Regensburg, 93053 Regensburg, Germany; charlotte.thiel@oth-regensburg.de (C.T.)

* Correspondence: sophie.albrecht@oth-regensburg.de

† This paper is an extended version of our paper published in Albrecht, S.V.; Hellerbrand, S.; Weininger, F.; Thiel, C. Possibilities for Reducing the Environmental Impact in the Construction Industry Using the Example of a 3D Printed Staircase. In Proceedings of the 4th International Conference on Sustainable Development in Civil, Urban and Transportation Eng. (CUTE 2024), Wroclaw, Poland, 14–17 October 2024.

Abstract: The construction industry faces significant challenges, including environmental sustainability, rising material costs, and a shortage of skilled labor. Digital fabrication technologies offer innovative solutions to address these issues by reducing raw material consumption and waste generation. Among these, 3D printing technologies offer distinct advantages over traditional construction methods, particularly in handling complex geometries. However, the significant environmental impact of cement in 3D printed concrete, due to its high rheological and printability requirements, remains a concern. This study introduces a novel application of 3D printed permanent formwork in the construction of a winder staircase, assessed through an Environmental Life Cycle Assessment (LCA) from cradle to gate. By comparing the environmental impacts of various construction materials and processes, this study highlights the comparative advantages and disadvantages of conventional methods versus 3D printing. The LCA results reveal that traditional production methods, particularly those using plywood formwork, exhibit higher environmental impacts. In contrast, timber formwork performs better than most 3D printed mixtures in terms of Global Warming Potential (GWP), Acidification Potential (AP), and abiotic depletion potential (ADP). The findings of this study underscore the potential of additive manufacturing for sustainable construction, particularly through the use of low-clinker cement in 3D printed formwork, offering a promising pathway towards reducing the environmental footprint of construction activities.

Keywords: additive manufacturing in construction (AMC); life cycle assessment (LCA); 3D concrete printing (3DCP); selective cement activation (SCA); permanent formwork

1. Introduction

The modern construction industry has to tackle the shortage of skilled workers [1], the reduction in negative environmental impact [2] and the housing shortage [3]. Sustainability and resource efficiency are crucial for the construction industry, which is one of the world's largest consumers of natural resources and energy [2]. The construction sector is responsible for a significant proportion of global carbon dioxide emissions, both through the production of building materials such as concrete and steel, and through the energy consumed during the construction and use of buildings [4]. In the face of climate change and resource scarcity, it is essential that the construction industry develops and implements more sustainable and resource-efficient practices. Through the use of efficient technologies, waste reduction and the conscious use of materials, the construction industry can significantly reduce its environmental footprint while ensuring the long-term availability of resources [5].

Comprehensive assessments like Life Cycle Analysis (LCA) are becoming increasingly important to assess and reduce potential environmental impacts [6,7].

An Environmental Life Cycle Assessment (E-LCA) is a method used to evaluate the environmental impacts associated with all stages of a product's life, from raw material extraction to disposal. In addition, even supplementary information beyond the building's life cycle can be considered as foreseen in EN 15804 phase D [8]. This provides a full picture of the product's environmental footprint, including emissions and waste produced along the way.

Recent advancements, such as additive manufacturing in construction, commonly known as 3D printing, offer potential solutions to the aforementioned challenges. The study by Garcia de Soto et al. demonstrates that additive manufacturing in construction has the potential to significantly enhance productivity in the fabrication of complex structures [9].

Agustí-Juan et al. [10] investigated the environmental impact of a digital fabrication method compared to traditional construction techniques for complex architectural forms. The LCA comparison between digital fabrication and conventional construction methods for various wall types shows that conventional methods have lower environmental impacts for simpler walls, but as structural complexity increases, the environmental performance of conventional walls deteriorates. For double-curved and complex walls, Mesh Mold offers significant reductions in climate change (38%) and human toxicity (31%) compared to conventional techniques using polystyrene formwork. In addition, the environmental impact of conventional methods increases with form complexity, whereas the impact of the Mesh Mold process remains constant, making it more advantageous for complex architectural designs [10].

Conventional one-time use wood formwork, particularly when utilizing plywood, exhibits significant potential for improvement in terms of environmental impact compared to plastic or steel formwork [11,12]. To investigate these aspects, this study utilizes a case study of a staircase, comparing conventional formwork with 3D printed permanent formwork, while also analyzing various databases to assess the impact of cement and researching different electricity mixes to evaluate their influence on the overall environmental performance. Through this comparison, the study aims to evaluate the potential of additive manufacturing in reducing the environmental impact of construction activities.

2. Materials and Methods

2.1. Goal and Scope

For the Life Cycle Assessment (LCA), the framework was developed in ISO 14040 and ISO 14044 was applied [13,14]. The software openLCA 2.2 and the database ecoinvent3.9.1 cut-off were used. The life cycle phases A1–A3, extraction of raw materials, transport, and processing, are analyzed in this study [8]. For the functional unit, a winder staircase with five steps and an angle of 90° was chosen, as in Figure 1.

The case study of a 3D printed lost formwork for a winding staircase illustrates the potential of additive manufacturing to achieve material efficiency and design adaptability in construction.

A promising approach to achieving efficiency improvement is the digital fabrication of building components, with a focus often on direct 3D printing with concrete. However, in this study, an alternative approach was pursued, focusing on indirect digital fabrication using a 3D printed integrated formwork. Staircase elements were produced using a 3D concrete printer, stacked, reinforced, and then concreted to achieve full load-bearing capacity

(Figure 2). As demonstrated in Figure 2, the additive manufacturing process achieved a high-resolution surface finish, eliminating the need for any additional surface treatment.

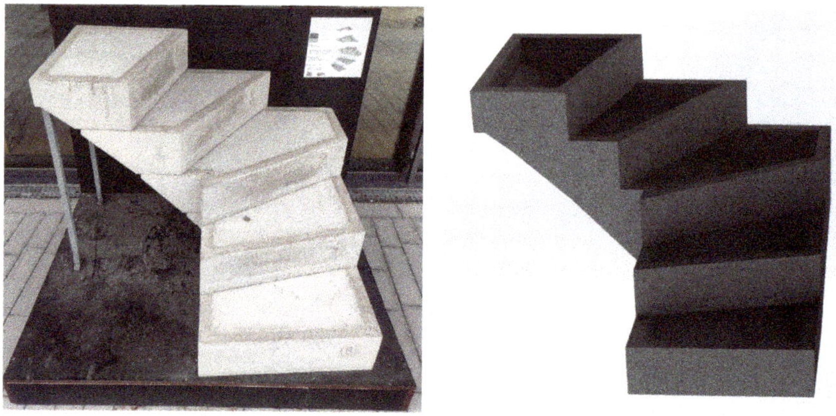

(a) (b)

Figure 1. A case study of a 3D printed lost formwork for a winding staircase, demonstrating material efficiency and adaptability: (**a**) the produced additively manufactured lost formwork staircase filled with concrete, serving as the final exhibition piece; (**b**) a rendering of the additively manufactured lost formwork staircase.

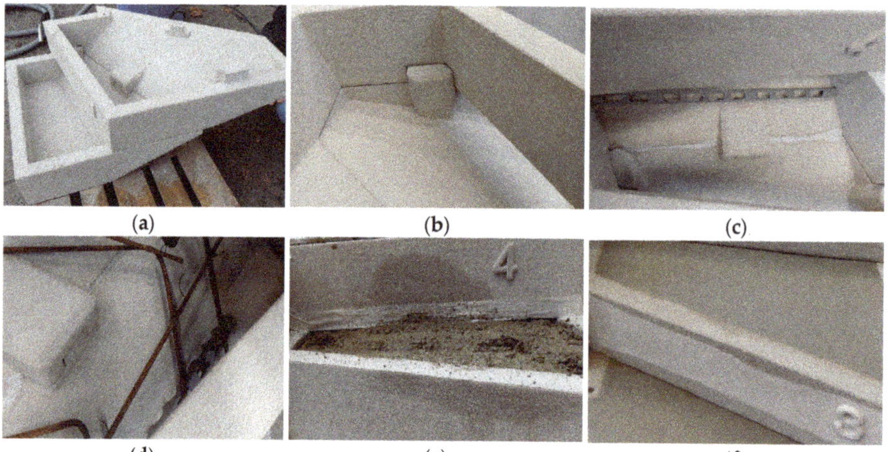

(a) (b) (c) (d) (e) (f)

Figure 2. Production process of staircase with additively manufactured lost formwork: (**a**) assembly of individually printed concrete throughs; (**b**) stacked throughs held together by joints; (**c**) sealing connections between elements; (**d**) insertion of reinforcement; (**e**) pouring of concrete infill; (**f**) smoothing of step surfaces.

A key advantage of the digital construction process employed is its ability to address the current skilled labor shortage in the construction industry. The lack of skilled workers, especially in traditional trades such as formwork construction and concrete processing, poses significant challenges for the industry. The use of 3D printing technologies offers a solution here, as it reduces the labor intensity and complexity of certain construction processes. This makes it possible to involve unskilled or less-qualified workers in the construction process. Since many of the work steps are automated or greatly simplified, the reliance on highly specialized personnel decreases.

The approach of 3D printed integrated formwork, as applied in the case of the trough staircase, offers further advantages compared to traditional formwork methods. The lost formwork was produced with the particle-bed 3D printing method, specifically Selective Cement Activation (SCA), known for its high resolution compared to other concrete AM methods [15]. SCA offers significant advantages, particularly in terms of design freedom, enabling the creation of complex, customized shapes such as overhangs, arches, and free-form structures without the need for costly formwork. With a high resolution and accuracy down to 0.1 mm, this method allows for the production of intricate and detailed components. However, it is limited by the size of the print area, which means that large components must be produced in sections. While design complexity does not affect production time, large-scale printing still requires significant processing time [16]. SCA enables complex geometries to be realized precisely and efficiently, a process that often requires significant effort using conventional techniques. This is evident, for example, in the view of the staircase from above. The surface created by the winding of the staircase is difficult to achieve with traditional formwork methods without significant effort. The staircase was parametrically designed in Rhinoceros and Grasshopper to be easily adjusted to different usage scenarios. The steps had dimensions of 71.00 cm in width and 20.00 cm in height.

Six different scenarios with varying production methods and materials used (Table 1) were investigated and assessed through a Life Cycle Assessment (LCA).

Table 1. Six scenarios with varying production methods and materials used, adapted from [17] and licensed by Springer Nature Customer Center GmbH.

Name	Type of Formwork	Material for Formwork
Scenario 1	Conventional one-time use wood formwork	plywood
Scenario 2	Conventional one-time use wood formwork	timber
Scenario 3	SCA 3D printed lost formwork	60/40 volumetric share of aggregate/cement [15] mixture with CEM I
Scenario 4	SCA 3D printed lost formwork	60/40 volumetric share of aggregate/cement [15] mixture with CEM III
Scenario 5	SCA 3D printed lost formwork	80/20 volumetric share of aggregate/cement [15] mixture with CEM I
Scenario 6	SCA 3D printed lost formwork	80/20 volumetric share of aggregate/cement [15] mixture with CEM III

The first scenario was conventional reinforced concrete production with a one-time use plywood formwork. The second scenario was the conventional manufacturing method as well, but using a one-time use untreated timber formwork. The formwork utilized in scenarios 1 and 2 closely resembled the design depicted in Figure 3.

Figure 3. Wooden formwork for a conventional winder staircase.

The choice of single-use formwork was dictated by the unique specifications of the staircase design, which required a tailored approach to meet its complex geometry. The use of a re-usable formwork was deemed unsuitable in this context, as it lacked the flexibility to meet the precise contours and dimensions required for this particular design. Instead, a specifically fabricated formwork was required to ensure that the architectural and functional features of the staircase were accurately realized. Conventional one-time use wood formwork performs best in terms of environmental impact compared to plastic or steel formwork [11,12]. Therefore, the materials plywood and timber were chosen to be analyzed.

For Scenarios 3 to 6, a permanent formwork with a thickness of 5 cm was printed by a SCA 3D printer, resulting in a smaller amount of filling with the conventional reinforced concrete mixture used in Scenarios 1 and 2. In Scenario 3 and 5 the formwork was printed with mixtures designed by Herding et al. [15]. To address the significant influence of cement dosage on the environmental impact of 3D printed concrete (3DPCC), the scenarios were deliberately designed with a reduced cement content. This approach considered the cement requirements necessary to achieve the printing properties, which were notably higher compared to conventional cast concrete [18]. These mixtures were changed for Scenarios 4 and 6 by replacing the Portland cement of the printed formwork with CEM III/B, a ground granulated blast furnace slag cement (GGBFS). Clinker substitutes such as GGBFS cements play a critical role in reducing greenhouse gas emissions in cement production by lowering the clinker content, which is the main source of emissions [19]. GGBFS, a by-product of the steel industry, exhibits hydraulic properties when mixed with clinker and gypsum, enabling its use as a partial replacement for clinker. Although these substitutes reduce process-, fuel-, and power-related CO_2-equivalent emissions, they may require additional energy for fine grinding, potentially offsetting gains in energy efficiency [19,20]. Despite its current limitations, such as not being printable for SCA, GGBFS cement was chosen for its high mitigation potential and role in optimizing resource use. While not yet applied to SCA, this study provides a foundation to encourage further exploration of GGBFS cement for future advancements.

2.2. Life Cycle Inventory (LCI)

The scenarios already described in [17] take into account both the input and output processes (materials and produced waste). The machines used in the processes and the electricity were not included due to lack of data on particle-bed 3D printers used in construction. To maintain comparability, the machines and electricity were therefore not included in the conventional scenarios either.

The datasets utilized in this LCA were exclusively sourced from the ecoinvent3.9.1 database, which was released on the 15th of December in 2022. To evaluate the impacts of stages A1 to A3, market datasets from ecoinvent3.9.1 were employed.

The selection of geographical locations followed a hierarchical approach: priority was given to datasets representing Europe as a whole. If no such dataset was available, datasets for "Europe without Switzerland" were used. In cases where neither of these options was applicable, datasets from specific European countries were selected. If no suitable European dataset was available, the global dataset was used as a last resort.

Scenarios 1 and 2 comprised two components: a one-time use wooden formwork and the corresponding concrete filling for the staircase. Scenarios 3 to 6 utilized a SCA 3D printed lost formwork in combination with the concrete filling.

The concrete used for the filling was a generic mixture of Ordinary Portland cement with an approximate density of 2400 kg/m^3. This concrete composition remained consistent across all six scenarios.

In Scenarios 1 and 2, the quantity of wood required for the one-time use formwork of the staircase was determined based on the surface area of the functional unit [21]. The nails needed for the formwork were not considered. The waste concrete was measured as 60 kg of unused material, which was disposed of. The waste wood was the one-time used wood formwork as the staircase was a non-standardized one and was only produced once. The total mass of the waste wood was calculated based on the density of the specific wood type used. Additionally, the wastewater from the concrete production was the water used to clean the equipment after the concrete work was finished. The LCI of Scenario 1 is seen in Table 2. For Scenario 2, the plywood formwork was replaced by timber using the dataset "market for cleft timber, measured as dry mass | cleft timber, measured as dry mass | Cutoff, U—Europe without Switzerland".

Table 2. LCI of Scenario 1—conventional reinforced concrete with plywood formwork, adapted from [17] and licensed by Springer Nature Customer Center GmbH.

Material	Process	Amount	Unit
	Concrete filling		
Cement, Portland	market for cement, Portland	cement, Portland	Cutoff, U—Europe without Switzerland	119.70	kg
Gravel, round	market for gravel, round	gravel, round	Cutoff, U—CH	349.30	kg
Reinforcing steel	market for reinforcing steel	reinforcing steel	Cutoff, U—GLO	10.63	kg
Sand	market for sand	sand	Cutoff, U—CH	285.95	kg
Tap water	market for tap water	tap water	Cutoff, U—Europe without Switzerland	63.00	kg
Waste concrete, not reinforced	market for waste concrete, not reinforced	waste concrete, not reinforced	Cutoff, U—Europe without Switzerland	60.00	kg
Wastewater from concrete production	market for wastewater from concrete production	wastewater from concrete production	Cutoff, U—CH	0.09	m^3
	One-time use wood formwork		
Plywood	market for plywood	plywood	Cutoff, U—RER	0.32904	m^3
Waste wood, untreated	market for waste wood, untreated	waste wood, untreated	Cutoff, U—AT	236.9088	kg

In Scenarios 3 to 6, the additively manufactured scenarios, the volume of concrete filling required for the staircase was reduced by 40% compared to Scenarios 1 and 2. This reduction was attributed to the presence of the SCA-printed lost formwork, which occupied a portion of the total volume, thereby decreasing the overall concrete demand. Scenario 3 utilized a mixture composition of 60/0/40_0.5, as described by Herding et al. [15]. The Life Cycle Inventory (LCI) for Scenario 5, calculated using the 80/0/20_0.5 mixture composition by Herding et al. [15], is presented in Table 3. In Scenarios 4 and 6, the Portland cement was replaced by CEM III/B "market for cement, CEM III/B | cement, CEM III/B | Cutoff, U—Europe without Switzerland".

Table 3. LCI of Scenario 5—80/20 mixture for SCA 3D printed formwork and reinforced concrete filling, adapted from [17] and licensed by Springer Nature Customer Center GmbH.

Material	Process	Amount	Unit
	Concrete filling		
Cement, Portland	market for cement, Portland	cement, Portland	Cutoff, U—Europe without Switzerland	71.82	kg
Gravel, round	market for gravel, round	gravel, round	Cutoff, U—CH	209.58	kg
Reinforcing steel	market for reinforcing steel	reinforcing steel	Cutoff, U—GLO	10.63	kg
Sand	market for sand	sand	Cutoff, U—CH	171.57	kg
Tap water	market for tap water	tap water	Cutoff, U—Europe without Switzerland	37.80	kg
Waste concrete, not reinforced	market for waste concrete, not reinforced	waste concrete, not reinforced	Cutoff, U—Europe without Switzerland	36.00	kg
Wastewater from concrete production	market for wastewater from concrete production	wastewater from concrete production	Cutoff, U—CH	0.09	m^3

Table 3. *Cont.*

Material	Process	Amount	Unit
	SCA 3D printed lost formwork		
Cement, Portland	market for cement, Portland ǀ cement, Portland ǀ Cutoff, U—Europe without Switzerland	83.72	kg
Silica sand	market for silica sand ǀ silica sand ǀ Cutoff, U—GLO	296.80	kg
Tap water	market for tap water ǀ tap water ǀ Cutoff, U—Europe without Switzerland	34.86	kg

Even though the mixture was not yet printable, it was modeled to show the environmental potential of low-clinker cementitious mixtures. The waste concrete was measured to be 36 kg, reflecting the overall reduction in the concrete filling compared to Scenarios 1 and 2. The same equipment was used for producing the concrete filling as the concrete in Scenarios 1 and 2, and, therefore, the same amount of wastewater was required for cleaning it. The LCI of Scenario 5 is seen in Table 3.

2.3. *Life Cycle Impact Assessment (LCIA)*

The impact categories Global Warming Potential (GWP); Ozone Depletion Potential (ODP); Acidification Potential (AP); eutrophication potential—aquatic freshwater (EP-freshwater); eutrophication potential—aquatic marine (EP-marine); eutrophication potential—terrestrial (EP-terrestrial); abiotic depletion potential for non-fossil resources (ADPE); abiotic potential for fossil resources (ADPF); water use (WDP); total use of renewable primary energy resources (PERT); and total use of non-renewable primary energy resources (PENRT) were determined according to DIN EN 15804 [8]. The results of ADPE, ADPF, and WDP should be used with care as the uncertainties of the results are high and as there is limited experience with the indicator.

The selection of impact categories in this study aligns with the core environmental indicators specified in DIN EN 15804 [8], reflecting their relevance and importance in LCAs. Global Warming Potential (GWP) and Ozone Depletion Potential (ODP) are classified as Type 1 impact categories, indicating that they are recommended and considered satisfactory due to their high environmental significance and widespread acceptance within both scientific and regulatory frameworks [8,22]. The alignment of the GWP and ODP categories with policy, their scientific credibility, and their adaptability to varying emission profiles make them a key tool for evaluating climate impacts in LCAs [22]. The Intergovernmental Panel on Climate Change (IPCC)'s 100-year timeframe model for calculating GWP and the World Meteorological Organization's (WMO)'s model for calculating ODP are the most recent and reliable midpoint models available for these categories [8,23]. Both indicators are fundamental to capturing critical aspects of environmental impact, as outlined in the standard.

Acidification Potential (AP); eutrophication potential—aquatic freshwater (EP-freshwater); eutrophication potential—aquatic marine (EP-marine); and eutrophication potential—terrestrial (EP-terrestrial) are classified as Type 2 categories, meaning they are also recommended, though some methodological improvements or refinements are necessary to enhance their robustness and applicability [8,22]. Acidification Potential (AP) and terrestrial eutrophication potential (EP-terrestrial) are calculated with the Accumulated Exceedance method and are highly applicable on a European scale, with strong environmental relevance for biodiversity and the natural environment. They provide a comprehensive assessment of atmospheric and soil impacts. However, they are sensitive to the emission scenario and current critical load, which may affect their reliability under varying conditions [8,22,23]. The aquatic eutrophication potentials (EP-freshwater and EP-marine), calculated with the recommended midpoint ReCiPe method, effectively assess aquatic ecosystem impacts with high environmental relevance. However, they do not cover terrestrial ecosystems and are parameterized for the US with spatial differentiation only at the state level, limiting their comprehensiveness. In summary, while these Type 2 categories are valuable, improvements in scope, spatial differentiation, and comprehensiveness are needed to enhance their scientific applicability.

Lastly, abiotic depletion potential for non-fossil resources (ADPE), abiotic potential for fossil resources (ADPF), and water use (WDP), are categorized as Type 3 indicators, suggesting that while they are part of the core set of indicators under DIN EN 15804, their application requires caution due to uncertainties or limitations in their assessment methodologies [10,22]. The CML method for finding the abiotic depletion potentials (ADPE and ADPF) is the recommended one, as it considers the extraction and the reserves of a resource. To assess WDP, the recently developed AWARE method was used. Due to the young age of the method, it lacks significance and needs further expertise and experience [8,24].

The uncertainties associated with the choice of the total use of renewable primary energy resources (PERT) and total use of non-renewable primary energy resources (PENRT) cumulative energy demand (CED) model is relatively low. The analysis of cumulative energy demand is valuable for obtaining a general overview of the energy-related environmental impacts throughout a product's life cycle and serves as an initial comparative tool for assessing individual products [25].

While Type 1 categories such as GWP and ODP are well-established and critical for Life Cycle Assessments, Type 2 categories like AP and EP offer valuable insights into other environmental impacts but require further refinement to improve their robustness and applicability. The Type 3 indicators provide useful data on resource depletion and water use but should be applied with caution due to uncertainties in their methodologies. The parameters for describing the use of resources (PERT and PENRT) are beneficial as they provide a comprehensive assessment of energy demand, helping to identify and compare energy-related environmental impacts across different product systems. Overall, continued advancements in the methodologies and broader scope of these indicators will enhance their effectiveness in supporting comprehensive Life Cycle Assessments, offering more accurate and reliable assessments of environmental sustainability.

3. Results and Interpretation

3.1. Results

In Section 2.1, during the selection of the scenarios, an optimization was already considered through the use of CEM III, the GGBFS cement. Additionally, the calculation of concrete volume and the wood consumption for the one-time use formwork were adjusted according to the specific scenarios. Consequently, the amount of waste was also adapted to each respective scenario.

Table 4 presents the results of the six scenarios, providing direct value comparison for each impact category. The analysis shows that the highest Global Warming Potential is observed in Scenario 1, with 276 kg CO_2 eq., followed closely by Scenario 3 at 246 kg CO_2 eq., underscoring the significant environmental impact of plywood. Moreover, altering the additive manufacturing mixture ratio from 60/40 to 80/20 results in a notable reduction of approximately 29% in CO_2 emissions, as evidenced by the comparison between Scenarios 3 and 5. The utilization of CEM III in Scenario 4 accounts for a reduction of 33% in CO_2 emissions compared to Scenario 3. Scenario 6 falls below 50% of the values of Scenario 1 in every single impact category. The biggest difference between these scenarios is for the PERT, where Scenario 6 only has 0.7% of the MJ eq. of Scenario 1. Considering combined primary energy resources (PEs), Scenario 6 consumes 7% MJ eq. in comparison to Scenario

1 due to the higher PENRT levels. The data indicates that the transition from Ordinary Portland cement to GGBFS cement is responsible for the observed increase in ADPE and PERT. The increased use of renewable primary energy resources is counterbalanced by a more significant reduction in non-renewable energy use. Consequently, although PERT increases, the overall primary energy demand decreases when CEM III/B is used.

Table 4. Results of six scenarios.

	GWP kg CO_2 eq.	ODP kg CFC-11 eq.	AP molc H+ eq.	EP-Freshwater kg P eq.	EP-Marine kg N eq.	EP-Terrestrial molc N eq.	ADPE kg Sb eq.	ADPF MJ	WDP m^3	PENRT MJ eq.	PERT MJ eq.
Scenario 1	276	1.88×10^{-4}	1.67	0.098	0.018	6.47	4.37×10^{-4}	2918	382	3979	9013
Scenario 2	144	4.19×10^{-5}	0.42	0.027	0.007	1.43	1.03×10^{-4}	944	72	811	3345
Scenario 3	246	1.78×10^{-5}	0.64	0.035	0.004	2.01	1.38×10^{-4}	1346	77	1134	80
Scenario 4	164	1.94×10^{-5}	0.57	0.032	0.003	1.55	1.53×10^{-4}	1252	75	1053	86
Scenario 5	176	1.52×10^{-5}	0.51	0.028	0.003	1.57	1.11×10^{-4}	1074	83	915	66
Scenario 6	134	1.56×10^{-5}	0.47	0.026	0.003	1.31	1.19×10^{-4}	1008	68	860	67

3.2. *Interpretation and Dominance Analysis*

The staircase serves as a conceptual framework that can take various forms, allowing for flexibility in its application. Rather than focusing on absolute data, the emphasis is on percentage-based results that highlight the potential impact and scalability of the approach. This perspective provides a more insightful understanding of relative gains, enabling users to appreciate how incremental improvements can add up. By concentrating on percentage outcomes, we can effectively illustrate the broader possibilities and adaptability of the concept without being confined to fixed data points or rigid structures.

Figure 4 depicts the overall results of every scenario. It is visible that Scenario 1 has the highest environmental impact in every category. The comparison of Scenarios 1 and 2 supports the findings that the change from plywood to timber is a severe improvement on the sustainability of the staircase. Balasbaneh et al. [11] determined that the GWP of single-use plywood formwork per square meter exceeds that of timber, a finding reaffirmed by the results of this study. Scenario 3, compared to Scenario 5, highlights the impact of a lower cement content for the permanent formwork. Scenarios 3 to 6 show that the substitution of the Portland cement with GGBFS cement lowers every category except for ODP, ADPE, and PERT, where the values are only negligibly higher than the original mixture compositions. Therefore, it can be stated that Scenario 6 performs comparably best and has the lowest values of GWP, EP-freshwater, EP-marine, EP-terrestrial, and WDP, and the second-lowest in the other categories except for ADPE.

Figure 4 also illustrates the division of the four main components of the staircase and their influence on each impact category. The change in dominance between Scenarios 1 and 2 is based on the change from plywood to timber formwork. Scenarios 3 to 6 have a certain similarity as the reinforcement, concrete, and waste are unchanged, and only the formwork with the different mixtures are altered. The waste is only a minor contributor for the 3D printed scenarios, and an increased one for the conventional productions, especially for ODP.

Figure 5 presents that the main contributor to each category of Scenario 1 is the plywood formwork, whose percentage share stays at almost 80% for every impact category except for GWP, EP-marine, and the PERT. Notably, 99% of the total PERT is attributed to the plywood formwork. The environmental improvement resulting from the shift from plywood to timber formwork is clearly highlighted in the visualization. Figure 5 highlights that cement is the highest contributor in all the scenarios, except when plywood is a part of them. The amount of reinforcement used in the scenarios is consistent across all cases. In conventional scenarios, waste plays a more significant role compared to additive manufacturing scenarios. The staircase built using the additive-manufactured optimized mixture and the low-clinker cement formwork outperformed the optimal single-use timber formwork (Scenario 2) across various environmental categories.

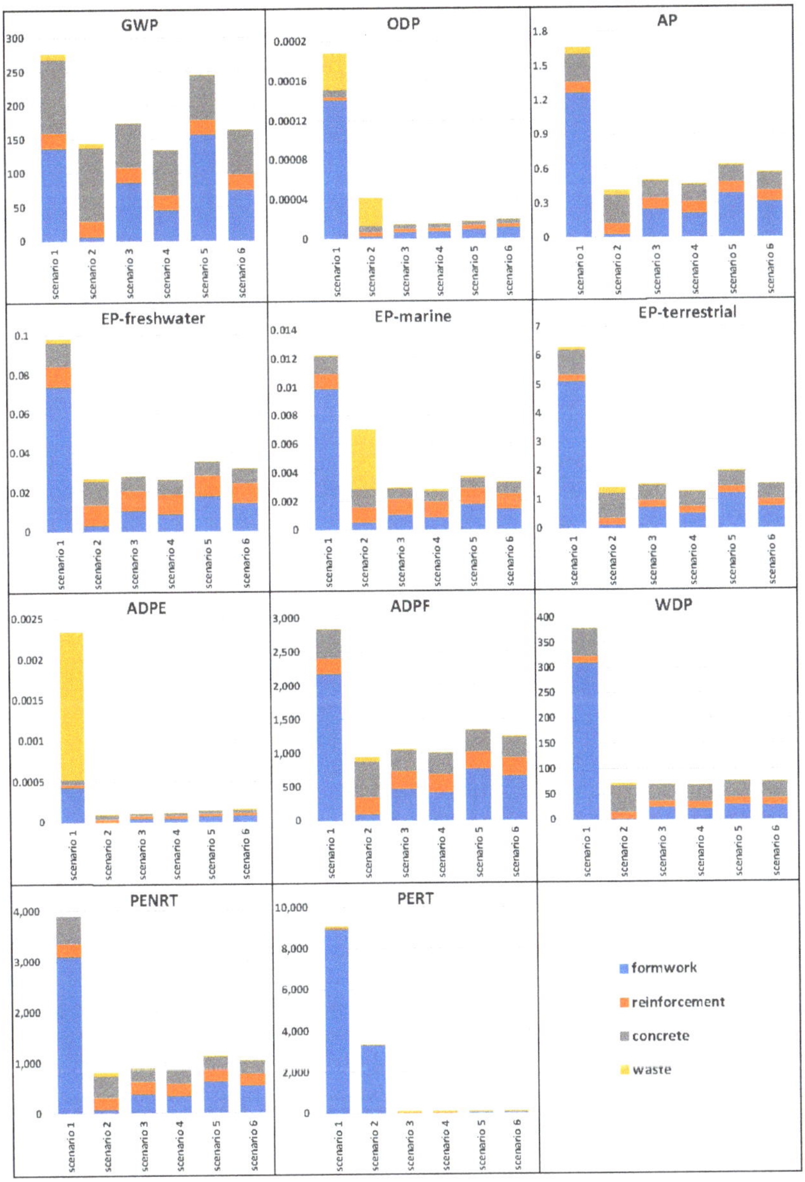

Figure 4. Results of Scenario 1 (conventional production with plywood formwork), Scenario 2 (conventional production with timber formwork), Scenario 3 (additively manufactured formwork with 60/40 mixture [15]), Scenario 4 (additively manufactured formwork with 60/40 mixture [15] and CEM III), Scenario 5 (additively manufactured formwork with 80/20 mixture [15]), and Scenario 6 (additively manufactured formwork with 80/20 mixture [15] and CEM III).

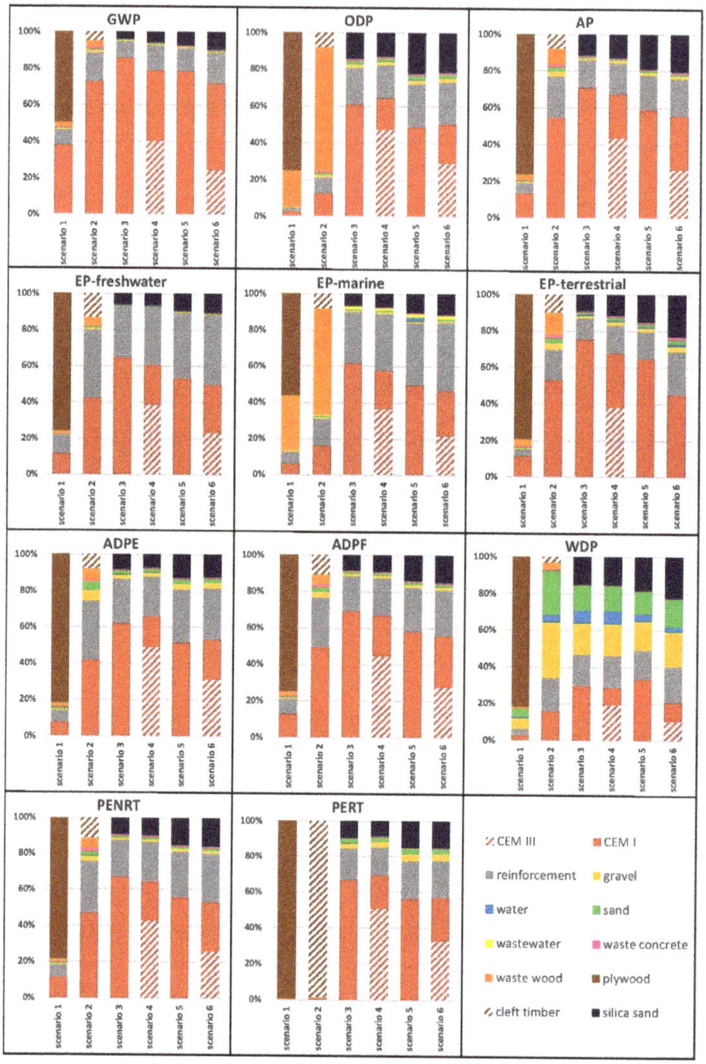

Figure 5. Analysis of the six scenarios. Dominance analysis separated by material.

4. Discussion

4.1. Cement Datasets

Concrete is the most used material in the construction industry, and it is estimated that the production of cement, its main binder, is responsible for approximately 7% of global carbon emissions [26]. Additionally, the location of cement production plays a critical role in determining the total emissions. Tables 5 and 6 highlight the geographical influence on GWP of the cement datasets. Across all regions analyzed, the vast majority of the GWP from cement production—ranging from 95.5% to 99.5%—originates from cement production, with transportation accounting for the remaining emissions. Switzerland has the highest share of emissions from cement production (99.5%), while India shows the lowest (95.5%), indicating that the transportation factor for Switzerland is very low compared to other regions. Other regions, including South Africa, Peru, the US, Europe, and Canada, report similar proportions, with cement production contributing approximately 97%.

Table 5. Comparison of 1 kg of Portland cement "market for cement, Portland | cement, Portland | Cutoff, U" from Switzerland, Canada, Europe without Switzerland, United States, Peru, India, and South Africa in ecoinvent3.9.1 database.

Market Process Medium Voltage	Switzerland	Canada	Europe/CH	US	Peru	India	South Africa
GWP [kg CO_2 eq.]	0.74008	0.86674	0.880346	0.90786	0.91638	0.93117	1.02766

Table 6. Comparison of 1 kg of ground granulated blast furnace slag (GGBFS) cement "market for cement, CEM III/B | cement, CEM III/B | Cutoff, U" from Switzerland, Europe without Switzerland, and Rest of World in ecoinvent3.9.1 database.

Market Process Medium Voltage	Switzerland	Europe/CH	Rest of World
GWP [kg CO_2 eq.]	0.31174	0.39590	0.44263

For the CEMIII/B dataset, there are only three geographical locations in the ecoinvent3.9.1 database available. For the "Rest of World" category, 93.6% of emissions are from cement production, while Europe without Switzerland shows a slightly higher share at 94.0%. Switzerland has the highest proportion, with 98.7% of emissions from cement production, leaving only a small fraction attributable to transportation.

The results indicate that factors such as the energy mix of the region (e.g., reliance on coal, natural gas, or renewable energy), transportation distances for raw materials, and the efficiency of local production technologies can significantly influence the carbon footprint of cement production.

The data in Tables 5 and 6 clearly illustrate that CEM III/B achieves more than a 50% reduction in GWP compared to Portland cement in the same geographical location. This demonstrates the effectiveness of CEM III/B in mitigating environmental impacts, making it a more sustainable alternative to traditional Portland cement.

Further reductions in conventional production scenarios can be achieved by replacing Portland cement with GGBFS cement in the concrete mixture, a strategy that also applies to the filling of the additive-manufactured lost formwork.

Reducing the clinker-to-cement ratio is a key strategy for lowering the GWP in cement production, but it is not the only solution. Electrifying production processes with renewable energy sources is another, and the scaling up of emerging technologies and alternative binders offer promising pathways for decarbonization [26]. A novel approach involves using captured CO_2 to produce nano-$CaCO_3$ directly within the plant, which is then mixed with cement clinkers to create a new generation of cement. This method can reduce CO_2 emissions by up to 70% (from 0.96 kg to 0.3 kg CO_2 eq.) while improving durability, extending the lifespan of structures, and lowering the overall environmental impact of cement production and construction [27,28].

In this study, ecoinvent3.9.1 was used as it was the most current and validated version available at the time the study began. During the research, the evoinvent3.10 database was released. Tables 7 and 8 show comparisons of the cement datasets used in the study, the updated one, and their deviation.

The values for Portland cement for the impact categories ADPE and PERT increase by 140.67% and 15.83%, respectively. While the PERT rises in evoincent3.10, the PENRT falls by 1.41%, which is positive overall as the primary energy has decreased and, within it, the amount of renewable energy has increased. For the GGBFS cement, the values in ecoinvent3.10 have increased for the impact categories GWP, AP, ADPE, and PERT. These results suggest an evolving understanding of the life cycle impacts of cement production, with potential shifts in energy sourcing, modeling assumptions, or updated input data in the ecoinvent3.10 database.

Table 7. A comparison of the impact categories of 1 kg "market for cement, Portland | cement, Portland | Cutoff, U—Europe without Switzerland" in the ecoinvent3.9.1 and ecoinvent3.10 databases and the deviation.

	GWP kg CO_2 eq.	ODP kg CFC-11 eq.	AP molc H+ eq.	ADPE kg Sb eq.	ADPF MJ	WDP m^3	PENRT MJ eq.	PERT MJ eq.
CEM I 3.9.1	0.880346	4.54×10^{-8}	0.001899	3.59×10^{-7}	3.191437	0.095574	3.906352	0.223288
CEM I 3.10	0.879083	4.42×10^{-8}	0.001875	8.64×10^{-7}	3.164876	0.087237	3.851083	0.258631
Deviation [%]	−0.14	−2.64	−1.26	+140.67	−0.83	−8.72	−1.41	+15.83

Table 8. A comparison of the impact categories of 1 kg "market for cement, CEM III/B | cement, CEM III/B | Cutoff, U—Europe without Switzerland" in the ecoinvent3.9.1 and ecoinvent3.10 databases and the deviation.

	GWP kg CO_2 eq.	ODP kg CFC-11 eq.	AP molc H+ eq.	ADPE kg Sb eq.	ADPF MJ	WDP m^3	PENRT MJ eq.	PERT MJ eq.
CEM III/B 3.9.1	0.396287	5.53×10^{-8}	0.001495	4.51×10^{-7}	2.706048	0.089025	3.343201	0.262794
CEM III/B 3.10	0.396783	5.20×10^{-8}	0.001514	1.11×10^{-6}	2.532155	0.059845	3.098149	0.276835
Deviation [%]	+0.13	−5.97	+1.27	+146.12	−6.43	−32.78	−7.33	+5.34

Tables 9 and 10 present a comparison of Portland cement and GGBFS cement data from the ecoinvent3.10 and oekobaudat databases, highlighting significant deviations across various environmental impact categories. The deviations between the databases range from 18% (GWP) to 149% (PERT), indicating substantial discrepancies in how environmental impact is assessed. These discrepancies are influenced not only by variations in data interpretation but also by the geographic scope of the databases: ecoinvent refers to Europe without Switzerland, while oekobaudat is a database that focuses specifically on Germany. Such large deviations emphasize the importance of carefully selecting and cross-referencing databases when conducting environmental impact assessments.

Table 9. A comparison of the impact categories of 1 kg "market for cement, Portland | cement, Portland | Cutoff, U—Europe without Switzerland" in the ecoinvent3.10 database and "Portland cement (CEM I)" phase A1–A3 in the oekobaudat database and the deviation.

	GWP kg CO_2 eq.	ODP kg CFC-11 eq.	AP molc H+ eq.	ADPE kg Sb eq.	ADPF MJ	WDP m^3	PENRT MJ eq.	PERT MJ eq.
CEM I 3.10	0.879083	4.42×10^{-8}	0.001875	8.64×10^{-7}	3.164876	0.087237	3.851083	0.258631
CEM I oekobaudat	0.665000	2.11×10^{-11}	0.000940	4.91×10^{-8}	2.060000	0.004690	2.060000	0.644000
Deviation [%]	−24.35	−99.95	−49.87	−94.32	−34.91	−94.62	−46.51	+149.00

Table 10. A comparison of the impact categories of 1 kg "market for cement, CEM III/B | cement, CEM III/B | Cutoff, U—Europe without Switzerland" in the ecoinvent3.10 database and "Blastfurnace cement CEM III/B 42.5 L-LH/SR" phase A1–A3 in the oekobaudat database.

	GWP kg CO_2 eq.	ODP kg CFC-11 eq.	AP molc H+ eq.	ADPE kg Sb eq.	ADPF MJ	WDP m^3	PENRT MJ eq.	PERT MJ eq.
CEM III/B 3.10	0.396783	5.20×10^{-8}	0.001514	1.11×10^{-6}	2.532155	0.059845	3.098149	0.276835
CEM III/B oekobaudat	0.327000	7.10×10^{-12}	0.000520	3.85×10^{-8}	1.410000	0.007850	1.410000	0.519000
Deviation [%]	−17.59	−99.99	−65.65	−96.53	−44.32	−86.88	−54.49	+87.48

Further research into advanced low-carbon binders, such as Celitement, could contribute significantly to reducing greenhouse gas emissions from cement. Celitement is an environmentally advantageous hydraulic binder that, when used as a replacement for Portland cement clinker, can reduce CO_2 emissions by up to 50% during production, making it a more sustainable alternative in cement manufacturing [29].

As the cement industry transitions towards more sustainable practices, the role of robotics and electricity consumption in production is expected to become increasingly influential in the overall environmental impact assessment. Moreover, thorough research into electricity consumption and its impact on the Life Cycle Assessment (LCA) of additive manufacturing products in construction will become increasingly essential. Understanding the energy demands and environmental implications of using electricity in these processes will be crucial for evaluating the overall sustainability and decarbonization potential of additive manufacturing technologies in the construction sector.

4.2. Energy Datasets

Considering the significance of electricity locations, it is therefore crucial to analyze the electricity mix used in the case of the SCA-printed lost formwork staircase. This study identifies plywood, Ordinary Portland cement, and GGBFS cement as the three largest contributors to GWP. On closer examination of the dataset "market for plywood | plywood | Cutoff, U—RER", 38.32% of the GWP is caused by the electricity used in the production process for the plywood. For plywood production, the electricity used in the market process in ecoinvent3.10 is medium voltage in Europe, while for both types of cement, the electricity mix is based on medium voltage in Europe without Switzerland. As observed in Table 11, there are significant variations in carbon dioxide emissions across different regions. These values show that Switzerland and France have notably lower carbon intensities due to their reliance on low-carbon energy resources (such as nuclear or hydroelectric power), while countries like China and Germany have higher carbon intensities due to a greater reliance on coal and other fossil fuels. Considering the values presented in Table 11, the overall sustainability of the SCA staircase is influenced by the electricity mix of the location.

Table 11. kg CO_2 eq. per kWh of electricity, medium voltage of different locations assessed with ecoinvent3.10.

Market Process Medium Voltage	Europe	Switzerland	Germany	France	China	Canada
GWP [kg CO_2 eq.]	0.32536	0.02332	0.42080	0.07711	0.95041	0.18963

While the cement industry holds significant potential for reducing its carbon footprint through various strategies, another critical factor in reducing the environmental impact is the energy mix used in the production processes of the product. The environmental impact of innovative technologies like robotic 3D printing in construction is closely tied to the energy sources that power these processes. As the cement industry continues to lower its Global Warming Potential, the relative impact of robotic processes and electricity consumption will become more significant in the overall environmental footprint of construction.

Kuzmenko [30] found that the environmental impact of the robotic 3D printing process is significantly influenced by both its lifespan and the local electricity mix, with the overall process-related impact being roughly evenly split between embodied and operational emissions. The supplementary burden associated with the six-axis extrusion robotic printing technology contributes approximately 13% to the overall climate change impact. Additionally, factors like printing resolution further influence the overall environmental footprint, particularly the GWP [30]. In contrast, direct emissions from conventional onsite construction contribute only 2.42% to the overall climate change impact [31]. The Environmental Product Declarations (EPDs) from Informationszentrum Beton GmbH indicate that the production processes of C12/15 and C20/25 concrete contribute approximately 3.0% to 4.0% of the total GWP in the A1–A3 life cycle stages [32]. These compressive strength categories closely resemble the compressive strength of the additive-manufactured mixtures in Scenario 3 and Scenario 5 [15]. Conventional construction methods exhibit a lower process-related GWP, contributing approximately 2.42–4.00% to the overall climate change impact, whereas additive manufacturing processes impose a higher environmental burden. However, these percentages of the different production processes can vary depending on regional energy sources, as shown in Table 11, with some regions experiencing a considerably higher process-related impact.

Overall, data on the impact of robotics and electricity consumption in additive manufacturing for construction remains very limited. Kuzmenko has conducted a thorough study on extrusion-based 3D printing, providing valuable insights into its environmental impact. However, the availability of data on particle-bed 3D printing, particularly regarding energy consumption, remains limited, constraining the scope of this study, which utilizes a particle-bed 3D printer. The GWP associated with the robotic printing process is notably high and can significantly influence the overall environmental impact of the 3D printing process. However, further investigations are necessary to determine whether the particle-bed printing process exhibits a similar level of impact as the extrusion-based process.

5. Conclusions and Outlook

This paper examines strategies to reduce the environmental impact of a 3D printed lost formwork for a winding staircase, highlighting the potential of additive manufacturing to enhance material efficiency and design flexibility in construction. Among the six analyzed scenarios, the additive manufacturing scenario utilizing an already optimized 60/40 mixture [15] for the lost formwork of the staircase demonstrates the best LCA results. In this scenario, the Portland Cement content was substituted with GGBFS cement for the SCA-printed lost formwork for the staircase. However, current mixtures with GGBFS for additive manufacturing processes are not yet printable. This study aims to highlight the significance of such substitutions, providing material researchers with insights into the environmental benefits and encouraging further development of printable, low-clinker cementitious materials. Therefore, further research is needed to lower the clinker content in additive manufacturing cement mixtures without compromising the performance.

A crucial limitation of this study is the exclusion of machinery and equipment, along with their respective electricity consumption, from the LCI. This decision was made due to the current lack of available data on particle-bed 3D printers for construction applications. To maintain comparability, the machines and electricity are therefore not included in the conventional scenarios either. Nevertheless, the environmental impact of robotic 3D printing processes is closely tied to the energy mix used. As such, the sustainability of robotic 3D printing processes cannot be fully evaluated without considering the variability in electricity mixes across different regions.

One of the primary contributions of this research is that it is among the first studies to analyze a particle-bed 3D printed lost formwork using an LCA approach. The findings illustrate both the opportunities and challenges of integrating additive manufacturing into construction. In essence, the example of a 3D printed staircase serves as a microcosm of broader trends shaping the construction industry's sustainability journey. By leveraging technology, creativity, and collaboration, stakeholders can pave the way for a greener, more sustainable built environment that meets the needs of present and future generations.

Further studies are needed to evaluate the environmental viability of 3D printed formwork based on its complexity. This research lays the groundwork for exploring the circularity potential, life cycle phases (B, C, and D), and cost efficiency of 3D printed lost formwork. Future research should also include robotic systems, their wear components, electricity consumption, and the overall lifespan of different robotic systems. The focus should be not only on cement-based materials, but also on alternative materials such as earth or steel. This research could help improve the sustainability and resource efficiency of additive manufacturing and identify new ways of using environmentally friendly materials in the construction industry. These topics are currently researched within the TRR277 [33], contributing to a deeper understanding of the sustainability and circularity of additive manufacturing in construction.

Author Contributions: Conceptualization, S.V.A. and C.T.; methodology, S.V.A. and C.T.; software, S.V.A.; validation, S.V.A. and C.T.; formal analysis, S.V.A. and S.H.; investigation, S.V.A. and S.H.; resources, S.V.A., S.H. and F.W.; data curation, S.V.A.; writing—original draft preparation, S.V.A.; writing—review and editing, C.T.; visualization, S.V.A., S.H. and F.W.; supervision, C.T.; project administration, C.T.; funding acquisition, C.T. All authors have read and agreed to the published version of the manuscript.

Funding: This research was funded by the Deutsche Forschungsgemeinschaft (DFG, German Research Foundation)—Project Number 414265976—TRR277. This publication was funded by Ostbayerische Technische Hochschule Regensburg, whose support is greatly appreciated.

Institutional Review Board Statement: Not applicable.

Informed Consent Statement: Not applicable.

Data Availability Statement: The original contributions presented in this study are included in the article. Further inquiries can be directed to the corresponding author.

Acknowledgments: This article is a revised and expanded version of a paper entitled Possibilities for Reducing the Environmental Impact in the Construction Industry Using the Example of a 3D Printed Staircase, which was presented at the 4th International Conference on Sustainable Development in Civil, Urban and Transportation Engineering (CUTE 2024), Wroclaw, Poland, 14–17 October 2024.

Conflicts of Interest: The authors declare no conflicts of interest. The funders had no role in the design of the study; in the collection, analyses, or interpretation of data; in the writing of the manuscript; or in the decision to publish the results.

Abbreviations

The following abbreviations are used in this manuscript:

3DPCC	3D printed concrete
AP	Acidification Potential
CED	cumulative energy demand
E-LCA	Environmental Life Cycle Assessment
EPD	Environmental Product Declaration
EP-freshwater	eutrophication potential—aquatic freshwater
EP-marine	eutrophication potential—aquatic marine
EP-terrestrial	eutrophication potential—terrestrial
GGBFS	ground granulated blast furnace slag
GWP	Global Warming Potential
IPCC	Intergovernmental Panel on Climate Change
LCA	Life Cycle Assessment
LCI	Life Cycle Inventory
LCIA	Life Cycle Impact Assessment
ODP	Ozone Depletion Potential
PE	primary energy resources
PENRT	total use of non-renewable primary energy resources
PERT	total use of renewable primary energy resources
POCP	Photochemical Ozone Creation Potential
SCA	Selective Cement Activation
WMO	World Meteorological Organization

References

1. Brucker Juricic, B.; Galic, M.; Marenjak, S. Review of the Construction Labour Demand and Shortages in the EU. *Buildings* **2021**, *11*, 17. [CrossRef]
2. United Nations Environment Programme. *Global Status Report for Buildings and Construction—Beyond Foundations: MainStreaming Sustainable Solutions to Cut Emissions from the Buildings Sector*; United Nations Environment Programme: Nairobi, Kenya, 2024.
3. Statista Research Department. Available online: https://www.statista.com/topics/5466/global-housing-market/#topicOverview (accessed on 25 October 2024).
4. Circle Economy Foundation. *The Circularity Gap Report*; Circle Economy Foundation: Amsterdam, The Netherlands, 2024.
5. United Nations Environment Programme. *Global Status Report for Buildings and Construction: Towards a Zero-Emissions, Efficient and Resilient Buildings and Construction Sector*; United Nations Environment Programme: Nairobi, Kenya, 2022.
6. UNEP/SETAC Life Cycle Initiative. In *Towards a Life Cycle Sustainability Assessment: Making Informed Choices on Products*; UNEP: Brussels, Belgium, 2011; pp. 1–5.
7. Hauschild, M.Z.; Rosenbaum, R.K.; Olsen, S.I. *Life Cycle Assessment: Theory and practice*, 1st ed.; Springer Nature: Cham, Switzerland, 2018. [CrossRef]
8. *DIN EN 15804:2022-03*; Sustainability of Construction Works—Environmental Product Declarations—Core Rules for the Product Category of Construction Products. Beuth: Berlin, Germany, 2022.
9. García de Soto, B.; Agustí-Juan, I.; Hunhevicz, J.; Joss, S.; Graser, K.; Habert, G.; Adey, B.T. Productivity of digital fabrication in construction: Cost and time analysis of a robotically built wall. *Autom. Constr.* **2018**, *92*, 297–311. [CrossRef]
10. Agustí-Juan, I.; Müller, F.; Hack, N.; Wangler, T.; Habert, G. Potential benefits of digital fabrication for complex structures: Environmental assessment of a robotically fabricated concrete wall. *J. Clean. Prod.* **2017**, *154*, 330–340. [CrossRef]
11. Balasbaneh, A.T.; Sher, W.; Ibrahim, M.H.W. Life cycle assessment and economic analysis of Reusable formwork materials considering the circular economy. *Ain Shams Eng. J.* **2024**, *15*, 102585. [CrossRef]
12. Bakırhan, E.K.; Kayılı, M.T. Evaluation of the Environmental Impact of Formwork Systems Depending on the Service Life and Cost Analysis. *CRPASE Trans. Civ. Environ. Eng.* **2023**, *9*, 1–10. [CrossRef]
13. *ISO 14040: 2006*; Environmental Management—Life Cycle Assessment—Principles and Framework. Beuth: Berlin, Germany, 2022.
14. *ISO 14044: 2006*; Environmental Management—Life Cycle Assessment— Requirements and Guidelines. Beuth: Berlin, Germany, 2022.
15. Herding, F.; Mai, I.; Lowke, D. Reduction of the cement content in Selective Cement Activation. *Ce/papers* **2023**, *6*, 790–796. [CrossRef]
16. Dirk Lowke, D.; Talke, D.; Dressler, I.; Weger, D.; Gehlen, C.; Ostertag, C.; Rael, R. Particle bed 3D printing by selective cement activation—Applications, material and process technology. *Cem. Concr. Res.* **2020**, *134*, 106077. [CrossRef]
17. Albrecht, S.V.; Hellerbrand, S.; Weininger, F.; Thiel, C. Possibilities for Reducing the Environmental Impact in the Construction Industry Using the Example of a 3D Printed Staircase. In Proceedings of the 4th International Conference on Sustainable Development in Civil, Urban and Transportation Engineering, CUTE 2024, Wrocław, Poland, 14–17 October 2024; Springer Nature: Singapore, 2024.
18. Fonseca, M.; Matos, A.M. 3D Construction Printing Standing for Sustainability and Circularity: Material-Level Opportunities. *Materials* **2023**, *16*, 2458. [CrossRef] [PubMed]
19. IEA; WBCSD. *Cement Technology Roadmap 2009 Carbon Emissions Reductions Up to 2050*; WBCSD: Paris, France; Conches: Geneva, Switzerland, 2009.
20. Bhagath Singh, G.V.P.; Durga Prasad, V. Environmental impact of concrete containing high volume fly ash and ground granulated blast furnace slag. *J. Clean. Prod.* **2024**, *448*, 141729. [CrossRef]
21. Krause, T.; Ulke, B. *Zahlentafeln für den Baubetrieb*, 9th ed.; Springer: Wiesbaden, Germany, 2016.
22. European Comission. *International Reference Life Cycle Data System (ILCD) Handbook: Recommendations for Life Cycle Impact Assessment in the European Context: Based on Existing Environmental Impact Assessment Models and Factors*, 1st ed.; Publications Office: Luxembourg, 2011; Available online: https://data.europa.eu/doi/10.2788/33030 (accessed on 25 November 2024).
23. Acero, A.P.; Rodríguez, C.; Ciroth, A. *LCIA methods: Impact Assessment Methods in Life Cycle Assessment and Their Impact Categories*; GreenDelta: Berlin, Germany, 2016.
24. Boulay, A.M.; Bare, J.; Benini, L.; Berger, M.; Lathuillière, M.J.; Manzardo, A.; Margni, M.; Motoshita, M.; Núnez, M.; Pastor, A.V.; et al. The WULCA consensus characterization model for water scarcity footprints: Assessing impacts of water consumption based on available water remaining (AWARE). *Int. J. Life Cycle Assess* **2018**, *23*, 368–378. [CrossRef]
25. Hischier, R.; Weidema, B.; Althaus, H.-J.; Bauer, C.; Doka, G.; Dones, R.; Frischknecht, R.; Hellweg, S.; Humbert, S.; Jungbluth, N.; et al. *Implementation of Life Cycle Impact Assessment Methods*; Ecoinvent Report No. 3, v2.2; Swiss Centre for Life Cycle Inventories: Dübendorf, Switzerland, 2010.

26. United Nations Environment Programme. *Building Materials and the Climate: Constructing a New Future*; United Nations Environment Programme: Nairobi, Kenya, 2023.
27. Poudyal, L.; Adhikari, K. Environmental sustainability in cement industry: An integrated approach for green and economical cement production. *Resour. Environ. Sustain.* **2021**, *4*, 100024. [CrossRef]
28. Batuecas, E.; Liendo, F.; Tommasi, T.; Bensaid, S.; Deorsola, F.A.; Fino, D. Recycling CO_2 from flue gas for $CaCO_3$ nanoparticles production as cement filler: A life cycle assessment. *J. CO2 Util.* **2021**, *45*, 101446. [CrossRef]
29. Stemmermann, P.; Schweike, U.; Garbev, K.; Beuchle, G. Celitement—A sustainable prospect for the cement industry. *Cem. Int.* **2010**, *8*, 52–67.
30. Kuzmenko, K. Environmental Performance in Construction: A Case-Study of 3D Concrete Printing Technology. Ph.D. Thesis, École des Ponts ParisTech, Champs-sur-Marne, France, 2021.
31. Hong, J.; Shen, G.Q.; Feng, Y.; Lau, W.S.; Mao, C. Greenhouse gas emissions during the construction phase of a building: A case study in China. *J. Clean. Prod.* **2015**, *103*, 249–259. [CrossRef]
32. Beton Umweltproduktdeklarationen. Available online: https://www.beton.org/betonbau/planungshilfen/umweltproduktdeklarationen/ (accessed on 25 November 2024).
33. AMC TRR 277. Available online: https://amc-trr277.de/projects/project-area-c/focus-area-c-09/ (accessed on 29 November 2024).

materials

MDPI

Article

Performance Evaluation of Shotcrete Mortar with Silicon Manganese Slag as Substitute for Fine Aggregate

Woo-Ri Kwon [1], Jung-Bin Lee [1], Bok-Mo Yoon [2] and Jang-Ho Jay Kim [1,*]

1 Department of Civil and Environmental Engineering, Yonsei University, Seoul 03722, Republic of Korea; uskwon@yonsei.ac.kr (W.-R.K.); jbiiee@naver.com (J.-B.L.)
2 Miju Ecorock Co., Ltd., 36 Seohanam-ro 43beon-gil, Hanam 12991, Gyeonggi-do, Republic of Korea; miju@ecorock.co.kr
* Correspondence: jjhkim@yonsei.ac.kr; Tel.: +82-(2)-2123-5802

Abstract: Shotcrete is a versatile construction material, yet its performance limitations, such as high rebound rates and poor adhesion, demand technological improvements to ensure structural reliability. Silicon manganese (SiMn) slag, a by-product of SiMn alloy production, has gained attention as a potential sustainable alternative to natural aggregates in construction materials, addressing both resource depletion and carbon reduction challenges in the industry. This study is conducted to develop and evaluate a new mix design of mortar incorporating SiMn slag as fine aggregate, focusing on enhancing performance. Mixtures with varying percentages (0%, 30%, 50%, 70%, and 100%) of SiMn slag as a fine aggregate replacement were evaluated for fresh properties (air content, slump), mechanical performance (compressive strength, flexural strength, splitting tensile strength), durability (chloride ion penetration resistance, freeze–thaw resistance, carbonation resistance), and constructability (rebound rate, free shrinkage) to assess suitability as mortar for shotcrete. The experimental results demonstrated that the mixture with 50% SiMn slag replacement demonstrated the most balanced performance, showing an increase of 12.33% in compressive strength, 8.97% in splitting tensile strength, and 18.4% in flexural strength compared to the control. Durability properties also improved by an average of 11.93%, while rebound rate and shrinkage were significantly reduced. The findings confirm that SiMn slag is a technically viable and advantageous substitute for fine aggregates in shotcrete. Further research is needed to refine its economic feasibility and broaden its implementation in sustainable construction.

Keywords: aggregate replacement; silicon manganese slag; shotcrete; durability; workability; constructability; sustainability

1. Introduction

The worsening climate crisis and global resource depletion have placed growing pressure on the construction industry to adopt sustainable and durable materials. As Pierre Humbert of the University of Oxford stated, "Let's get this on the table right away, without mincing words. With regard to the climate crisis, yes, it's time to panic" [1]. This urgent message highlights the responsibility of material engineers and researchers to contribute to low-carbon, resource-efficient construction practices. In this context, improving both the durability and sustainability of construction materials such as shotcrete is no longer optional but essential.

Shotcrete is broadly adopted in many different construction works, such as underground mining, tunnel lining, slope stabilization, and concrete repair strengthening due to its implementation flexibility, rapid application, and excellent bonding performance [2]. In the tunnel

https://doi.org/10.3390/ma18081754

industry, shotcrete is primarily employed to form thin and flexible self-supporting layers inside tunnels [3] or as a tunnel lining serving as permanent support without the need for additional concrete members [4,5]. This application significantly enhances the safety of underground construction [6,7] while ensuring that projects are executed without unnecessary delays [8]. In addition, shotcrete offers significant advantages in performance and cost-effectiveness, particularly when applied to relatively light-load-bearing structures, such as landscaping tunnels, artificial rock formations, underground parking lots, and small-scale underground storage facilities. Shotcrete can achieve rapid construction as it does not require formwork, demonstrate adaptability to various structural design criteria, and result in reduced overall costs due to its minimal equipment needs. Despite its advantages, shotcrete encounters significant obstacles, such as poor adhesion [9], high rebound rate [10], and interface generation [11]. In 2022, a worker died at a railway construction site in Goyang, South Korea, due to shotcrete detachment. Also, a construction accident occurred in 2023, where a shotcrete wall collapsed at a condominium construction site in Coquitlam, Canada. Fortunately, no casualties were reported, but such incidents show the necessity of improving the performance of shotcrete to prevent recurring accidents.

The modern construction industry faces challenges, such as natural resource depletion and strict environmental regulations. For example, net-zero carbon emission goals emphasize the importance of sustainable construction [12]. In particular, the depletion of natural aggregates has driven the development of alternative aggregates through recycling waste resources, including disused aggregates, agricultural residues, and industrial wastes [13,14]. Recycling industrial waste, which places a significant burden on waste management systems due to the globally increasing volume of waste, not only reduces environmental pollution but also offers a sustainable solution by converting waste into valuable resources, thereby contributing to the transition toward a circular economy [15]. SiMn slag is a byproduct of silicon manganese production, which is generated in high-carbon ferro-manganese blast furnaces using manganese ore, silica stone, and coke [16]. SiMn slag is a waste resource frequently investigated for its potential reuse as aggregates in construction materials, along with blast furnace slag, steel slag, and fly ash [17]. The production process of SiMn slag is shown in Figure 1. The process involves producing silicon manganese in an electric furnace [18]. The remaining slag is cooled using high-pressure water and crushed for potential use [19]. As the global market for SiMn alloy production expands, the accumulation of SiMn slag has also increased, which poses environmental challenges due to its large volume [20–22]. Recycling SiMn slag provides solutions to the environmental and economic challenges caused by the production of silicon manganese, namely the need for large-scale treatment plants for the disposal of slag and generated fugitive dust and leachate [23]. To address this problem, many efforts have been made to recycle SiMn slag as a means of promoting resource circulation.

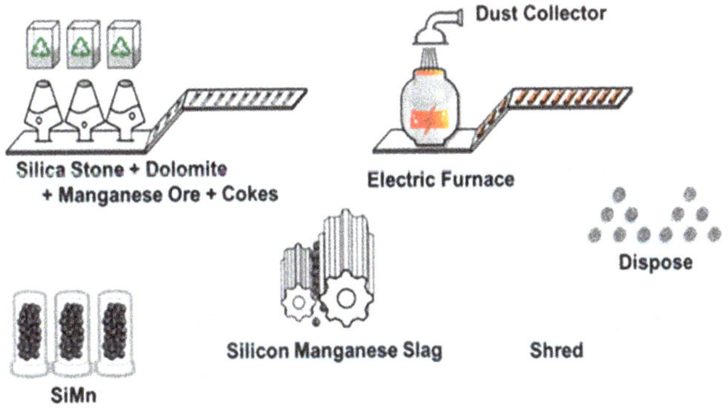

Figure 1. Production of silicon manganese (SiMn) slag.

Several studies have positively evaluated the potential of using SiMn slag as a sustainable next-generation aggregate substitute. Wong et al. highlighted the potential for recycling SiMn slag as construction material [24]. Patil et al. further expanded its application, noting that SiMn slag is suitable for road pavement and railway tracks due to its high resistance to wear, abrasion, and degradation [25]. Similarly, Choi et al. demonstrated that mortar produced using rapid-cooled SiMn slag aggregate exhibits proper engineering properties, confirming its viability as a lightweight aggregate for construction [19]. Furthermore, Nath et al. suggested that SiMn slag can be used as both a binder and an aggregate in construction to enhance concrete properties [26].

The potential of SiMn slag as a sustainable substitute for fine aggregates in mortar for shotcrete was investigated to evaluate its performance. A new mix proportion incorporating SiMn slag as fine aggregates was developed through experimental research. Mix proportions were designed with varying SiMn slag replacement levels (0%, 30%, 50%, 70%, and 100%), and each mixture was evaluated to determine the most effective and practical formulation for shotcrete application. Fresh properties of the mortar were assessed through air content and slump tests. Mechanical properties of the mixtures were assessed through compressive strength, flexural strength, and splitting tensile strength tests, while durability was evaluated using chloride ion penetration resistance, freeze–thaw resistance, and carbonation resistance tests. Additionally, workability was examined through rebound rate and drying shrinkage tests to further understand the practical application. These evaluations identified the optimal mix proportion, ensuring superior performance and efficiency for practical shotcrete applications. The findings are expected to contribute to the development of high-performance, eco-friendly construction materials and provide a novel solution for sustainable construction and recycling waste.

2. Materials and Methods

2.1. Materials

A hazardous substance test conducted in accordance with waste process standards (No. 2015-50 from the Korean Ministry of Environment) indicates that SiMn slag is a harmless and eco-friendly material. As shown in Table 1, the only items detected in significant amounts were Pb, Cu, and Cd, and all 11 test items were below threshold values. The result of the test supports the suitability of SiMn slag as a harmless and eco-friendly substitute for fine aggregate.

Table 1. Waste process test.

Item	Unit	Detection Limit	Detection Criterion	Test Result
Pb	mg/L	0.040	3	0.086
Cu	mg/L	0.008	3	0.090
Cd	mg/L	0.002	0.3	0.008
Hg	mg/L	0.0005	0.005	-
Cr^{6+}	mg/L	0.01	1.5	-
CN^{-}	mg/L	0.01	1	-
As	mg/L	0.004	1.5	-
OPC	mg/L	0.0005	1	-
C_2HCl_3	mg/L	0.008	0.3	-
C_2Cl_4	mg/L	0.002	0.1	-
Cr	mg/L	0.01	-	-

The physical property test was carried out according to KS F 2504 (Standard Test Method for Density and Water Absorption Rate of Fine Aggregates) [27]. Table 2 shows that SiMn slag has a saturated surface-dry density of 3.00 g/cm^3, an absolute dry density of 2.97 g/cm^3, and a water absorption rate of 0.79%. The particle size of the SiMn slag used in this study ranged from 0.2 to 0.6 mm, which is included in the range defined for natural sand in KS F 2527 (Aggregates for Concrete) [28]. Compared to general fine aggregate, which has a specific gravity of 2.65 and a water absorption of 1%, SiMn slag exhibits higher specific gravity and a lower water absorption rate, indicating greater strength and enhanced resistance to freezing and thawing. This aspect suggests that SiMn slag has potential as a fine aggregate replacement for concrete and cement mortar.

Table 2. Physical properties of SiMn slag.

Property		Unit	Test Result
Density	Saturated Surface-Dry (SSD) Density	g/cm^3	3.00
	Absolute Dry Density	g/cm^3	2.97
Water Absorption		%	0.79

To examine the chemical composition of SiMn slag, X-ray fluorescence spectrometry (XRF) was performed. XRF uses X-rays to emit electrons and identifies the atoms by filling the vacated electron spots with other electrons from X-rays to measure the energy of each atom's unique value. In Table 3, a comparison of the chemical composition of SiMn slag and fine aggregate is shown. The results from the XRF analysis showed that SiO_2, CaO, MnO, and Al_2O_3 are dominant elements in SiMn slag, with low levels of MgO, Fe_2O_3, and NaCl. However, fine aggregate is dominated by 77.2% of SiO_2 and 10.2% of Al_2O_3, with low levels of CaO, MnO, MgO, Fe_2O_3, and NaCl. Therefore, SiMn slag can be defined as a latent hydraulic material, preventing alkaline aggregate reaction and reducing concrete expansion to minimize cracking. Overall, the findings suggest that SiMn slag has significant potential as an environmentally friendly and effective alternative to traditional fine aggregate in concrete and cement mortar.

Table 3. Chemical composition of SiMn slag and fine aggregate.

Material	SiO_2	Al_2O_3	Fe_2O_3	CaO	MgO	MnO	NaCl
Silicon Manganese Slag	37.1	12.1	0.73	23.5	4.69	17.1	0.007
Fine Aggregate	77.2	10.2	2.87	1.12	0.65	0.09	0.008

2.2. Preliminary Test

Preliminary tests were conducted to determine the water–cement ratio for the mix and the required amount of plasticizer. In the initial assessment of the water–cement ratio for the mix, the results of the compressive strength test for various ratios were examined, as shown in Table 4. A mix proportion was formulated by adjusting the water–cement ratio within the range of 30% to 50% and focusing on the complete replacement of aggregate with SiMn slag. Compressive strength tests were conducted according to KS F 2405 (Test method for compressive strength of concrete) [29] on specimens cured for 3, 7, 14, and 28 days. The test results show that the highest compressive strength of 36.8 MPa at 28 days of age was obtained at the water–cement ratio of 35%. Also, a water–cement ratio of 40% showed a similar compressive strength of 35.8 MPa at 28 days of age.

Table 4. Compressive strength results based on water–cement ratio conditions.

W/C (%)	Cement (kg/m³)	Water (kg/m³)	SiMn Slag (kg/m³)	Fine Aggregate (kg/m³)	Compressive Strength (MPa)		
					3 Days	7 Days	28 Days
30	626	188	1461	-	23.35	26.24	32.1
35	626	219	1461	-	27.8	30.7	36.8
40	626	251	1461	-	27.14	30.1	35.8
45	626	282	1461	-	25.2	28.2	33.1
50	626	313	1461	-	22.48	25.23	30.4

The preliminary test to determine the required plasticizer amount for sufficient workability was carried out. The target slump value ranged between 12 mm and 15 mm. The tests were conducted with a plasticizer amount ranging from 0.3% to 0.9% of the cement weight, with 100% replacement of fine aggregate with SiMn slag. The slump test results are shown in Table 5. The results showed that the slump value of 13 mm, which corresponds to 0.5% plasticizer, meets the target slump value range. Thus, 0.5% plasticizer of the cement weight was selected as a final amount for the mix proportion.

Table 5. Slump test results based on plasticizer conditions.

Plasticizer (C*%)	W/C (%)	Cement (kg/m³)	Water (kg/m³)	SiMn Slag (kg/m³)	Fine Aggregate (kg/m³)	Slump (mm)
0	40	626	251	1461	-	5
0.9	40	626	251	1461	-	FLOW
0.8	40	626	251	1461	-	27
0.7	40	626	251	1461	-	21
0.6	40	626	251	1461	-	16
0.5	40	626	251	1461	-	13
0.4	40	626	251	1461	-	11
0.3	40	626	251	1461	-	9

2.3. Mix Proportion

To assess the performance of shotcrete with SiMn slag as the fine aggregate, preliminary compressive strength tests were conducted to determine a mix design. The mix designs listed in Table 6 are proposed to obtain a target compressive strength exceeding 21 MPa, with a slump range of 12–15 mm. Five cases with varying percentages of SiMn slag (0%, 30%, 50%, 70%, and 100%) as fine aggregate were considered. The water–cement ratio was set to 40%, and a plasticizer amount of 0.50% of the cement weight was determined from the previous preliminary tests. Ordinary Portland Cement (KS L 5201 Type 1) [30] and a polycarboxylate-based superplasticizer with air-entraining properties were used in this study. A reference specimen with 0% SiMn slag served as the control specimen (S-0), with the other specimens labeled as S-30, S-50, S-70, and S-100 based on the respective slag replacement percentages.

Table 6. Mortar mix proportion based on SiMn slag replacement rates.

Specimen	W/C (%)	Cement (kg/m³)	Water (kg/m³)	SiMn Slag (kg/m³)	Fine Aggregate (kg/m³)	Plasticizer (C*%)
S-0 (Control)	40	626	251	-	1462	0.50
S-30	40	626	251	439	1023	0.50
S-50	40	626	251	731	731	0.50
S-70	40	626	251	1023	439	0.50
S-100	40	626	251	1462	-	0.50

2.4. Test Methods

Air content and slump of the fresh mortar were evaluated for 0%, 30%, 50%, 70%, and 100% replacement of SiMn slag as the fine aggregate.

2.4.1. Air Content Test

To determine the air content of the mortar, the air content measurement test was conducted following the guidelines of KS F 2421 (Standard Test Method for Air Content of Fresh Concrete by the Pressure Method) [31]. Figure 2 illustrates the air content test setup, including an air content meter and a cylindrical metal container with a minimum volume of 5 L. The air content of the mortar was calculated using Equation (1):

$$A = A_1 - G \quad (1)$$

where A (%) is the air content of fresh concrete; A_1 (%) is the apparent content; G is the aggregate correction factor.

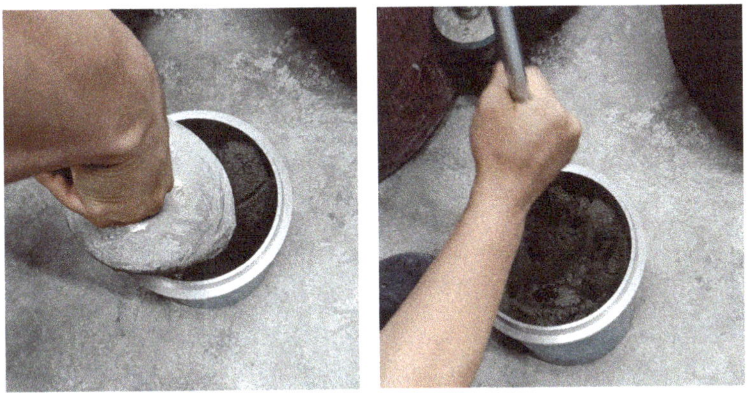

(**a**) (**b**)

Figure 2. Air content test: (**a**) pouring mortar mix; (**b**) pounding mortar mix.

2.4.2. Slump Test

The slump test for the mortar was conducted following the guidelines outlined in KS F 2402 (Test Method for Concrete Slump) [32]. The slump of the concrete was measured by the difference in height between the original height of the cone and the highest point of the slumped concrete, as shown in Figure 3.

Figure 3. Slump test based on different plasticizer content.

To evaluate the mechanical properties of the hardened mortar, compressive, flexural, and splitting tensile strength tests were conducted.

2.4.3. Compressive Strength Test

The compressive strength test for the mortar was conducted following the guidelines of KS F 2405. Cylindrical specimens were prepared according to the specifications outlined in KS F 2403 (Standard Test Method for Making Concrete Specimens) [33]. Mixes of the specimens contained various fine aggregate replacement ratios of 0%, 30%, 50%, 70%, and 100%. Specimens were wet-cured at a temperature of (20 ± 2) °C until the strength test. Before the compressive strength test, the top and bottom surfaces of the specimens were ground to be perpendicular to the compressive loading. A compressive strength test was performed using a Universal Testing Machine (UTM) with a maximum capacity of 1000 kN and precision of 500 N, under a constant loading rate of (0.6 ± 0.2) MPa/s. The compressive strength tests were conducted at 3, 7, 14, and 28 days of curing age. The compressive strength was calculated using Equation (2):

$$f_c = P/\pi(d/2)^2 \tag{2}$$

where f_c (MPa) is the compressive strength of concrete; P (N) is the load; d (mm) is the diameter of the specimen.

2.4.4. Flexural Strength Test

To assess the flexural strength of the mortar with SiMn slag as fine aggregate, the test was carried out following the guidelines of KS F 2408 (Standard Test Method for Flexural Strength of Concrete) [34]. Rectangular specimens with dimensions of 100 × 100 × 400 mm were prepared according to KS F 2403. The specimens were made with various aggregate replacement ratios of 0%, 30%, 50%, 70%, and 100%. The 3-point loading was used for the flexural strength test. The loading rate was maintained at (0.06 ± 0.04) MPa/s. Specimens were wet-cured at a temperature of (20 ± 2) °C until testing. Testing was conducted at 14 and 28 days of age, and flexural strength values were determined using Equation (3):

$$f_b = PL/bh^2 \tag{3}$$

where f_b (MPa) is the flexural strength of concrete; P (N) is the maximum load from the testing machine; L (mm) is the span; b (mm) is the width of the specimen; h (mm) is the height of the specimen.

2.4.5. Splitting Tensile Strength Test

The splitting tensile strength test was conducted following the guidelines of KS F 2423 (Standard Test Method for Splitting Tensile Strength of Concrete) [35]. Cylindrical specimens with a 100 mm diameter and 200 mm length were prepared in accordance with KS F 2403. Specimens were prepared with various aggregate replacement ratios of 0%, 30%, 50%, 70%, and 100%. Specimens were wet-cured at a temperature of (20 ± 2) °C until testing. A load with a rate of (0.06 ± 0.04) MPa/s was applied until the maximum load was reached. Splitting tensile strength tests were conducted at 14 and 28 days of age. The splitting tensile strength was calculated according to Equation (4):

$$f_{sp} = 2P/\pi dl \tag{4}$$

where f_{sp} (MPa) is the splitting tensile strength of concrete; P (N) is the maximum load, as indicated by the testing machine; l (mm) is the span; d (mm) is the diameter of the specimen.

For durability, chloride ion penetration resistance, freezing, and thawing as well as accelerated carbonation tests were conducted.

2.4.6. Chloride Ion Penetration Resistance Test

To assess the chloride ion penetration resistance of the mortar, the test was conducted following the guidelines of KS F 2711 (Standard Test Method for Resistance of Concrete to Chloride Ion Penetration by Electrical Conductance) [36]. Three cylindrical specimens were prepared for each SiMn slag replacement ratio, which were cured for 28 days. Table 7 outlines the evaluation standards for the total charge passage specified by KS F 2711. The total charge passage value was calculated using Equation (5):

$$Q = 900(I_0 + 2I_{30} + 2I_{60} + \cdots + 2I_{330} + I_{360}) \quad (5)$$

where Q (C) is the charge passed in coulombs; I_0 (A) is the current immediately after the voltage is applied; I_t (A) is the current at t minutes after the voltage is applied.

Table 7. Evaluation standard of total charge passage.

Total Passed Charge (Coulombs)	Chloride Ion Permeability
>4000	High
2000–4000	Moderate
1000–2000	Low
100–1000	Very low
<100	Negligible

2.4.7. Freezing and Thawing Test

To evaluate the freezing and thawing resistance of the mortar with aggregate replacement, the test was conducted following the guidelines of KS F 2456 (Test Method for Resistance of Concrete to Rapid Freezing and Thawing) [37]. Freezing and thawing resistance was quantified by calculating the relative dynamic modulus using Equation (6):

$$P_c = \left(n_c^2 / n_0^2\right) \times 100 \quad (6)$$

where P_c (%) is the relative dynamic modulus of elasticity after freeze–thaw cycle C; n_c (Hz) is the primary resonance frequency of the strain vibration after freeze–thaw cycle C; n_0 (Hz) is the primary resonance of strain vibration in freeze–thaw cycle 0.

2.4.8. Accelerated Carbonation Resistance Test

To assess the carbonation of the mortar, the test was conducted following KS F 2584 (Standard Test Method for Accelerated Carbonation of concrete) [38], as shown in Figure 4. Carbonization depth was measured after 4 weeks of exposure to carbon dioxide-filled test equipment by spraying phenolphthalein solution as a reagent. While typical times for carbonization depth measurement are 1, 4, 8, 13, and 26 weeks, the measurements were taken at a 4-week incubation period to correspond to the curing duration.

To evaluate the constructability of the mix proportion, rebound and drying shrinkage crack tests were conducted.

2.4.9. Rebound Test

The rebound test was conducted to evaluate workability of the mortar. The mortar was shotcreted over a steel system consisting of steel bars and steel mesh, arranged in a grid pattern of 250 × 250 mm. The mortar was applied to the system over an area covering four grids with a thickness of 30 mm. To minimize the variability caused by operator skill, all spraying work was performed by a single skilled worker. The rebound amount was measured by collecting and weighing the mixture that fell to the ground after shotcreting was completed. The mixtures with water–cement ratios of 35% and 40% were tested, and rebound percentages were calculated using Equation (7):

$$Rebound\ rate\ (\%) = \frac{Weight\ of\ rebound\ mixture}{Total\ weight\ of\ mixture} \times 100 \quad (7)$$

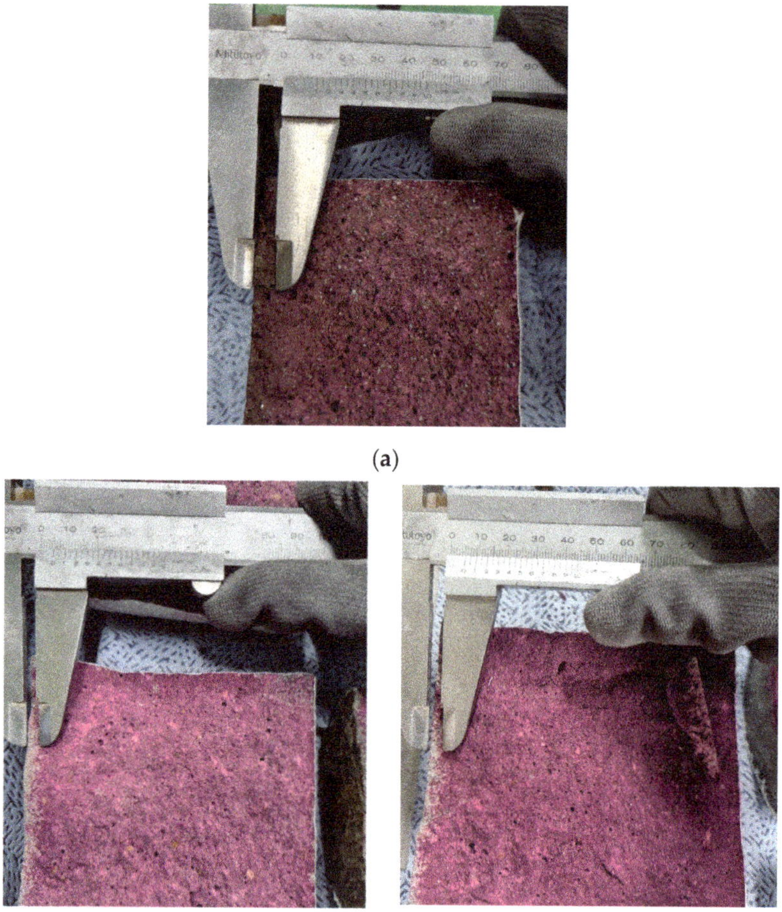

(a)

(b) (c)

Figure 4. Carbonation depth measurement: (**a**) S-0 (control); (**b**) S-50; (**c**) S-100.

2.4.10. Drying Shrinkage Crack Test

To evaluate the initial drying shrinkage performance of the mortar and to assess the initial shrinkage crack occurrence, the free shrinkage strain measurement test was conducted in accordance with KS F 2595 (Standard Test Method for Drying Shrinkage Crack in Concrete) [39]. As shown in Figure 5, the specimens were prepared with dimensions of 40 × 40 × 160 mm, which were attached with strain gauges to monitor strain over 30 days in a controlled environment of 23 °C temperature and 50% relative humidity.

Figure 5. Free drying shrinkage crack test.

3. Results

3.1. Fresh Properties

The air content and slump test results were tabulated and are given in Table 8. The air content and slump test results for all the specimens ranged between 5.7–6.6% and 130–220 mm, respectively, as shown in Figure 6. The control specimens without slag replacement exhibited the lowest air content average of 5.7%. As the SiMn slag replacement ratio increased incrementally from 30% to 100%, the air content correspondingly increased. More specifically, the S-30, S-50, S-70, and S-100 slag specimens had air contents of 6.1%, 6.0%, 6.3%, and 6.5% respectively. The full slag replacement S-100 specimens had the highest air content. All of the specimens displayed slump values exceeding 120 mm. The control specimens without slag substitution had the highest average slump of 210 mm. The S-30 specimens with a 30% fine aggregate replacement ratio exhibited a slightly lower but still high slump of 205 mm. Slump values progressively declined as higher SiMn slag contents were substituted. More specifically, the S-50, S-70, and S-100 specimens demonstrated slumps of 185 mm, 160 mm, and 135 mm, respectively. Therefore, an inverse correlation was observed between the slag replacement ratio and slump measurement, with the S-100 mixture having the lowest workability. Overall, greater SiMn slag substitution correlated with higher air content values and lower flowability and slump.

Table 8. Average of air content test and slump test results.

Specimen	SiMn Slag (%)	Fine Aggregate (%)	Air Content (%)	Slump (mm)
Control	0	100	5.7	210
S-30	30	70	6.1	205
S-50	50	50	6	185
S-70	70	30	6.5	160
S-100	100	0	6.5	135

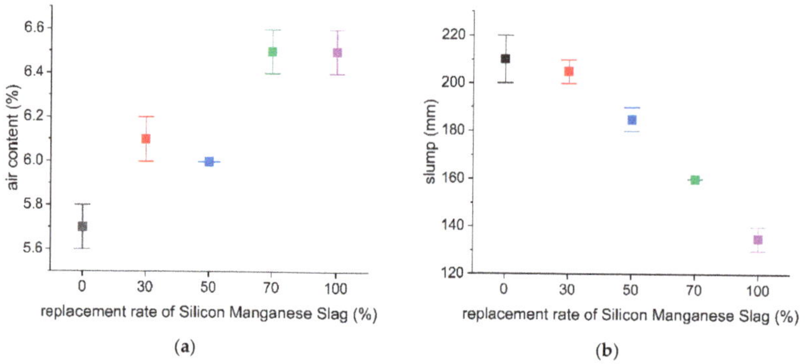

Figure 6. Air content and slump based on replacement rate of SiMn slag: (**a**) air content test; (**b**) slump test.

3.2. Mechanical Properties

The compressive and splitting tensile strength test results are summarized in Table 9. Also, the flexural strength test results are summarized in Table 10. In all of the strength test results, the strengths increased as SiMn slag replacement ratios increased. As shown in Table 9, the control specimens without SiMn slag exceeded the minimum specified 21 MPa compressive strength at 28 days, averaging 24.9 MPa. Figure 7 illustrates superior strength development in the specimens with higher SiMn slag replacement ratios for all curing ages (3–28 days). In Figure 7a, the S-30, S-50, S-70, and S-100 specimens displayed 8%, 12%, 19%, and 43% higher 28-day compressive strengths than the strength of the control specimen, respectively. With respect to the splitting tensile strength results shown in Figure 7b, the same trend as that of the compressive strength results is demonstrated, with a 41.3% increase in the S-100 specimen compared to the control specimen. Likewise, the flexural strength results showed a similar trend with increased slag content, with S-100 specimens having the highest strength. At 28 days, the mortar with 30%, 50%, 70%, and 100% replacement ratios exhibited a 14%, 18%, 22%, and 37% flexural strength increase compared to the control specimen strength, respectively, as shown in Table 10 and Figure 7c. Thus, all of the investigated mechanical strengths were enhanced with increasing SiMn slag contents, verifying its feasibility and benefits as a shotcrete constituent for improved material performance.

Table 9. Average of compressive strength test and splitting tensile strength test results.

Specimen	SiMn Slag (%)	Fine Aggregate (%)	Compressive Strength (MPa)				Splitting Tensile Strength (MPa)	
			3 Days	7 Days	14 Days	28 Days	14 Days	28 Days
Control	0	100	17.85	20.53	22.15	24.9	2.7	2.9
S-30	30	70	19.58	21.78	24.67	26.87	3.06	3.23
S-50	50	50	21.03	23.51	25.85	27.97	3	3.16
S-70	70	30	22.43	26.38	27.54	29.74	3.13	3.26
S-100	100	0	26.9	29.95	33.07	35.54	3.63	4.1

Table 10. Average of flexural strength test results.

Specimen	SiMn Slag (%)	Fine Aggregate (%)	Flexural Strength (MPa)	
			14 Days	28 Days
Control	0	100	5.55	6.25
S-30	30	70	6.3	7.15
S-50	50	50	6.75	7.4
S-70	70	30	7.4	7.65
S-100	100	0	8.2	8.55

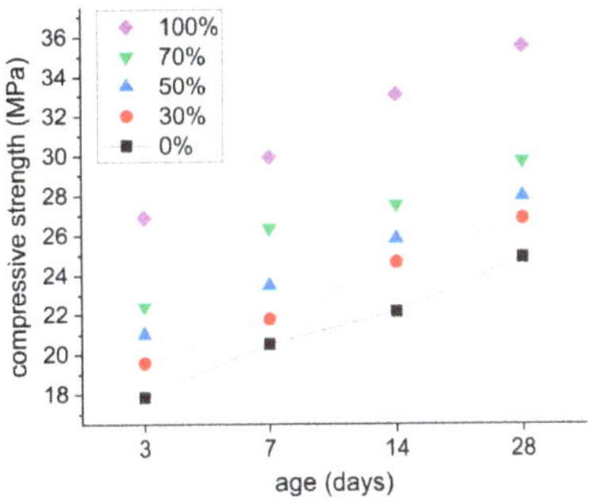

(**a**)

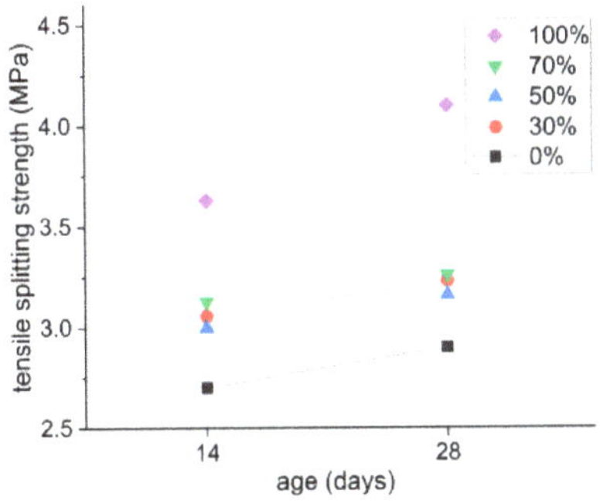

(**b**)

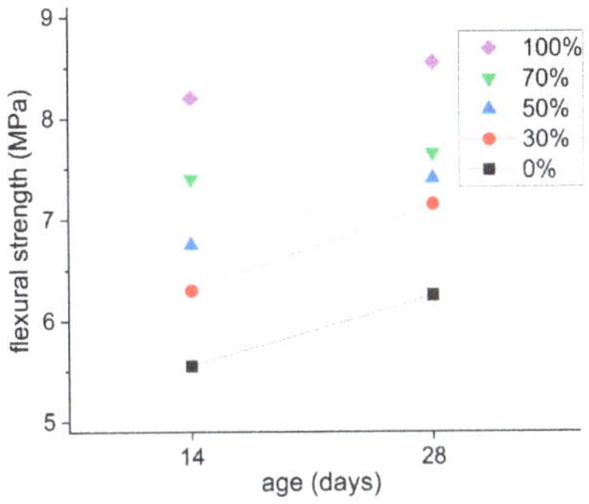

(**c**)

Figure 7. Compressive, splitting tensile, and flexural strength based on replacement rate of SiMn slag: (**a**) compressive strength; (**b**) splitting tensile strength; (**c**) flexural strength.

3.3. Durability Properties

Durability performance of the mortar was evaluated under simulated environmental conditions. The durability tests of chloride ion penetration, freeze–thaw resistance, and accelerated carbonation were conducted. The results are tabulated in Table 11 and shown in Figure 8. In the accelerated chloride ion penetration test, the control specimen (0% SiMn slag) was applied with an average charge of 4546 coulombs for 28 days. As demonstrated in Figure 8a, the S-30, S-50, S-70, and S-100 specimens showed a 10%, 4%, 19%, and 27% decrease in chloride ion penetration, respectively, compared to the control specimen. The resulting trends showed that the chloride ion penetration resistance improved as the slag replacement ratio increased.

Table 11. Average of chloride ion penetration, freezing and thawing, and accelerated carbonation test results.

Specimen	SiMn Slag (%)	Fine Aggregate (%)	Total Charge Passage (C)	Relative Dynamic Modulus of Elasticity (%)		Carbonation Depth (mm)
				0 Cycle	300 Cycles	
Control	0	100	4546	100	88.67	6.06
S-30	30	70	4086	100	90.36	4.45
S-50	50	50	4338	100	90	4.26
S-70	70	30	3667	100	90.77	3.93
S-100	100	0	3326	100	91.07	3.86

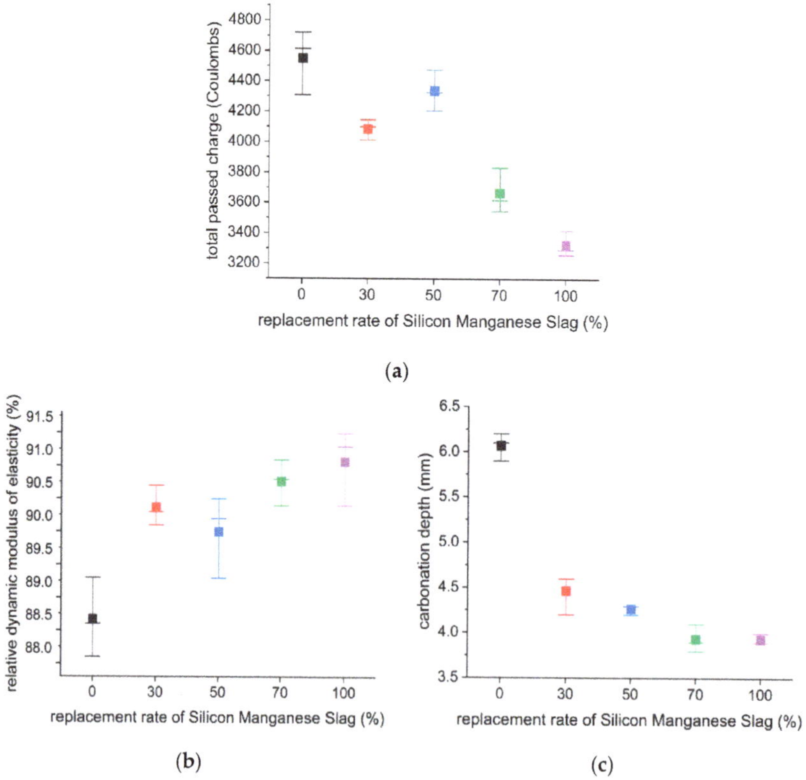

Figure 8. Chloride ion penetration resistance, freezing and thawing, and accelerated carbonation test based on replacement rate of SiMn slag: (**a**) chloride ion penetration resistance test; (**b**) freezing and thawing test; (**c**) accelerated carbonation test.

For freezing–thawing resistance evaluation of the mortar, the relative dynamic modulus of elasticity (RDME) was measured after 300 freeze–thaw cycles. In Table 11 and Figure 8b, the RDME results are shown as a percentage of the initial RDME measured from the specimens without a freeze–thaw cycle (0 cycle). The results showed that all of the measured RDMEs from the specimens after 300 cycles exceeded 90%, except the control specimen (0% SiMn slag). The increasing RDME results coincide with the SiMn slag content increase, with the largest average RDME of 91.07% obtained from the S-100 specimen. The results indicate that SiMn slag mortar has better freeze–thaw resistance than ordinary silica sand mortar when used in the shotcrete mortar mix.

For carbonation resistance evaluation, the specimens are exposed to a carbon dioxide environment using the accelerated carbonation test machine to measure carbonation depth. The carbonation depth results are tabulated in Table 11 and shown in Figure 8c. The results showed that the carbonation depth drastically decreased as the SiMn replacement ratio increased. The mortar with the replacement ratios of 0, 30, 50, 70, and 100% had average carbonation depths of 6.06, 4.45, 4.26, 3.93, and 3.86 mm, respectively. Even with 30% SiMn slag replacement of silica sand, an approximate 36.18% reduction of carbonation depth was achieved, showing a dramatic enhancement of carbonation resistance when using SiMn slag over silica sand. Thus, replacing fine aggregate with SiMn slag enhanced the resistance of the mortar to chloride intrusion, freezing damage, and carbonation in harsh exposures.

3.4. Constructability

To evaluate the constructability of the mix, a shotcrete rebound test was conducted. The results indicate that only the mixtures with SiMn slag replacement ratios of 30% (S-30) and 50% (S-50) were feasible, with S-50 being identified as the most appropriate mix. The mixtures with SiMn slag replacement ratios of 70% (S-70) and 100% (S-100) showed the highest strength, but shotcrete construction application was not possible due to very low workability and the inability to be sprayed as shotcrete. It was observed that at the higher SiMn slag content, the mortar failed to pass through the hose during spraying, leading to backlogging. This can be attributed to the relatively heavier weight of the SiMn slag particles (e.g., a saturated surface-dry density of SiMn slag of 3 g/cm^3), which is heavier than the commonly used silica sand (e.g., a saturated surface-dry density of 2.60 g/cm^3). The rebound amounts for S-30 and S-50 mixes appeared visually similar, as shown in Figure 9. However, quantitative analysis revealed notable differences between the two mixtures of S-30 and S-50, demonstrating rebound rates of 51.2% and 18.5%, respectively. The higher rebound rate in S-30 was attributed to its relatively higher water content compared to the SiMn slag content, which resulted in inadequate adhesion of the mortar to the structure and a greater amount of material detachment during spraying. Consequently, the S-50 mix was determined to be the most suitable, as it exhibited lower material loss and ensured both physical and durability properties along with adequate workability.

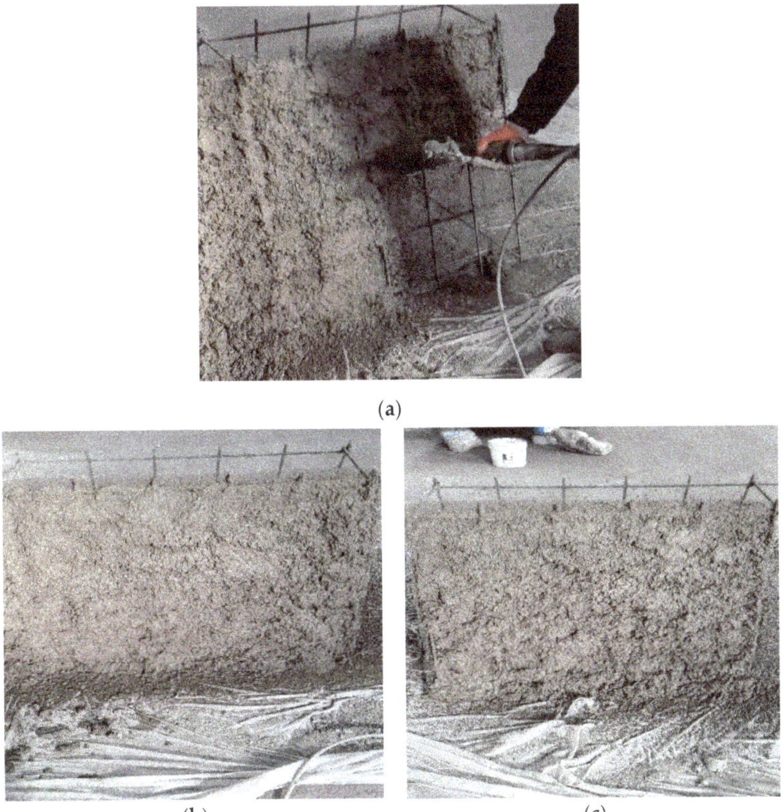

(a) (b) (c)

Figure 9. Shotcrete sprayed with different replacement ratios: (**a**) shotcrete spraying; (**b**) S-30; (**c**) S-50.

For serviceability and stability evaluation, a 30-day free drying shrinkage test was conducted. The shrinkage measurement results for S-0 and S-50 mixes are shown in Figure 10. The results demonstrated the significant effect of SiMn slag replacement on the shrinkage behavior. The S-0 mixture exhibited a free shrinkage strain of approximately 600 micro-strain, exceeding the criterion of 500 micro-strain, which is based on standard guidelines to prevent cracking and ensure durability. In contrast, the S-50 mixture stabilized at approximately 500 micro-strain, satisfying the criterion and demonstrating the effectiveness of replacing 50% of the fine aggregate with SiMn slag in reducing shrinkage strain. Both mixtures exhibited rapid shrinkage during the first 7 days, followed by gradual stabilization, indicating that most shrinkage occurs in the early curing phase, as expected. The results confirmed that the replacement of fine aggregate with SiMn slag enhances curing stability, likely due to the heavier density of the slag particles, which contributes to better internal restraint against shrinkage.

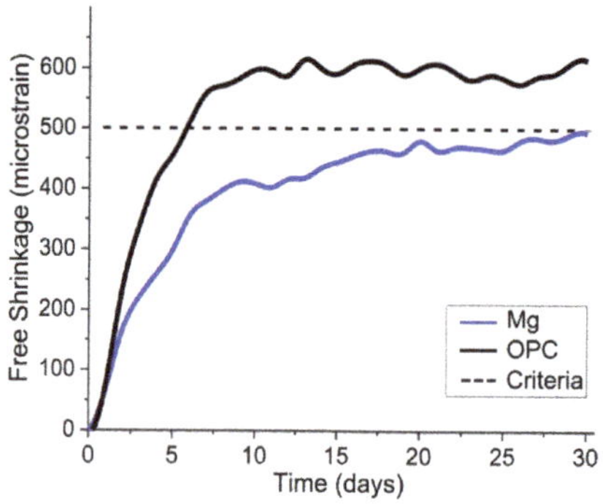

Figure 10. Free shrinkage strains of S-0 and S-50.

Based on the combined results of the rebound test and free drying shrinkage test, replacing 50% of the fine aggregate with SiMn slag is determined to be a suitable mixture for practical shotcrete construction applications where durability and constructability are critical. This replacement ratio not only ensures both mechanical and durability performance but also maintains sufficient constructability, making it an optimal mix.

4. Discussion

This study aimed to apply SiMn slag as a fine aggregate replacement in shotcrete mortar by adjusting replacement ratios and derived the optimal mix proportion suitable for the purpose. In addition, varying mix proportions using SiMn slag in shotcrete mortar were evaluated by assessing its influence on fresh properties, mechanical performance, durability, and constructability. To evaluate the practical applicability of SiMn slag, the study examined not only inherent material performances but also durability properties and constructability. The experimental findings confirmed that increasing SiMn slag content enhances mechanical strength and durability properties, making it a promising alternative material for sustainable construction applications. Among the evaluated durability indicators, enhanced resistance to freeze–thaw cycles and chloride ion penetration are especially important for shotcrete, which is primarily applied in structures exposed to external environments, such as landscape elements, tunnel linings, and bridge substructures. Such exposure demands high durability against cyclic freezing and chloride ingress to ensure long-term structural integrity.

Despite its beneficial effect on strength and durability, excessive SiMn slag replacement negatively impacted workability. Particularly, for the mixes with 70% and 100% replacement ratios, significant reductions in slump and spraying feasibility were observed. The rebound test results identified the S-50 mixture as the most suitable one, balancing mechanical performance with practical constructability while also exhibiting reduced shrinkage strain and improved adhesion during shotcrete application.

The findings of this study suggest that replacing 50% of fine aggregate with SiMn slag provides an optimal mix that ensures structural reliability while addressing sustainability concerns by reducing reliance on natural resources. However, this study does not provide a comprehensive theoretical assessment of potential chemical interactions between SiMn slag and cementitious components. Additionally, the risk of harmful substance leaching, especially under long-term exposure to moisture or aggressive environments, remains unexamined. These limitations should be addressed in future studies to ensure environmental safety and long-term durability.

Although the current production cost of SiMn slag mortar may exceed that of conventional natural fine aggregates, the overall life-cycle cost (LCC) could be comparable when considering environmental factors such as reduced carbon emissions, the conservation of natural resources, and the avoidance of environmental degradation. From a resource recycling perspective, the use of SiMn slag offers additional sustainability benefits, reinforcing its potential as a practical and eco-efficient alternative. Building upon this, future research should also aim to discover chemical interactions between SiMn slag and cementitious components, evaluate the economic viability of large-scale SiMn slag utilization, and validate its practical effectiveness through field implementation and long-term monitoring.

5. Conclusions

For industrial waste recycling purposes, an optimum mix proportion for shotcrete mortar was studied by varying replacement ratios (0%, 30%, 50%, 70%, and 100%) of SiMn slag as fine aggregate. Fresh and hardened performances of the mixture with various replacement ratios were evaluated to test the potential usage of SiMn slag as a sustainable construction material. The study results are as follows.

(1) The new mix proportions determined from preliminary tests have a water–cement ratio of 40% and a plasticizer content of 0.50% by weight of cement. Additionally, the unit weights of cement and water are 626 kg/m^3 and 251 kg/m^3, respectively. The total unit weight of aggregates including SiMn slag and fine aggregate is 1462 kg/m^3.

(2) Fresh properties of the mortar were assessed by the air content and slump tests. As the SiMn slag replacement ratio increased incrementally from 0% to 100%, the average air content increased from 5.7% to 6.5% and the average slump decreased from 210 mm to 135 mm. The increase in air content and decrease in slump with higher SiMn slag replacement in fresh mortar performance suggest the need for mix adjustments to ensure practical applicability.

(3) Mechanical performances of the mortar were evaluated by the compressive, splitting tensile, and flexural strength tests. As the SiMn slag replacement ratio increased incrementally from 0% to 100%, the average compressive strength at 28 days increased from 24.9 MPa to 35.54 MPa. The splitting tensile and flexural strength at 28 days increased from 2.9 MPa to 4.1 MPa and from 6.25 MPa to 8.55 MPa, respectively. The results suggest that SiMn slag can be effectively utilized to enhance the structural capacity of mortar.

(4) Durability properties of the mortar were evaluated by the tests of chloride ion penetration, freeze–thaw resistance, and accelerated carbonation. As the SiMn slag replacement ratio increased incrementally from 0% to 100%, the chloride ion penetration resistance showed improvement as indicated by the test data of average charge decreasing from 4546 to 3326 coulombs for 28 days. From the freeze–thaw resistance test, the average of relative dynamic modulus of elasticity (RDME) after 300 cycles

increased from 88.67% to 91.07%. From the carbonation test, the average carbonation depth decreased from 6.06 mm to 3.86 mm. The improved durability performance demonstrates the potential of SiMn slag for use in shotcrete structures exposed to aggressive environments.

(5) For the evaluation of constructability and serviceability of the mortar, the shotcrete rebound test and 30-day free drying shrinkage test were conducted. The rebound rates of the mortar with 30% and 50% replacement ratios were 51.2% and 18.5%, respectively, but the mortar with other replacement ratios was unsuitable for spraying. From the 30-day free drying shrinkage test, the mortar with a 50% replacement ratio exhibited below 500 micro-strain, satisfying the criterion of 500 micro-strain, whereas the mortar with a 0% replacement ratio exhibited approximately 600 micro-strain. The mortar with 50% SiMn slag replacement showed that SiMn slag can enhance both the efficiency of application and long-term performance in field conditions.

(6) The optimal shotcrete mix for sustainable construction was determined to be a mix consisting of 1462 kg/m^3 of fine aggregate with 50% SiMn replacement (731 kg/m^3), 626 kg/m^3 of Ordinary Portland Cement (ASTM C150 Type I, equivalent to KS L 5201 Type 1), 251 kg/m^3 of water, and 3.13 kg/m^3 of polycarboxylate-based superplasticizer with air-entraining properties. The findings demonstrate the positive contribution of SiMn slag to the development and utilization of durable and sustainable recycled materials.

Author Contributions: Conceptualization, W.-R.K., J.-B.L., and J.-H.J.K.; validation, W.-R.K. and B.-M.Y.; formal analysis, W.-R.K. and J.-B.L.; investigation, W.-R.K. and J.-B.L.; resources, B.-M.Y.; writing—original draft preparation, W.-R.K. and J.-B.L.; writing—review and editing, W.-R.K. and J.-H.J.K.; visualization, W.-R.K.; supervision, J.-H.J.K.; project administration, W.-R.K.; funding acquisition, J.-H.J.K. All authors have read and agreed to the published version of the manuscript.

Funding: This work was supported by the Korea Institute of Energy Technology Evaluation and Planning (KETEP), granted financial resources from the Ministry of Trade, Industry & Energy, Republic of Korea (RS-2020-KP002217).

Institutional Review Board Statement: Not applicable.

Informed Consent Statement: Not applicable.

Data Availability Statement: The original contributions presented in the study are included in the article, further inquiries can be directed to the corresponding author.

Conflicts of Interest: Author Bok-Mo Yoon was employed by the company Miju Ecorock Co., Ltd. The remaining authors declare that the research was conducted in the absence of any commercial or financial relationships that could be construed as a potential conflict of interest.

Abbreviations

The following abbreviations are used in this manuscript:

SiMn	Silicon Manganese
RDME	relative dynamic modulus of elasticity

References

1. Pierrehumbert, R. There is no Plan B for dealing with the climate crisis. *Bull. At. Sci.* **2019**, *75*, 6–7. [CrossRef]
2. Bernardo, G.; Guida, A.; Mecca, I. Advancements in Shotcrete Technology. In Proceedings of the Structural Studies, Repairs and Maintenance of Heritage Architecture XIV, A Coruna, Spain, 13–15 July 2015; WIT Press: Southampton, UK, 2015. [CrossRef]
3. Ghiasi, V.; Omar, H. Analysis of Shotcrete Lining of Underground Tunnels. *Pertanika J. Sci. Technol.* **2011**, *19*, 249–257.
4. Lackner, R.; Macht, J.; Hellmich, C.; Mang, H.A. Hybrid Method for Analysis of Segmented Shotcrete Tunnel Linings. *J. Geotech. Geoenviron. Eng.* **2002**, *128*, 298–308. [CrossRef]

5. Rispin, M.; Kleven, O.; Dimmock, R.; Myrdal, R. Shotcrete: Early Strength and Re-Entry Revisited—Practices and Technology. In Proceedings of the First International Conference on Underground Mining Technology, Sudbury, ON, Canada, 11–13 October 2017. [CrossRef]
6. Moon, T.; Wang, E.; Oh, J.; Li, K.; Ramandi, H.L.; Zhang, C.; Saydam, S. Evaluation of Shotcrete Performance from Full-Scale Load-Deflection Tests Using Numerical Analysis. In Proceedings of the ARMA US Rock Mechanics/Geomechanics Symposium, Atlanta, GA, USA, 25–28 June 2023. [CrossRef]
7. Oreste, P.P. A Procedure for Determining the Reaction Curve of Shotcrete Lining Considering Transient Conditions. *Rock Mech. Rock Eng.* **2003**, *36*, 209–236. [CrossRef]
8. Naseri, S.; Bahrani, N. Design of Initial Shotcrete Lining for a Mine Shaft Using Two-Dimensional Finite Element Models Considering Excavation Advance Rate. *Geotech. Geol. Eng.* **2021**, *39*, 4709–4732. [CrossRef]
9. Bajwa, D.S.; Pourhashem, G.; Ullah, A.H.; Bajwa, S.G. A concise review of current lignin production, applications, products and their environmental impact. *Ind. Crop. Prod.* **2019**, *139*, 111526. [CrossRef]
10. Dong, W.; Kong, F.; He, T.; Liu, M.; Wu, J.; Zhang, L. Enhancing performance and reducing carbon emissions in sprayed concrete using green silica-lignin admixture and recycled aggregates. *Low-Carbon Mater. Green Constr.* **2024**, *2*, 28. [CrossRef]
11. Garba, M.J.; Tian, Y.; Xie, Z.; Yu, C.; Hu, C.; Chen, L.; Yuan, Q. Effect of accelerators on the long-term performance of shotcrete and its improvement strategies: A review. *J. Build. Eng.* **2024**, *89*, 109364. [CrossRef]
12. Chen, L.; Huang, L.; Hua, J.; Chen, Z.; Wei, L.; Osman, A.I.; Fawzy, S.; Rooney, D.W.; Dong, L.; Yap, P.-S. Green construction for low-carbon cities: A review. *Environ. Chem. Lett.* **2023**, *21*, 1627–1657. [CrossRef]
13. Mohanta, N.R.; Murmu, M. Alternative coarse aggregate for sustainable and eco-friendly concrete—A review. *J. Build. Eng.* **2022**, *59*, 105079. [CrossRef]
14. Tayeh, B.A.; Saffar, D.M.A.; Alyousef, R. The Utilization of Recycled Aggregate in High Performance Concrete: A Review. *J. Mater. Res. Technol.* **2020**, *9*, 8469–8481. [CrossRef]
15. Akter, S.; Muniruzzaman, M. Industrial Waste Management and Environment: A Study in Kamrangirchar (Raised Land), Dhaka. *Environ. Manag. Sustain. Dev.* **2021**, *10*, 24–44. [CrossRef]
16. Horii, K.; Tsutsumi, N.; Kitano, Y.; Kato, T. Processing and reusing technologies for steelmaking slag. *Nippon. Steel Tech. Rep.* **2013**, *104*, 123–129.
17. Ma, Y.; Moosavi-Khoonsari, E.; Kero, I.T.; Tranell, G.M. Recycling of Industrial Byproducts in Construction: Metallurgical Slag as a Sustainable Aggregate. *Metall. Mater. Trans. B* **2018**, *49*, 2444–2457. [CrossRef]
18. Olsen, S.E.; Tangstad, M. Silicomanganese production—Process Understanding. In Proceedings of the Infacon X, Cape Town, South Africa, 1–4 February 2004.
19. Choi, H.B.; Kim, J.M. Properties of silicon manganese slag as an aggregate for concrete depending on cooling conditions. *J. Mater. Cycles Waste Manag.* **2020**, *22*, 1067–1080. [CrossRef]
20. Buruiana, D.L.; Obreja, C.-D.; Herbei, E.E.; Ghisman, V. Re-Use of Silico-Manganese Slag. *Sustainability* **2021**, *13*, 11771. [CrossRef]
21. Ting, J.J.; Low, W.W.; Wong, K.S.; Ting, T.Z.H.; Abdul-Rahman, H. Feasibility of recycling silicomanganese slag as a cementitious material. *J. Eng. Technol. Adv.* **2023**, *8*, 97–115. [CrossRef]
22. Makhambetov, Y.; Myngzhassar, Y.; Abdirashit, A.; Onuralp, Y. Analysis of the silicomanganese market worldwide: A review. *Eng. J. Satbayev Univ.* **2024**, *146*, 9–15. [CrossRef]
23. Saha, A.K.; Khan, M.; Sarker, P.K. Properties of Concrete Incorporating Recycled Materials. *Resour. Conserv. Recycl.* **2018**, *134*, 10–24. [CrossRef]
24. Wong, N.H.; Kong, Z.Y.; Sambo, R.; Chai, C.S.; Khoso, A.R.; Bamgbade, J.A.; Sunarso, J. Physicochemical Characteristics of Silicomanganese Slag as a Recycling Construction Material: An Overview. *Min. Metall. Explor.* **2024**, *41*, 1891–1907. [CrossRef]
25. Patil, A.V.; Pande, A.M. Strength Characteristics of Concrete Using Recycled Aggregates. *Adv. Mater. Res.* **2011**, *255*, 3258–3262. [CrossRef]
26. Nath, S.K.; Randhawa, N.S.; Kumar, S. Environmental Impact of Concrete Incorporating Recycled Materials: A Comparative Study. *Resour. Conserv. Recycl.* **2022**, *176*, 105946. [CrossRef]
27. *KS F 2504*; Korean Industrial Standard: Standard Test Method for Density and Water Absorption Rate of Fine Aggregates. Korean Standards Association: Seoul, Republic of Korea, 2024.
28. *KS F 2527*; Korean Industrial Standard: Aggregates for Concrete. Korean Standards Association: Seoul, Republic of Korea, 2024.
29. *KS F 2405*; Korean Industrial Standard: Test Method for Compressive Strength of Concrete. Korean Standards Association: Seoul, Republic of Korea, 2020.
30. *KS L 5201*; Korean Industrial Standard: Portland Cement. Korean Standards Association: Seoul, Republic of Korea, 2021.
31. *KS F 2421*; Korean Industrial Standard: Standard Test Method for Air Content of Fresh Concrete by the Pressure Method. Korean Standards Association: Seoul, Republic of Korea, 2016.
32. *KS F 2402*; Korean Industrial Standard: Test Method for Concrete Slump. Korean Standards Association: Seoul, Republic of Korea, 2022.

33. *KS F 2403;* Korean Industrial Standard: Standard Test Method for Making Concrete Specimens. Korean Standards Association: Seoul, Republic of Korea, 2019.
34. *KS F 2408;* Korean Industrial Standard: Standard Test Method for Flexural Strength of Concrete. Korean Standards Association: Seoul, Republic of Korea, 2016.
35. *KS F 2423;* Korean Industrial Standard: Standard Test Method for Splitting Tensile Strength of Concrete. Korean Standards Association: Seoul, Republic of Korea, 2021.
36. *KS F 2711;* Korean Industrial Standard: Standard Test Method for Resistance of Concrete to Chloride Ion Penetration by Electrical Conductance. Korean Standards Association: Seoul, Republic of Korea, 2016.
37. *KS F 2456;* Korean Industrial Standard: Test Method for Resistance of Concrete to Rapid Freezing and Thawing. Korean Standards Association: Seoul, Republic of Korea, 2023.
38. *KS F 2584;* Korean Industrial Standard: Standard Test Method for Accelerated Carbonation of Concrete. Korean Standards Association: Seoul, Republic of Korea, 2016.
39. *KS F 2595;* Korean Industrial Standard: Standard Test Method for Drying Shrinkage Crack in Concrete. Korean Standards Association: Seoul, Republic of Korea, 2021.

materials

MDPI

Article

Experimental Investigation on the Fracture Behavior of PET-Modified Engineered High-Ductility Concrete: Effects of PET Powder and Precursor Composition

Fei Meng [1], Shen Luo [1], Jingxian Sun [1], Cheng Zhang [1], Leilei Xu [1], Liqun Zhang [1], Fumin Qing [2], Junfeng Zeng [3], Ruihao Luo [3,*] and Yongchang Guo [4]

1. Zhongshan Power Supply Bureau of Guangdong Power Grid Co., Ltd., Zhongshan 528400, China; blackstar_002@sina.cn (F.M.); zsluozong@163.com (S.L.); 1290025427@163.com (J.S.); 314355@163.com (C.Z.); 13924971515@139.com (L.X.); 13560600262@163.com (L.Z.)
2. Fujian Southeast Design Group Co., Ltd., Sanming 365000, China; supervisionr@outlook.com
3. School of Civil and Transportation Engineering, Guangdong University of Technology, Guangzhou 510006, China; 13823893571@139.com
4. School of Transportation and Civil Engineering and Architecture, Foshan University, Foshan 528225, China; guoyc@fosu.edu.cn

* Correspondence: luoryann@outlook.com

Abstract: The utilization of polyethylene terephthalate (PET) powder as aggregate in the development of environmentally friendly high-ductility composites (P-EHDC) offers a promising pathway for advancing sustainable and high-performance concrete materials. Despite its potential, the fracture behavior of P-EHDC—particularly under the influence of alkali-activated precursors—remains insufficiently explored. In this study, the fracture performance of P-EHDC was evaluated by varying the precursor composition ratios (GGBS:FA = 4:6, 3:7, and 2:8) and PET powder replacement ratios (0%, 15%, 30%, and 45% by volume). Fracture modes, Mode I fracture energy (G_F), and crack propagation behavior were analyzed using the *J*-integral method. All specimens exhibited ductile fracture characteristics, a clear contrast to the brittle failure observed in conventional concrete. The replacement of 15 vol% PET powder significantly increased G_F in precursor systems with higher GGBS content (4:6 and 3:7), and 30 vol% was more effective in fly ash-rich systems (2:8). The *J*-integral method, which offers broader applicability compared to conventional methods such as the double-K fracture model, provided a more comprehensive understanding of the fracture behavior. The results showed that PET powder reduced the matrix fracture toughness, promoted matrix cracking, and weakened the fiber-bridging effect, leading to enhanced energy absorption via fiber pull-out. At low PET powder replacement ratios (e.g., 15 vol%), the cracking threshold of the matrix was not significantly reduced, while more fibers engaged during the crack instability stage to absorb fracture energy through pull-out. This behavior highlights the synergistic toughening effect between PET powder and fibers in the P-EHDC system. The effect became more pronounced when the PET content was below 45 vol% and the precursor matrix contained a higher proportion of GGBS, leading to enhanced ductility. This study introduces a novel approach to fracture behavior analysis in PET-modified alkali-activated composites and provides theoretical support for the toughening design of high-performance, low-carbon concrete materials.

Keywords: PET-modified environmentally friendly high-ductility (P-EHDC); fracture performance; *J*-integral method

https://doi.org/10.3390/ma18092132

1. Introduction

High-ductility concrete (HDC), such as engineered cementitious composites (ECC) [1–3], aims to address the inherent brittleness of traditional concrete. Through micromechanical mechanisms governing the interaction between the cementitious matrix and fibers, HDC demonstrates significant tensile strain-hardening behavior and multiple cracking characteristics [4–6], enhancing material toughness. ECC has gained attention for its high ductility and crack control capabilities [7–10].

The production of ECC typically requires large amounts of traditional cementitious materials, such as Portland cement. However, in light of the increasing global climate challenges, countries worldwide are setting carbon reduction goals. China has specifically committed to reaching peak carbon emissions by 2030 and achieving carbon neutrality by 2060. The production of Portland cement generates harmful substances, including CO_2, nitrogen oxides, sulfur oxides, and dust. For every ton of cement produced, 0.7 to 1.1 tons of CO_2 are emitted, contributing to 5–7% of global carbon emissions annually [11–15]. As a result, developing alternative green binders has become a research focus. Alkali-activated composites, which use industrial waste materials like slag (GGBS) and fly ash (FA) with volcanic ash activity, combined with alkaline activators, are considered promising green binders [16–18]. Compared to Portland cement, alkali-activated materials reduce CO_2 emissions by 55–75% and energy consumption by 36%, while offering superior properties like high compressive strength and low permeability [19–23]. Thus, these materials are seen as promising substitutes for Portland cement [24,25]. Alkali-activated materials are also viewed as a viable alternative for producing high-ductility alkali-activated concrete (HDAAC), marking a new direction for high-performance concrete development.

In the past few decades, plastic has accounted for approximately 13% of the global annual municipal solid waste (MSW), and 19.2% and 16.4% of the total MSW sent to landfills and incineration, respectively. However, plastic recycling accounts for only 4.4% [26]. The inadequate management and disposal of these plastics have led to severe soil and water pollution [27]. Polyethylene terephthalate (PET) is commonly used to manufacture plastic bottles and transparent packaging, which are discarded after single use. Studies have shown that using plastic waste as a replacement for natural aggregates in concrete can reduce the extraction of natural resources and mitigate the negative environmental impacts of plastic [28–31].

Existing research indicates that the application of waste plastic in cement-based composites presents multidimensional technical characteristics. In fiber-reinforced systems, PET fibers have been shown to effectively enhance the mechanical properties of engineered cementitious composites (ECC) and engineered geopolymer composites (EGC). Their reinforcement effect competes with conventional polyvinyl alcohol (PVA) and polypropylene (PP) fibers in terms of tensile strength, compressive strength, and multiple crack control [32]. Particularly, the interface bonding characteristics of PET fibers provide significant advantages in matrix stress transfer and crack propagation suppression. Studies on plastic as a substitute for aggregates show significant differences in the performance impact depending on the type of plastic. When PET and PE are used as fine aggregate replacements, both lead to a reduction in compressive, tensile, and flexural strengths, with PE having a more pronounced negative effect [33]. However, it is noteworthy that while the incorporation of plastic aggregates weakens strength parameters, it does not significantly deteriorate the material's toughness, indicating its potential value in specific engineering scenarios. Mechanism studies suggest that the interfacial transition zone (ITZ) between plastic particles and the cement matrix is a key factor affecting material strength. When the plastic content exceeds a critical threshold, the significant reduction in interfacial bonding strength and adhesion becomes the primary limiting factor [34]. On the other hand, experiments replacing quartz powder with PET powder show an optimal replacement threshold. When the replacement ratio is controlled around 10%, the compressive and tensile strengths of geopolymer concrete can be increased by approximately 5.8% and 24%, respectively. This enhancement is attributed to the densification effect of PET powder on the matrix microstructure [35]. However, this enhancement effect is highly sensitive to the dosage, as excessive incorporation leads to performance degradation due to interfacial bonding deterioration. Scanning electron microscopy (SEM) analysis further reveals a negative correlation between plastic content and ITZ quality. A high content of plastic particles forms weak bonding regions at the matrix-plastic interface, thereby reducing the overall performance of the composite material [35].

The fracture behavior of conventional concrete is typically characterized by brittle or quasi-brittle fracture. Under the framework of fracture mechanics, its failure can be analyzed using linear elastic fracture mechanics methods [36–38], such as the traditional dual-K fracture criterion [39–41], as the double-K fracture model characterizes the fracture behavior of concrete by introducing two distinct fracture toughness values for crack initiation and unstable propagation. However, high-ductility concretes, such as HDAAC, exhibit distinct fracture characteristics compared to conventional fiber-reinforced concretes; they display tough fracture behavior and cannot satisfy the basic assumptions of the dual-K fracture model, specifically the linear superposition assumption [42,43]. As a result, some studies suggest that the *J*-integral method, based on nonlinear principles, is more suitable for assessing the fracture behavior of high-ductility concretes. The *J*-integral method's path-independent nature avoids the issues associated with the linear superposition assumption [44].

In summary, current research on PET-modified concrete primarily focuses on concrete using Portland cement as the binder, and those studies mainly investigate basic mechanical properties such as compressive, tensile, and flexural strength, while research on fracture mechanics is relatively limited. Moreover, there is a significant lack of studies on the application of PET powder in alkali-activated high-ductility composites, which are designed using alkali-activated binders and synthetic fibers as reinforcement materials. On the other hand, conventional analytical models such as the double-K fracture model are only applicable to ordinary concrete materials exhibiting brittle failure.

Therefore, this study utilized PET powder to modify alkali-activated high-ductility composites, thereby developing PET-modified engineered high-ductility concrete (P-EHDC). Based on the *J*-integral fracture model, the study investigated the effects of different precursor material ratios (GGBS:FA = 4:6, 3:7, and 2:8) and various PET powder replacement ratios for quartz powder (0%, 15%, 30%, and 45% by volume) on the fracture behavior of P-EHDC. Through three-point bending tests, the failure modes of P-EHDC were analyzed, and the material's performance was comprehensively evaluated from both localized crack propagation and overall deformation perspectives using *P-CMOD* and *P-δ* curves. The Model-I fracture energy (G_F) and fracture energy calculated using the *J*-integral method were also determined, with a thorough analysis of the material's energy dissipation capacity. Additionally, scanning electron microscopy (SEM) is employed for microstructural analysis.

2. Materials and Methods

2.1. Materials

The materials used in this study include ground granulated blast-furnace slag (GGBS), class F fly ash (FA), quartz powder (QP), plastic raw material polyethylene terephthalate powder (PET powder), alkali activators, ultra-high molecular weight polyethylene fibers (PE fibers), retarders (barium chloride), and water. The specific parameters of the materials used in this study are shown in Tables 1 and 2 and Figure 1. GGBS, a white solid powder, was sourced from Longze Water Purification Materials Co., Ltd. (Gongyi, China). FA, a gray-black solid powder, was also sourced from Longze Water Purification Materials Co., Ltd. (Gongyi, China). GGBS and FA compose the binder material. PET powder, a white powder with a density of 1.68×10^3 kg/cm^3, was obtained from Guangyuan Plastic Co., Ltd. (Dongguan, China). Quartz powder was used with a particle size range of 76 μm to 150 μm and a density of 2.65×10^3 kg/cm^3. The appearance of the two aggregates is shown in Figure 2. Sodium silicate solution was sourced from Yourui Refractory Materials Co., Ltd. (Jiaxing, China). Sodium hydroxide, a white solid particle with a purity of 96%, was purchased from Xilong Science Co., Ltd. (Jiaxing, China). PE fibers, white bundled monofilaments, were provided by Suotite Special Line Co., Ltd. in (Dongguan, China). Barium chloride ($BaCl_2$), used as a retarder to delay the initial setting time of P-EHDC, at a dosage of 1% by mass of the precursors, was sourced from Xilong Science Co., Ltd. (Jiaxing, China) [45]. The reaction between $BaCl_2$ and sodium silicate solution can produce a dense cladding layer, thus preventing the GGBS from coming into direct contact with the sodium silicate solution and achieving a certain retarding effect [46].

Table 1. Property Parameters of cementitious materials and aggregate.

Property Parameters	CaO	SiO_2	Al_2O_3	SO_3	Fe_2O_3	MgO	TiO_2	Other	Loss on Ignition (%)	Compressive Strength (MPa)	Elastic Modulus (GPa)
	wt%										
GGBS	34.0	34.5	17.7	1.64	1.03	6.01	/	5.12	0.840	/	/
FA	4.01	54.8	31.2	2.20	4.16	1.01	1.13	2.37	4.60	/	/
QP	0.09	89.2	5.97	0.23	0.67	0.09	0.20	0.65	/	/	/
PET	/	/	/	/	/	/	/	/	/	130	1.10

Table 2. Properties of PE.

Fiber Type	Length (mm)	Diameter (um)	Strength (MPa)	Aspect Ratio (L/d)	Density (g/cm^3)	Elastic Modulus (GPa)	Elongation (%)
PE	12.0 (Mean value)	20.0	2500	600	0.970	120	3.70

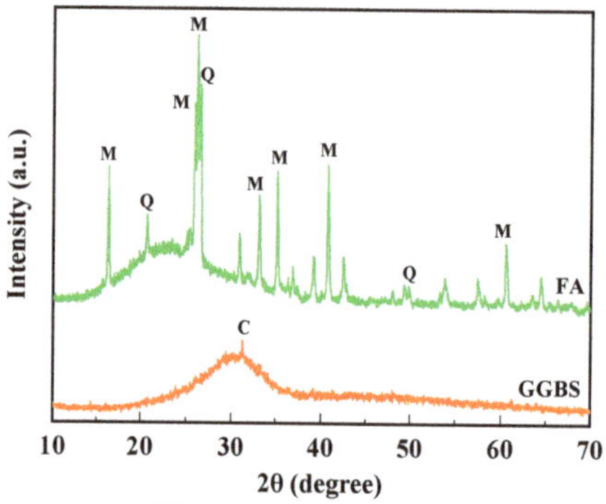

(**a**) XRD patterns of FA and GGBS

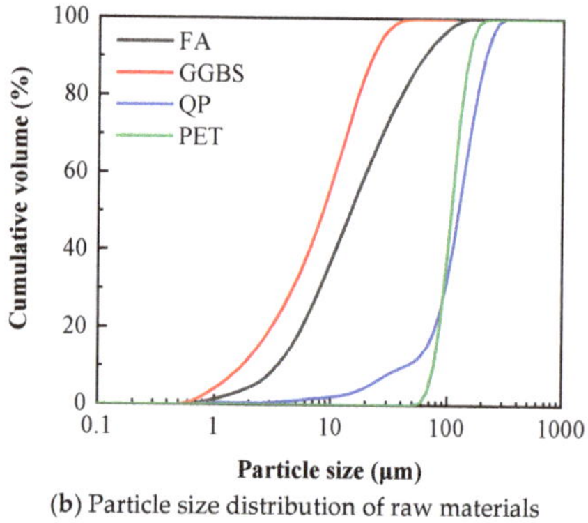

(**b**) Particle size distribution of raw materials

Figure 1. Physical and chemical property parameters of raw materials.

(**a**) PET powder, macroscopic image and SEM micrograph

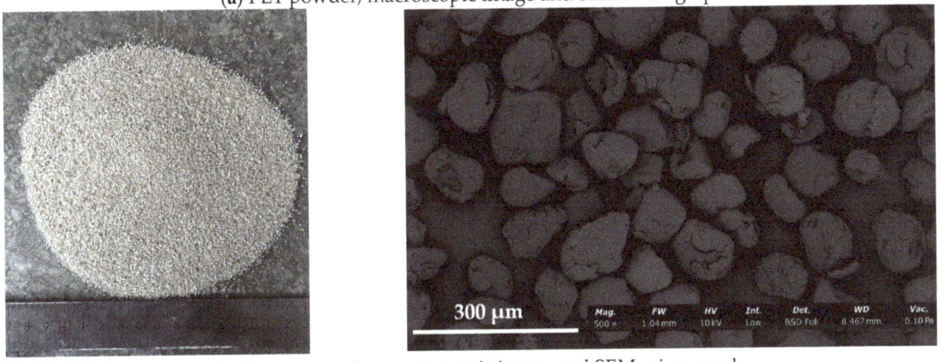

(**b**) Quartz powder, macroscopic image and SEM micrograph

Figure 2. Appearance of aggregates.

The mix proportions used in this study are shown in Table 3. Three precursor matrix types, B1, B2, and B3, were prepared based on different GGBS and FA ratios. These ratios primarily affect the strength of the material matrix, with strength decreasing as the proportion of GGBS decreases [47,48]. Similarly, when determining the final proportions of GGBS and FA, primary consideration was given to the effect of precursor ratios on the matrix strength. Preliminary experiments were also conducted to guide the decision, resulting in the selection of three ratios as shown in Table 3: 4:6, 3:7, and 2:8. Based on these three different binder ratios, PET powder was added to replace quartz powder at volumes of 0%, 15%, 30%, and 45% by volume. Thus, this study involves 12 mix designs of P-EHDC, with a water-to-binder ratio (w/b) of 0.32 for all binder compositions.

The specimen preparation process is divided into three main stages. First, a sodium hydroxide solution with a molarity of 14 mol/L is prepared using solid sodium hydroxide 24 h prior to casting. After allowing it to cool to room temperature naturally, the solution is mixed evenly with sodium silicate solution, then left to cool to room temperature again. Second, all dry materials, except for the fibers, are added to a planetary mixer as shown in Table 3 and mixed for 3 min to ensure uniform blending. The alkali activator and additional water are then mixed and added to the mixer, where the mixture is stirred for another 3 min. In the final stage, PE fibers are added, and the mixing is completed within 2 min, as shown in Figure 3. Finally, the fully mixed P-EHDC is poured into a prismatic mold. After initial setting, the specimens are covered with plastic film for curing. After 24 h, the specimens are removed from the molds and immersed in a water tank for continuous curing until 28 days [17].

Table 3. Mix design.

Cementitious Material ID	Mix ID	Mix Design (kg/m^3)							
		GGBS	FA	QP	Alkaline Solution	Water	Retarder	PET	PE
B1	B1-P0	491	736	245	491	73.6	12.3	0.00	19.4
	B1-P15	491	736	209	491	73.6	12.3	23.3	19.4
	B1-P30	491	736	172	491	73.6	12.3	46.7	19.4
	B1-P45	491	736	135	491	73.6	12.3	70.0	19.4
B2	B2-P0	364	850	243	486	72.8	12.1	0.00	19.4
	B2-P15	364	850	206	486	72.8	12.1	23.1	19.4
	B2-P30	364	850	170	486	72.8	12.1	46.2	19.4
	B2-P45	364	850	134	486	72.8	12.1	69.3	19.4
B3	B2-P0	239	956	239	478	71.7	12.0	0.00	19.4
	B2-P15	239	956	203	478	71.7	12.0	22.7	19.4
	B2-P30	239	956	167	478	71.7	12.0	45.5	19.4
	B2-P45	239	956	132	478	71.7	12.0	68.2	19.4

Note: B1-P15 refers to a cementitious material with a GGBS to FA ratio of 4:6, and PET replacing 15% of the quartz powder by volume.

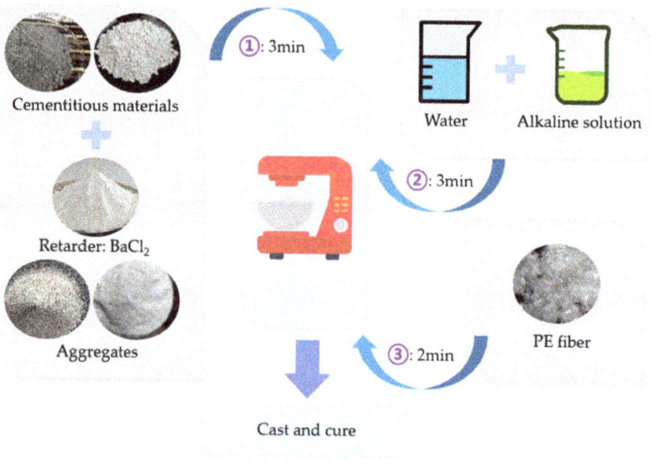

Figure 3. Preparation procedure for P-EHDC.

The schematic diagram of the fracture specimen used in this study is shown in Figure 4. The specimen dimensions are 40 mm × 40 mm × 160 mm, with a span of 120 mm. The preformed notch was manually cut with a cutting machine after the specimen had dried and completed curing. The notch depth is 16 mm, with a width within 1 mm, and the depth error is controlled within 0.5 mm.

2.2. Testing Setup

2.2.1. Fracture Test

The fracture testing setup used in this study is shown in Figure 5. Following the JCI-S-001 recommendation [49], a three-point bending fracture test was conducted to assess the fracture properties of concrete based on the *P-CMOD* curve. Referencing JCI-S-001 and existing research on the fracture performance of high-ductility concrete [36,42,44,50], a three-point bending test was employed to investigate the fracture behavior of P-EHDC. During the three-point bending loading process, the testing machine used a displacement control mode with a rate of 0.1 mm/min. Two displacement gauges with a range of 25 mm were placed on both sides of the specimen to measure the deflection at the midpoint. A clip-on extensometer was used at the bottom of the specimen to measure the crack opening displacement (*CMOD*). Additionally, the crack initiation load of P-EHDC was measured using two 10 mm strain gauges horizontally arranged at both ends of the preformed notch.

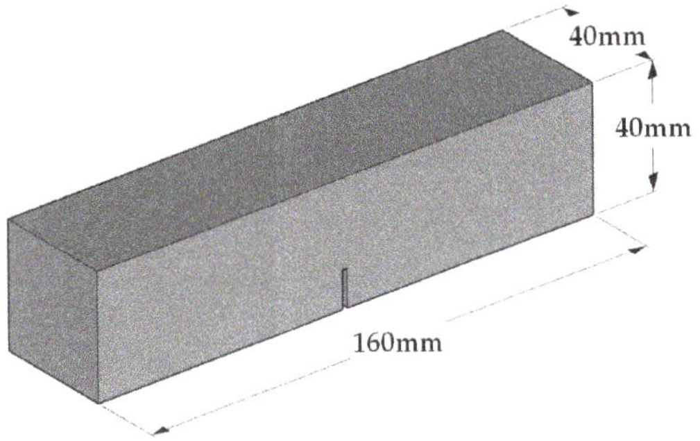

Figure 4. Schematic dimension of fracture specimen.

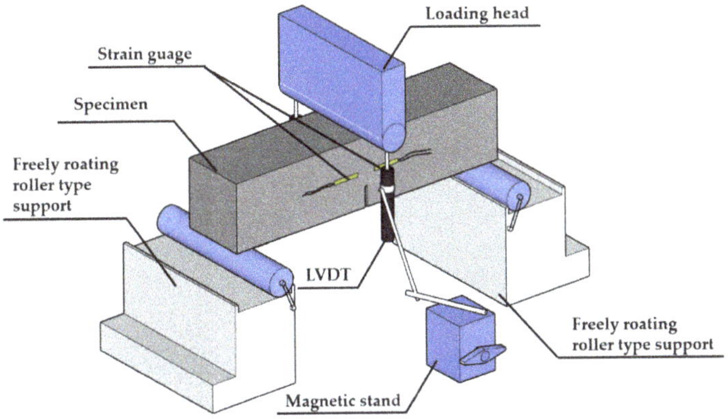

Figure 5. Schematic diagram of the fracture experimental setup.

2.2.2. SEM Analysis

The scanning electron microscopy was conducted using an S-3400N-II scanning electron microscope, produced by Hitachi (Tokyo, Japan). The SEM test mainly characterized the morphology of the interface between the aggregates and the matrix in the fractured cross-sections.

3. Results and Discussion

3.1. Failure Mode

The typical failure mode of the P-EHDC fracture specimen is shown in Figure 6. During the three-point bending test, microcracks were first observed at the tip of the preformed notch, indicating that the applied load had reached the crack initiation load of P-EHDC. As the loading continued, macrocracks gradually propagated from the notch tip (refer to the red curve in Figure 6), accompanied by an increasing opening angle at the notch. Ultimately, the macrocracks continued to develop until the specimen became unstable. Fibers were progressively pulled out or fractured from the matrix, and the specimen lost all its load-bearing capacity. As shown in Figure 6, the change in PET powder replacement ratio affects the initiation of microcracks near the main crack of the specimen (refer to the white curve in Figure 6). For example, in the B1 and B2 series test groups, microcracks were still observable on the surface of specimens with 0 vol% and 15 vol% PET powder replacement. However, as the PET powder replacement ratio increased, microcracks were no longer visible. On the other hand, the increase in PET powder replacement ratio made the propagation path of the main crack more tortuous, exhibiting the so-called "ductile fracture" mode. The change in failure mode indicates that replacing quartz powder with PET powder as aggregate increased the fracture surface area. Furthermore, PET replacement altered the matrix properties, and the tortuous main crack path suggests unevenly distributed defects in the matrix. The figure also shows that the reduction in matrix strength, as indicated by the GGBS-to-FA ratio, caused the macrocrack propagation path to become more distorted.

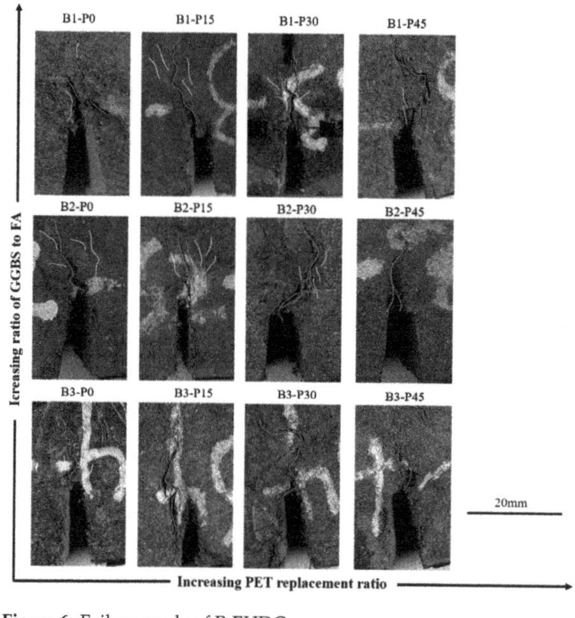

Figure 6. Failure mode of P-EHDC.

3.2. *P-δ Curves and P-CMOD Curves*

Figure 7 shows the typical *P-CMOD* curve of ductile materials. The P-δ and *P-CMOD* curves for all P-EHDC specimens are shown in Figures 8 and 9. It is observed that the curve propagation process shares common characteristics across all test groups. During the initial loading phase, the load increases rapidly at a nearly constant rate. After reaching the first inflection point, the load develops to the first peak value, then fluctuates around the peak load, forming a noticeable platform segment. Based on the fracture failure mode analysis, the appearance of the platform segment coincides with the time when the main crack in the fracture surface develops tortuously and when numerous microcracks form. More specifically, the *P-CMOD* curves for P-EHDC specimens exhibiting "ductile fracture" go through the following six stages, is also shown in Figure 7.

Stage I is the linear elastic stage, where the load is primarily carried by the P-EHDC matrix. In this stage, PE fibers exist as embedded defects within the matrix, which is due to the significant difference in elastic modulus between the PE fibers and the matrix [51,52]. Stage II is the inelastic deformation stage, where the load reaches the crack initiation load, and microcracks begin to form at the edges of the preformed notch. In Stage III, the load continues to increase with minor fluctuations. Notably, the slope of the curve decreases in the first three stages, meaning the rate of load increase slows down, and the stiffness of the P-EHDC decreases. In Stage IV, the *P-CMOD* curve begins to show noticeable oscillations, with the load continuing to increase amidst the fluctuations and accompanied by the formation of new cracks. This indicates that the specimen has entered the multi-crack stable propagation stage. In Stage V, a small platform section appears on the curve, with the load no longer increasing significantly. Although some microcracks still form at the notch tip, the matrix is almost no longer active, and the PE fibers begin to carry the majority of the load. This signifies that the width of the most prominent main crack at the notch is continuously increasing, and the specimen is about to enter the instability stage. Stage VI is the crack instability and propagation stage, where fibers begin to pull out or fracture, and the load drops significantly, accompanied by rapid expansion of the main crack, ultimately leading to specimen failure.

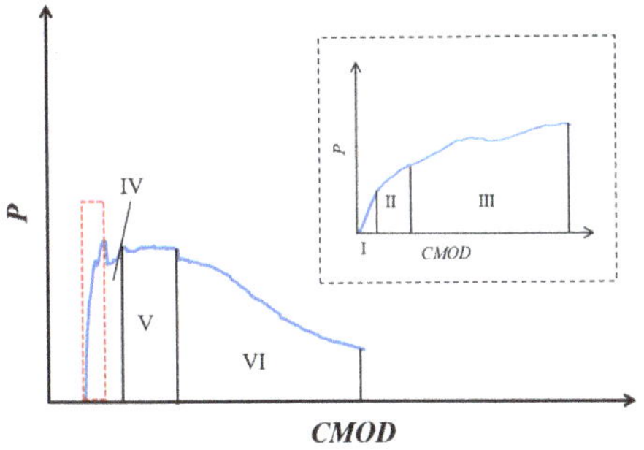

Figure 7. Schematic diagram of the stages in the *P-CMOD* curve for ductile fracture materials.

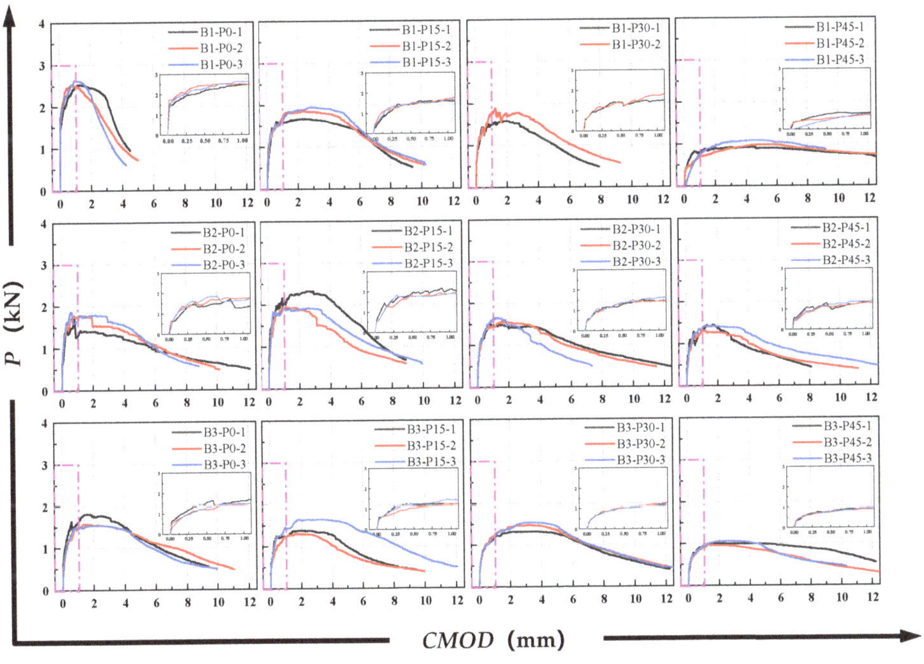

Figure 8. *P-CMOD* curves of P-EDHC.

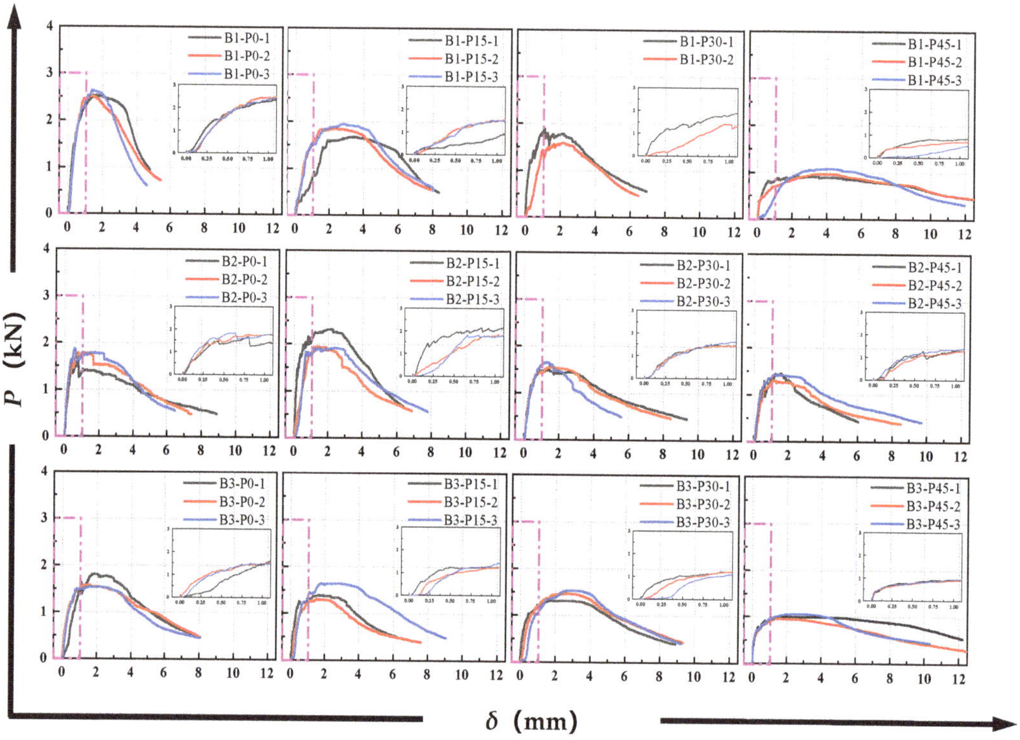

Figure 9. *P*-*δ* curves of P-EDHC.

From Figures 8 and 9, it is evident that, compared to B1-P0, B2-P0, and B3-P0, the slope of Stage I in the *P-CMOD* curve significantly decreased as the precursor matrix of P-EHDC changed. This indicates that the reduction in the GGBS ratio in the precursor weakens the stiffness of the linear elastic stage of P-EHDC. Since the load is primarily carried by the P-EHDC matrix in this stage, it suggests that the alkali-activated binder system with GGBS and FA as precursors shows a positive effect of GGBS on the strength of the cementitious products, and this is consistent with existing research findings [47,48]. On the other hand, as the PET replacement ratio increased, the curves of the first three stages in the *P-CMOD* curve become more gradual, with the slope significantly decreasing. The crack initiation load at the end of Stage II also gradually decreases. Moreover, the peak load that the specimen can reach decreased as the PET content increased.

3.3. Mode I Fracture Energy

The fracture energy G_F of concrete is one of the key parameters for evaluating its fracture performance and energy dissipation capacity. Although there are some differences in the definition and measurement methods of fracture energy in different standards and codes, in this study, despite the formation of some microcracks during the fracture process of P-EHDC specimens, the primary crack still plays a crucial role in determining the fracture energy of the specimen. Therefore, the G_F of P-EHDC is measured based on the JCI-S-001-2003 standard. According to the definition in the JCI-S-001-2003 standard, fracture energy G_F refers to the energy absorbed by the specimen per unit area when a crack is formed along a pre-set path. The specific calculation formulas can be referred to in Equations (1) and (2).

$$G_F = \frac{0.75W_0 + W_1}{A_{lig}} \tag{1}$$

$$W_1 = 0.75\left(\frac{S}{L}m_1 + 2m_2\right)g \cdot CMOD_0 \tag{2}$$

$$A_{lig} = B(W - a_0) \tag{3}$$

where W_0 represents the area under the *P-CMOD* curve; W_1 is the work done by the self-weight of the specimen and the fixtures; A_{lig} is the ligament area; B, W, and a_0 are the specimen's thickness, height, and notch depth, respectively; S/L is the ratio of the span to the total length of the specimen; m_1 and m_2 are the weights of the specimen and the fixtures not fully fixed in the test machine, with m_2 typically taken as 0; g = 9.81 m/s^2 is the acceleration due to gravity; and $CMOD_0$ is the *CMOD* value corresponding to 30% of the maximum load P_{max} on the *P-CMOD* curve.

Table 4 presents the results of the Model I fracture energy calculations, and Figure 10 shows the variation in fracture energy with the PET powder replacement ratio and precursor ratio. It is evident that the incorporation of PET powder has a positive impact on fracture energy. For instance, compared to B1P0, the fracture energy of B1P15, B1P30, and B1P45 increased by 60.3%, 19.5%, and 25.0%, respectively. Similarly, in the B2 and B3 series, the incorporation of PET powder also resulted in increased Model I fracture energy. This can be explained as follows: First, the alkali-activated high-ductility concrete made with GGBS and FA as precursors already exhibits ductile fracture characteristics [53,54], with microcracks surrounding the main crack. Introducing materials with hydrophobic properties or lower elastic moduli reduces the strength of the matrix and facilitates crack propagation [50], with PET powder serving this function. Under the influence of PET powder, the main crack path in P-EHDC becomes more tortuous, extending the *P-CMOD* curve. Even though the peak fracture load significantly decreases, the fracture energy is still enhanced.

Table 4. Fracture energy calculation results.

Mix IDs	*Pini* (kN)	*Pmax* (kN)	m_1 (kg)	$CMOD_0$ (mm)	W_0 (kN·mm)	W_1 (kN·mm)	G_F (kJ·m^{-2})
B1-P0	0.705 (0.022)	2.56 (0.025)	0.521 (0.001)	4.58 (0.39)	8.48 (0.77)	0.0244 (0.0022)	6.55 (0.61)
B1-P15	0.773 (0.015)	1.83 (0.171)	0.509 (0.001)	10.2 (0.5)	13.4 (1.2)	0.0376 (0.0033)	10.5 (0.9)
B1-P30	0.605 (0.025)	1.59 (0.60)	0.489 (0.007)	8.53 (0.91)	9.99 (1.72)	0.0270 (0.0046)	7.83 (1.35)
B1-P45	0.372 (0.025)	1.29 (0.77)	0.482 (0.003)	13.2 (0.1)	10.6 (0.0)	0.0282 (0.0000)	8.31 (0.00)
B2-P0	0.873 (0.101)	1.79 (0.02)	0.494 (0.006)	10.2 (1.6)	11.6 (0.23)	0.0316 (0.0006)	9.07 (0.18)
B2-P15	0.804 (0.110)	2.07 (0.33)	0.488 (0.001)	9.13 (0.67)	13.1 (1.8)	0.0353 (0.0049)	10.3 (1.4)
B2-P30	1.04 (0.09)	1.58 (0.06)	0.463 (0.002)	10.5 (2.8)	10.6 (2.5)	0.0271 (0.0063)	8.31 (1.93)
B2-P45	0.642 (0.044)	1.40 (0.14)	0.458 (0.002)	10.7 (2.5)	9.40 (2.30)	0.0238 (0.0058)	7.37 (1.80)
B3-P0	0.642 (0.044)	1.65 (0.83)	0.462 (0.004)	10.1 (0.8)	11.4 (0.8)	0.0291 (0.0021)	8.91 (0.63)
B3-P15	0.923 (0.031)	1.45 (0.50)	0.450 (0.011)	10.5 (1.3)	10.5 (3.3)	0.0261 (0.0082)	8.27 (2.58)

Table 4. *Cont.*

Mix IDs	*Pini* (kN)	*Pmax* (kN)	m_1 (kg)	$CMOD_0$ (mm)	W_0 (kN·mm)	W_1 (kN·mm)	G_F (kJ·m^{-2})
B3-P30	0.988 (0.022)	1.45 (0.30)	0.433 (0.003)	12.0 (0.3)	12.2 (0.6)	0.0292 (0.0015)	9.56 (0.48)
B3-P45	0.733 (0.009)	1.01 (0.19)	0.427 (0.002)	11.7 (1.2)	8.91 (1.38)	0.0210 (0.0032)	6.99 (1.08)

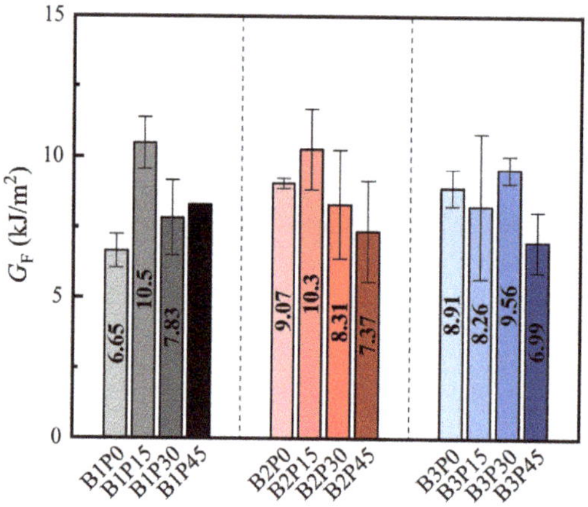

Figure 10. Effect of precursor matrix and PET powder replacement ratio on fracture energy.

3.4. Fracture Energy Based on J Integral Method

Various methods are used in the industry to evaluate the fracture behavior of materials, such as the dual-K criterion [39–41]. However, the dual-K criterion can only be applied when the fracture behavior of the material satisfies the important assumption of linear elastic fracture mechanics (as shown in Figure 11). Clearly, the ductile fracture behavior exhibited by P-EHDC in this study does not meet this assumption. Therefore, following the approach of Liu et al. [44] for assessing the fracture performance of strain-hardening concrete, this study employs a *J*-integral-based method to evaluate the ductile fracture behavior of P-EHDC. Based on previous research [38,43], this study introduces J_{IC}, defined as the energy flowing into the plastic zone when macrocracks appear, which can be considered the material's crack initiation fracture energy. Additionally, J_{IF} is introduced, defined as the energy localized in the plastic zone at the crack tip, marking the point of stable crack growth. The specific relationships are as follows:

$J = J_{IC}$, The formation of first macroscopic crack,

$J_{IC} < J < J_{IF}$, The stage of stable crack growth,

$J = J_{IF}$, Termination point of stable crack growth,

$J > J_{IF}$, Unstable crack propagation stage of the material.

J_{IC} and J_{IF} are calculated by Equations (4) and (5):

$$J_{IC} = \frac{2S_{IC}}{A_{lig}} \tag{4}$$

$$J_{IF} = \frac{2S_{IF}}{A_{lig}} \tag{5}$$

where S_{IC} and S_{IF} represent the areas under the P-δ curve of P-EHDC from the origin to the crack initiation load P_{ini}, and from the origin to the failure load P_{IF}, respectively.

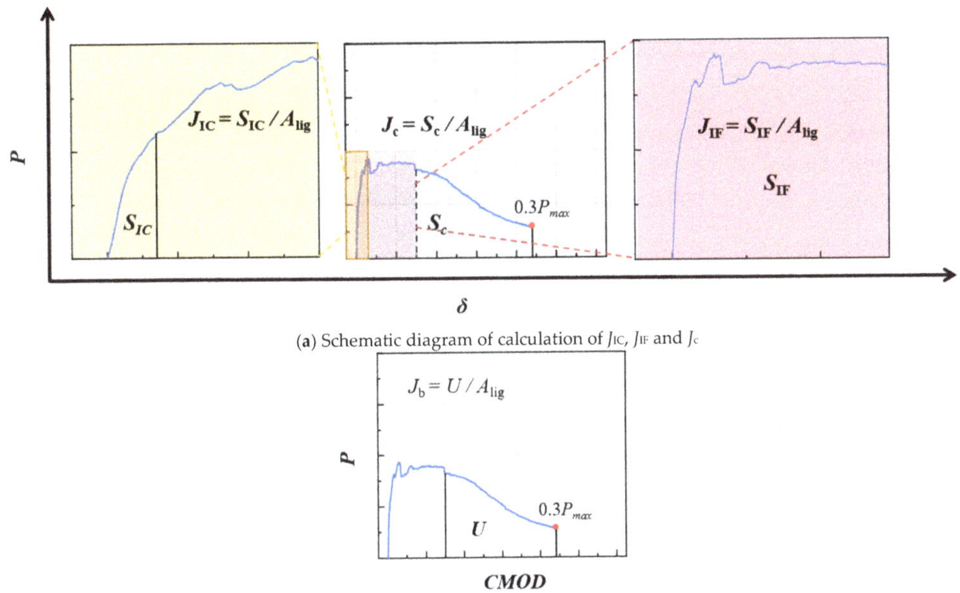

(**a**) Schematic diagram of calculation of J_{IC}, J_{IF} and J_c

(**b**) Schematic diagram of calculation of J_b

Figure 11. Schematic diagram of calculation of fracture energy based on *J*-integral method.

Given that P-EHDC is essentially a composite material—particularly due to the significant difference in properties between its fiber and matrix components—this study introduces a *J*-integral-based fracture energy analysis method to better investigate the respective contributions of the matrix and fibers to fracture behavior and energy dissipation throughout the entire cracking process. This approach aims to provide deeper insights into the underlying toughening mechanisms and enrich the understanding of the material's fracture characteristics.

Fiber-reinforced high-ductility composites can transfer stress at the crack tip to the surrounding area through fibers, leading to the propagation of microcracks. As a result, the total energy absorbed during fracture *J* consists of two parts: the energy from the fiber pull-out process on the main fracture surface (fiber-bridging fracture energy, J_b) and the energy from the microcracking process (energy associated with crack surface out-of-plane fracture, J_m) [55]. To evaluate the toughening mechanism of PET powder in P-EHDC, this study calculates the composite total fracture energy J_c, the total out-of-plane fracture energy J_m, and the energy consumed by fiber bridging within the crack surface J_b. Figure 11 provides a schematic of the fracture energy calculations based on the *J*-integral. It is particularly noted that, due to the good ductility of P-EHDC, it is difficult to load the specimen to complete fracture (i.e., to zero load). Therefore, $0.3P_{max}$ is used as the calculation endpoint, which is consistent with the calculation rule for Mode I fracture energy G_F. The relationships and calculation methods for the composite total fracture energy J_c, total out-of-plane fracture energy J_m, and energy consumed by fiber bridging within the crack surface J_b are as follows: [55]

$$J_c = J_m + J_b \tag{6}$$

$$J_c = \frac{2S_c}{A_{lig}} \tag{7}$$

$$J_b = \frac{U}{A_{lig}} \tag{8}$$

The determination of Equation (4) is based on the research conducted by the ASTM-24 committee [44,56] on three-point bending fracture specimens, where S_c represents the area under the *P-δ* curve of the specimen. Equation (5) calculates the fracture energy J_b consumed by fiber bridging using the post-peak area U from the *P-CMOD* curve.

Figure 12 and Table 5 present the variation of J_{IC} and J_{IF} for P-EHDC calculated based on the *J*-integral method, along with the specific data. As shown in Figure 12, the crack initiation fracture energy J_{IC} is relatively small compared to the failure fracture energy J_{IF} for all specimen groups, indicating that the first stage of fracture behavior in all groups was not fully developed. As the PET powder content increases, the overall trend for J_{IC} in the B1 series is a decrease. However, for the B2 and B3 series, the J_{IC} values initially increase and then decrease, with the peak values observed for B2P15 and B3P30. Notably, B2P15 exhibits the highest J_{IC} among all test groups with the same precursor ratio and PET powder replacement ratio, reaching 0.194 kN/m^2, which is similar to B1P0. The failure fracture energy J_{IF} is similar to the Mode I fracture energy, as both are determined by the area under the *P-δ* curve. In this study, J_{IF} shows a significant response to PET powder replacement rates. A small amount of PET powder increases J_{IF}, as seen in B1P15, B2P15, and B3P30. However, when the PET powder replacement ratio exceeds 30%, J_{IF} decreases significantly. Specifically, the lowest J_{IF} values across different precursor ratio groups are observed at a 45 vol% PET powder replacement ratio, with values of 4.83 kN/m^2, 4.49 kN/m^2, and 4.44 kN/m^2, respectively, representing reductions of 39.8%, 45.4%, and 42.0% compared to the failure fracture energies of B1P15, B2P15, and B3P30. On the other hand, changes in matrix strength affected the trend of J_{IF} variation. In groups with different matrix strengths, the specimen with the highest J_{IF} does not have the same PET powder replacement ratio. The highest J_{IF} = 8.24 kN/m^2 was also observed in B2P15.

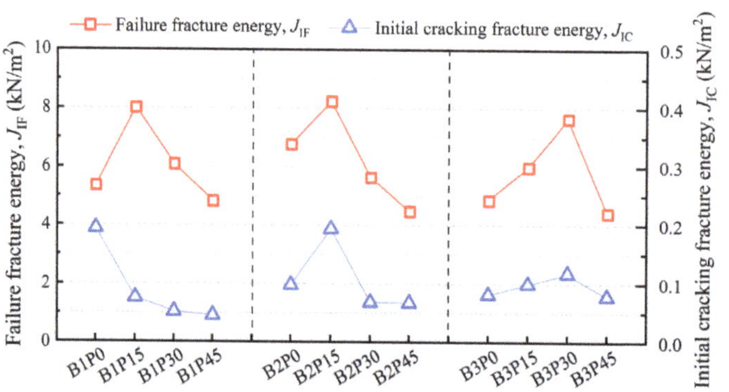

Figure 12. Effect of precursor matrix and PET powder replacement ratio on J_{IC} and J_{IF}.

J_{IC} and J_{IF} represent the initiation and failure critical points of P-EHDC, respectively. In Figure 12, the curve of the initiation fracture energy J_{IC} shows visible fluctuations, indicating that increasing the PET powder replacement ratio to 45 vol% can reduce the initiation fracture energy J_{IC} to a lower level. On the other hand, changing the GGBS content in the precursor (from B1 to B3) at the same PET powder replacement ratio does not show a clear trend regarding its impact on J_{IC}. From the values and the range of change in J_{IC}, it can be seen that lower GGBS content and higher PET powder content are detrimental to the development of J_{IC}. Su et al. [38,52] pointed out that the load before cracking is primarily borne by the matrix in fiber-reinforced concrete. Therefore, a low GGBS content and high PET powder content in the precursor are not conducive to the crack initiation strength of the matrix. Carpinteri et al. [57] indicated that the role of fibers in fiber-reinforced concrete is to bridge cracks in the matrix, preventing crack propagation until fiber debonding, pullout, or failure occurs. The mechanism of fiber action in P-EHDC is similar in this regard.

Table 5. Calculation results of fracture energy based on J integral method.

Mix IDs	J_{IC} (kJ/m^2)	J_{IF} (kJ/m^2)	J_c (kJ/m^2)	J_b (kJ/m^2)	J_m (kJ/m^2)
B1-P0	0.194 (0.041)	5.35 (0.99)	18.3 (1.6)	6.48 (0.68)	11.8 (0.9)
B1-P15	0.0757 (0.0455)	8.02 (1.19)	21.7 (1.1)	9.44 (0.99)	12.3 (0.9)
B1-P30	0.0524 (0.0031)	6.09 (1.22)	15.9 (2.5)	6.10 (1.61)	9.85 (0.88)
B1-P45	0.0456 (0.0209)	4.83 (0.088)	20.2 (1.6)	5.77 (0.15)	15.9 (2.4)
B2-P0	0.0985 (0.0476)	6.77 (1.88)	17.8 (0.6)	7.61 (1.40)	10.2 (0.9)
B2-P15	0.194 (0.068)	8.24 (2.32)	20.2 (2.3)	8.49 (0.69)	11.7 (2.0)
B2-P30	0.0689 (0.0217)	5.64 (1.63)	16.4 (3.7)	8.29 (0.15)	9.04 (2.14)
B2-P45	0.0677 (0.0107)	4.48 (0.82)	14.6 (3.6)	6.73 (1.43)	7.84 (2.21)
B3-P0	0.0816 (0.0180)	4.86 (1.32)	18.0 (0.8)	8.93 (1.15)	9.08 (1.09)
B3-P15	0.100 (0.034)	5.99 (1.81)	16.3 (4.8)	7.23 (2.11)	9.11 (2.11)
B3-P30	0.118 (0.022)	7.65 (0.18)	18.7 (1.0)	7.16 (0.80)	11.5 (0.3)
B3-P45	0.0783 (0.0301)	4.44 (1.16)	14.6 (2.2)	6.27 (0.48)	8.35 (2.43)

Note: the values are standard deviations in parentheses.

On the other hand, the failure fracture energy J_{IF} is considered to represent the end-point of crack stable development and the onset of material failure. Moreover, the level of J_{IF} reflects the energy borne by fibers in bridging cracks effectively in the matrix. The trend shown in Figure 12 indicates that J_{IF} initially increases and then decreases as the PET powder replacement ratio increases. Hou et al. [58] pointed out that rubber particles not only reduce the toughness of the matrix but also play a bridging role to some extent. The synergistic effect of both factors leads to a reduction in the crack width of HDC. The role of PET powder should be similar to that of rubber particles. The increase in J_{IF} with PET powder substitution is mainly due to its own bridging effect and the creation of more microcracks as the matrix fracture toughness decreases. This also helps fibers bridge more effectively and absorb more energy. However, this positive synergistic effect is limited to a certain range of PET powder replacement ratios, which, in this study, ranges from 15 vol% to 30 vol%.

Figure 13 shows the effect of PET powder substitution ratio on J_m, J_b, and J_c in different precursor matrices. According to the calculation assumptions of the J-integral and the method presented in Equation (7), J_c represents the critical point of P-EHDC's final failure. It includes the total fracture energy from the first four stages of P-EHDC's fracture behavior and the energy consumed during crack instability and development. It can be directly observed that the trend of J_c across all test groups is consistent with the trend of G_F shown in Figure 10. The highest values of J_c in the three precursor series were found at PET powder replacement ratios of 15 vol%, 15 vol%, and 30 vol%, with values of 21.7 kN/m^2, 20.2 kN/m^2, and 18.7 kN/m^2, respectively. On the other hand, the J_c variation in the B1 and B3 series exhibited two peaks, which is different from the B2 series. The lowest values of the composite fracture energy J_c for each series were found in B1P30, B2P45, and B3P45, at 15.9 kN/m^2, 14.6 kN/m^2, and 14.6 kN/m^2, respectively. These are 26.7%, 25.8%, and 21.9% lower than the highest values. Moreover, the change in precursor ratio did not significantly affect the composite fracture energy J_c in the B1 and B3 series.

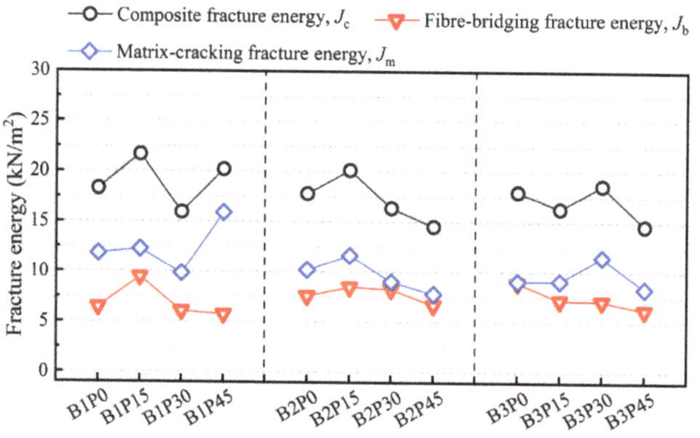

Figure 13. Effect of precursor matrix and PET powder replacement ratio on J_m, J_b and J_c.

J_b and J_m represent the energy consumed by fiber bridging within the crack surface and the total fracture energy outside the crack surface, respectively. J_b is calculated using Equation (8) and is primarily related to the crack instability and expansion phase. As shown in Figure 13, the energy consumed by fiber bridging within the crack surface (J_b) is the smallest among the three physical quantities. In the B1 and B2 series, J_b initially increases and then decreases with the increase in PET powder replacement ratio, while in the B3 series, J_b decreases progressively. The variation of total fracture energy outside the crack surface (J_m) with the PET powder replacement ratio follows the same trend as the composite fracture energy J_c.

The calculated results for J_b above indicate that when the precursor ratio of GGBS:FA is 4:6 and 3:7, a 15 vol% PET powder replacement enhances the energy dissipated by fiber pull-out and fracture within the crack surface. The energy consumed by fiber bridging (J_b) is related to the bond between the fibers and the matrix. The bonding condition influences whether the fiber will pull out or fracture during the failure process. Fibers that pull out from the matrix dissipate more energy than those that fracture. In the B1 series, the B1P0 specimen, which contains no PET powder, has a more compact matrix interface. When the specimen enters stage VI of the fracture behavior phase, most of the fibers fail due to strong bonding with the matrix, as evidenced by the steep slope of the descending section of the *P-CMOD* curve in Figure 8. After replacing quartz powder with 15 vol% PET powder, the surface energy of the PET powder (40–50 mN/m) is lower than the surface tension of water (72 mN/m), and the hydrophobic properties of PET powder weaken the compactness of the matrix interface. This reduces the bonding between the matrix and fibers, allowing more fibers to pull out from the matrix.

However, in the B3 series, the low GGBS content in the precursor results in a weak matrix [47,48]. Even without PET powder (0 vol%), the matrix cannot form enough fiber bridging. As a result, J_b decreases with increasing PET powder substitution. On the other hand, the incorporation of PET powder makes the matrix more prone to cracking. When the fibers can no longer absorb more fracture energy within the crack surface, the easily cracked matrix contributes to an increase in J_m, which in turn enhances the composite fracture energy. This is reflected in the decrease of J_b and the increase of J_m in both B1P45 and B3P30, leading to an overall increase in J_c. Similar findings were also supported by Su et al. [38] in their study of hydrophobic aggregates. Similarly, the reduction in matrix strength also plays a similar role.

3.5. The Analysis of Microstructure

Considering the significant effect of PET powder on the fracture energy of P-EHDC, Figure 14 presents SEM images of the microstructure at the interface between the aggregate and the matrix for both PET powder-free and PET powder-incorporated P-EHDC materials, focusing on the interface density. From Figure 14a, it is evident that the boundary between the quartz powder particles and the surrounding matrix is rather indistinct, indicating a dense bond between them. In contrast, Figure 14b clearly shows an obvious transition zone between the PET particles and the surrounding matrix, resembling a trench. This further suggests the presence of a distinct interfacial transition zone (ITZ) between the PET powder and the cement matrix, which appears more readily identifiable at the same magnification level than that between QP and the matrix. This phenomenon occurs because PET, as a hydrophobic material, naturally creates a less cohesive interface with the matrix. Notably, by comparing the overall interface density of the materials at the same observation scale, it is clear that the interface in the PET-free specimen is much denser. In Figure 14b, numerous fine cracks can be observed around the PET particles in the matrix, indicating that the incorporation of PET weakens the matrix strength and promotes cracking. This observation helps explain the increase in Jm with higher PET powder replacement rates, as seen in B1P45 and B3P30 in Section 3.4. Even at relatively low levels, PET powder can enhance the synergy with fibers, contributing to the material's toughness.

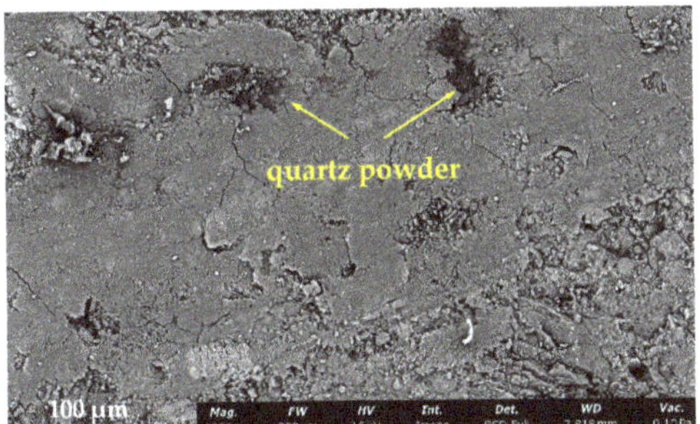

(**a**)

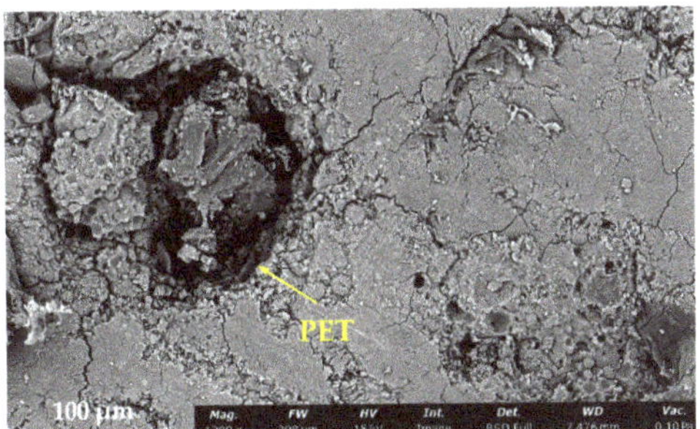

(**b**)

Figure 14. Microstructure of P-EHDC: (**a**) microstructure without incorporation of the PET powder; (**b**) microstructure with incorporation of the PET powder.

4. Conclusions

This study experimentally investigated the fracture behavior of P-EHDC under different precursor ratios and PET powder replacement ratios. The Mode I fracture energy was calculated to evaluate its energy absorption capacity, and the double-J fracture criterion was used to calculate J_{IC} and J_{IF} for assessing the fracture performance of P-EHDC. Based on the J-integral method, the total composite fracture energy J_c, the energy consumed by fiber bridging within the crack J_b, and the total fracture energy outside the crack J_m were calculated. Additionally, through microstructural observations, the influence mechanisms of precursor ratio and PET powder replacement ratio on the fracture performance of P-

EHDC were comprehensively studied, along with the toughening principles. The following conclusions were drawn:

(1) All P-EHDC specimens in this study exhibited "ductile fracture" behavior, with the final failure mode characterized by a main crack accompanied by some microcracks. The increase in PET powder replacement ratio led to a more tortuous path for the main crack but also reduced the initiation of surrounding microcracks to some extent. PET replacement altered the matrix properties, and the tortuous main crack path suggests unevenly distributed defects in the matrix. This should not be regarded as a negative outcome, as the tortuous crack path of the matrix suggests greater fracture energy absorption.

(2) Compared to the precursor ratio, the PET powder replacement ratio has a more significant effect on Mode I fracture energy (G_F). The presence of PET powder promotes cracking in the material, leading to the formation of more fine microcracks around the main crack. The reduction in matrix strength has the same effect, this is also the reason why the main crack path propagates more tortuously.

(3) The hydrophobic PET powder weakens the overall density of the matrix, promoting material cracking. While PET powder itself has a certain bridging effect, excessive usage can still weaken the fiber-matrix bridging. The proportion of GGBS in the precursor affects the matrix strength—higher matrix strength tends to anchor fibers, leading to fiber breakage, while lower matrix strength allows more fibers to pull out and absorb fracture energy. In the B1 series, with a high GGBS proportion, replacing 15 vol% of quartz powder with PET powder lowers the crack initiation fracture energy (J_{IC}). However, the PET powder's bridging effect and reduced fracture toughness create more microcracks, which increases the fracture energy at failure (J_{IF}). Despite the relatively low replacement ratio, the matrix's crack initiation threshold does not significantly decrease, but it allows more fibers to pull out and absorb fracture energy during the crack instability phase. Therefore, compared to B1P0, B1P15 shows a 45.7% increase in J_b and a 4.24% increase in J_m.

(4) Based on the findings of this study, it is recommended to use a higher matrix strength with no more than 45 vol% PET powder as a replacement for quartz powder to prepare P-EHDC. This approach can improve the ductile fracture behavior and fracture energy of the material while enhancing the recycling of waste PET powder and promoting the green and low-carbon development of high-performance new materials. Admittedly, the study did not fully integrate the surface crack propagation process of the specimens with the J-integral-based fracture analysis. Future research could employ high-resolution imaging techniques to capture the detailed evolution of surface cracks, enabling correlation with fracture energy parameters derived from the J-integral method. This approach may contribute to the development of a more suitable framework for analyzing and evaluating the fracture behavior of high-ductility materials.

Author Contributions: Conceptualization, F.M. and S.L.; methodology, J.S.; software, C.Z.; validation, C.Z., L.X. and L.Z.; formal analysis, F.Q.; investigation, J.Z.; resources, R.L.; data curation, R.L.; writing—original draft preparation, F.M.; writing—review and editing, S.L. and R.L.; visualization, R.L.; supervision, Y.G. and R.L.; project administration, Y.G. and R.L.; funding acquisition, Y.G. All authors have read and agreed to the published version of the manuscript.

Funding: This research was funded by National Natural Science Foundation of China, grant number 12272099.

Institutional Review Board Statement: Not applicable.

Informed Consent Statement: Not applicable.

Data Availability Statement: The data presented in this study are available on request from the corresponding author. The data are not publicly available due to confidentiality issues.

Conflicts of Interest: Fei Meng, Shen Luo, Jingxian Sun, Cheng Zhang, Leilei Xu, and Liqun Zhang are employed by the Zhongshan Power Supply Bureau of Guangdong Power Grid Co., Ltd., Fumin Qing is employed by the Fujian Southeast Design Group Co., Ltd. The authors declare that this study received funding from the National Natural Science Foundation of China. The research was conducted autonomously, ensuring the integrity and objectivity of the findings.

References

1. Wang, B.; Li, Q.; Liu, F.; Wang, J.; Xu, S. Shear bond assessment of UHTCC repair using push-out test. *Constr. Build. Mater.* **2018**, *164*, 206–216. [CrossRef]
2. Yu, J.; Li, H.; Leung, C.K.Y.; Lin, X.; Lam, J.Y.K.; Sham, I.M.L.; Shih, K. Matrix design for waterproof Engineered Cementitious Composites (ECCs). *Constr. Build. Mater.* **2017**, *139*, 438–446. [CrossRef]
3. Deng, M.; Ma, F.; Ye, W.; Li, F. Flexural behavior of reinforced concrete beams strengthened by HDC and RPC. *Constr. Build. Mater.* **2018**, *188*, 995–1006. [CrossRef]
4. Redon, C.; Li Victor, C.; Wu, C.; Hoshiro, H.; Saito, T.; Ogawa, A. Measuring and Modifying Interface Properties of PVA Fibers in ECC Matrix. *J. Mater. Civ. Eng.* **2001**, *13*, 399–406. [CrossRef]
5. Ding, Y.; Yu, K.-Q.; Yu, J.-t.; Xu, S.-l. Structural behaviors of ultra-high performance engineered cementitious composites (UHP-ECC) beams subjected to bending-experimental study. *Constr. Build. Mater.* **2018**, *177*, 102–115. [CrossRef]
6. Zhou, Y.; Zhong, Q.; Xing, F.; Sui, L.; Huang, Z.; Guo, M. Influence of cyclic loading on the tensile fracture characteristics of ultra-high performance engineered cementitious composites. *Constr. Build. Mater.* **2020**, *240*, 117937. [CrossRef]
7. Ouyang, J.; Guo, R.; Wang, X.-Y.; Fu, C.; Wan, F.; Pan, T. Effects of interface agent and cooling methods on the interfacial bonding performance of engineered cementitious composites (ECC) and existing concrete exposed to high temperature. *Constr. Build. Mater.* **2023**, *376*, 131054. [CrossRef]
8. Jiang, B.; Qian, Z.; Gu, D.; Pan, J. Repair concrete structures with high-early-strength engineered cementitious composites (HES-ECC): Material design and interfacial behavior. *J. Build. Eng.* **2023**, *68*, 106060. [CrossRef]
9. Hou, M.; Zhang, D.; Li, V.C. Crack width control and mechanical properties of low carbon engineered cementitious composites (ECC). *Constr. Build. Mater.* **2022**, *348*, 128692. [CrossRef]
10. Luo, J.; Wu, G.; Zhao, G.; Ma, Y.; Fang, Z.; Fang, S. Experimental and numerical analysis on shear performance of single embedded nut bolted shear connectors in prefabricated steel-UHPC composite structures under cyclic loading. *Structures* **2025**, *73*, 108446. [CrossRef]
11. Feiz, R.; Ammenberg, J.; Baas, L.; Eklund, M.; Helgstrand, A.; Marshall, R. Improving the CO_2 performance of cement, part I: Utilizing life-cycle assessment and key performance indicators to assess development within the cement industry. *J. Clean. Prod.* **2015**, *98*, 272–281. [CrossRef]
12. Chen, G.; Zhuo, K.-X.; Luo, R.-H.; Lai, H.-M.; Cai, Y.-J.; Xie, B.-X.; Lin, J.-X. Fracture behavior of environmentally friendly high-strength concrete using recycled rubber powder and steel fibers: Experiment and modeling. *Case Stud. Constr. Mater.* **2024**, *21*, e03501. [CrossRef]
13. Xiong, Z.; He, S.H.; Kwan, A.K.H.; Li, L.G.; Zeng, Y. Compressive behaviour of seawater sea-sand concrete containing glass fibres and expansive agents. *Constr. Build. Mater.* **2021**, *292*, 123309. [CrossRef]
14. Cardinale, T.; D'Amato, M.; Sulla, R.; Cardinale, N. Mechanical and Physical Characterization of Papercrete as New Eco-Friendly Construction Material. *Appl. Sci.* **2021**, *11*, 1011. [CrossRef]
15. Sharma, B.; Gatóo, A.; Bock, M.; Ramage, M. Engineered bamboo for structural applications. *Constr. Build. Mater.* **2015**, *81*, 66–73. [CrossRef]
16. Ding, Y.; Dai, J.-G.; Shi, C.-J. Mechanical properties of alkali-activated concrete: A state-of-the-art review. *Constr. Build. Mater.* **2016**, *127*, 68–79. [CrossRef]
17. Chen, G.; Liu, L.-Y.; Li, J.-F.; Lin, J.-X.; Guo, Y.-C. In-situ tensile behavior of hybrid PE/PP fibers reinforced strain-hardening alkali-activated composites (H-SHAAC) subjected to different temperatures. *J. Build. Eng.* **2024**, *98*, 111332. [CrossRef]
18. Chen, G.; Zheng, D.-p.; Chen, Y.-w.; Lin, J.-X.; Lao, W.-j.; Guo, Y.-c.; Chen, Z.-b.; Lan, X.-w. Development of high performance geopolymer concrete with waste rubber and recycle steel fiber: A study on compressive behavior, carbon emissions and economical performance. *Constr. Build. Mater.* **2023**, *393*, 131988. [CrossRef]
19. Amran, M.; Al-Fakih, A.; Chu, S.H.; Fediuk, R.; Haruna, S.; Azevedo, A.; Vatin, N. Long-term durability properties of geopolymer concrete: An in-depth review. *Case Stud. Constr. Mater.* **2021**, *15*, e00661. [CrossRef]

20. Ganesan, N.; Abraham, R.; Deepa Raj, S. Durability characteristics of steel fibre reinforced geopolymer concrete. *Constr. Build. Mater.* **2015**, *93*, 471–476. [CrossRef]
21. Ariffin, M.A.M.; Bhutta, M.A.R.; Hussin, M.W.; Mohd Tahir, M.; Aziah, N. Sulfuric acid resistance of blended ash geopolymer concrete. *Constr. Build. Mater.* **2013**, *43*, 80–86. [CrossRef]
22. Pan, H.; Xie, Z.; Chen, G.; Su, J.; Zhuo, K.; Chen, Z.; Lin, J.; Feng, C.; Guo, Y. Dynamic compressive behavior of high-strength engineered geopolymer composites. *J. Build. Eng.* **2023**, *80*, 108036. [CrossRef]
23. Chen, Z.; Pan, H.; Zhuo, K.; Su, J.; Xie, B.; Lin, J.; Guo, Y. Dynamic compressive behavior of environmentally friendly high-strength concrete: Experimental investigation and modelling. *Constr. Build. Mater.* **2024**, *427*, 136259. [CrossRef]
24. Nematollahi, B.; Sanjayan, J.; Qiu, J.; Yang, E.-H. Micromechanics-based investigation of a sustainable ambient temperature cured one-part strain hardening geopolymer composite. *Constr. Build. Mater.* **2017**, *131*, 552–563. [CrossRef]
25. Yang, K.-H.; Song, J.-K.; Song, K.-I. Assessment of CO2 reduction of alkali-activated concrete. *J. Clean. Prod.* **2013**, *39*, 265–272. [CrossRef]
26. Bamigboye, G.; Tarverdi, K.; Adigun, D.; Daniel, B.; Okorie, U.; Adediran, J. An appraisal of the mechanical, microstructural, and thermal characteristics of concrete containing waste PET as coarse aggregate. *Clean. Waste Syst.* **2022**, *1*, 100001. [CrossRef]
27. Bamigboye, G.O.; Ngene, B.U.; Ademola, D.; Jolayemi, J.K. Experimental Study on the Use of Waste Polyethylene Terephthalate (PET) and River Sand in Roof Tile Production. *J. Phys. Conf. Ser.* **2019**, *1378*, 042105. [CrossRef]
28. Alqahtani, F.; Ghataora, G.; Iqbal, K.; Dirar, S.; Azzedine, K.; Mansour, A.-O. Lightweight Concrete Containing Recycled Plastic Aggregates. In Proceedings of the 2015 International Conference on Electromechanical Control Technology and Transportation, Zhuhai, China, 31 October—1 November 2015.
29. Alqahtani, F.K.; Sherif, M.A.; Ghanem, A.M. Green lightweight concrete utilizing sustainable processed recycled plastic aggregates: Technical, economic and environmental assessment. *Constr. Build. Mater.* **2023**, *393*, 132027. [CrossRef]
30. Alqahtani, F.K.; Ghataora, G.; Khan, M.I.; Dirar, S. Novel lightweight concrete containing manufactured plastic aggregate. *Constr. Build. Mater.* **2017**, *148*, 386–397. [CrossRef]
31. Yang, S.; Yue, X.; Liu, X.; Tong, Y. Properties of self-compacting lightweight concrete containing recycled plastic particles. *Constr. Build. Mater.* **2015**, *84*, 444–453. [CrossRef]
32. Khan, S.U.; Ayub, T. PET Fiber–Reinforced Engineered Geopolymer and Cementitious Composites. *J. Mater. Civ. Eng.* **2022**, *34*, 06021010. [CrossRef]
33. Tayeh, B.A.; Almeshal, I.; Magbool, H.M.; Alabduljabbar, H.; Alyousef, R. Performance of sustainable concrete containing different types of recycled plastic. *J. Clean. Prod.* **2021**, *328*, 129517. [CrossRef]
34. Olofinnade, O.; Chandra, S.; Chakraborty, P. Recycling of high impact polystyrene and low-density polyethylene plastic wastes in lightweight based concrete for sustainable construction. *Mater. Today Proc.* **2021**, *38*, 2151–2156. [CrossRef]
35. Chithambar Ganesh, A.; Deepak, N.; Deepak, V.; Ajay, S.; Pandian, A.; Karthik. Utilization of PET bottles and plastic granules in geopolymer concrete. *Mater. Today Proc.* **2021**, *42*, 444–449. [CrossRef]
36. Gao, S.; Qi, L.; Zhu, Y.; Wang, W. Effect of notch depth ratio on mode I and mixed mode I-II fracture properties of engineered cementitious composites (ECC). *Int. J. Solids Struct.* **2022**, *236–237*, 111363. [CrossRef]
37. Cheng, Z.; Wang, P.; Jia, X.; Li, Z.; Pan, J.; Zhou, J. Crack development and fracture performance of Recycled Powder Engineered Cementitious Composites (ECC): Experimental investigation. *Constr. Build. Mater.* **2024**, *429*, 136388. [CrossRef]
38. Su, J.-Y.; Luo, R.-H.; Chen, Z.-B.; Lin, J.-X.; Huang, P.-Y.; Guo, Y.-C. Experimental study on the fracture performance of rubberized high strength-high ductility concrete with J-integral method. *Constr. Build. Mater.* **2024**, *421*, 135668. [CrossRef]
39. Xu, S.; Reinhardt, H.W. A simplified method for determining double-K fracture parameters for three-point bending tests. *Int. J. Fract.* **2000**, *104*, 181–209. [CrossRef]
40. Xu, S.; Reinhardt, H.W. Determination of double-K criterion for crack propagation in quasi-brittle fracture, Part II: Analytical evaluating and practical measuring methods for three-point bending notched beams. *Int. J. Fract.* **1999**, *98*, 151–177. [CrossRef]
41. Xu, S.; Reinhardt, H.W. Determination of double-Determination of double-K criterion for crack propagation in quasi-brittle fracture Part I: Experimental investigation of crack propagation. *Int. J. Fract.* **1999**, *98*, 111–149. [CrossRef]
42. Xu, S.; Lyu, Y.; Xu, S.; Li, Q. Enhancing the initial cracking fracture toughness of steel-polyvinyl alcohol hybrid fibers ultra high toughness cementitious composites by incorporating multi-walled carbon nanotubes. *Constr. Build. Mater.* **2019**, *195*, 269–282. [CrossRef]
43. Chen, Z.-B.; Su, J.-Y.; Luo, R.-H.; Lin, J.-X.; Chen, X.-F.; Huang, P.-Y.; Guo, Y.-C. Experimental investigation on fracture behavior of rubberized engineered cementitious composites under thermo-mechanical loadings. *Eng. Fract. Mech.* **2025**, *315*, 110838. [CrossRef]
44. Liu, W.; Xu, S.; Li, Q. Experimental study on fracture performance of ultra-high toughness cementitious composites with J-integral. *Eng. Fract. Mech.* **2012**, *96*, 656–666. [CrossRef]

45. Peng, Y.-Q.; Zheng, D.-P.; Pan, H.-S.; Yang, J.-L.; Lin, J.-X.; Lai, H.-M.; Wu, P.-Z.; Zhu, H.-Y. Strain hardening geopolymer composites with hybrid POM and UHMWPE fibers: Analysis of static mechanical properties, economic benefits, and environmental impact. *J. Build. Eng.* **2023**, *76*, 107315. [CrossRef]
46. Sanyin, Z.; Qijun, Y.; Fei, Q.; Jiaqi, H.; Suhong, Y.; Ziyun, W.; Guobang, G. Setting and strength characteristics of alkali-activated carbonatite cementitious materials with ground slag replacement. *J. Wuhan Univ. Technol. Mater. Sci. Ed.* **2006**, *21*, 125–128. [CrossRef]
47. Swathi, B.; Vidjeapriya, R. Influence of precursor materials and molar ratios on normal, high, and ultra-high performance geopolymer concrete—A state of art review. *Constr. Build. Mater.* **2023**, *392*, 132006. [CrossRef]
48. Zhang, H.Y.; Qiu, G.H.; Kodur, V.; Yuan, Z.S. Spalling behavior of metakaolin-fly ash based geopolymer concrete under elevated temperature exposure. *Cem. Concr. Compos.* **2020**, *106*, 103483. [CrossRef]
49. *JCI-S-001-2003*; Method of Test for Fracture Energy of Concrete by Use of Notched Beam, Materials and Structures. JCI Standard: Tokyo, Japan, 2003.
50. Su, J.-Y.; Chen, G.; Pan, H.-S.; Lin, J.-X.; Zhang, J.; Zhuo, K.-X.; Chen, Z.-B.; Guo, Y.-C. Rubber modified high strength-high ductility concrete: Effect of rubber replacement ratio and fiber length. *Constr. Build. Mater.* **2023**, *404*, 133243. [CrossRef]
51. Lin, J.-X.; Luo, R.-H.; Su, J.-Y.; Guo, Y.-C.; Chen, W.-S. Coarse synthetic fibers (PP and POM) as a replacement to steel fibers in UHPC: Tensile behavior, environmental and economic assessment. *Constr. Build. Mater.* **2024**, *412*, 134654. [CrossRef]
52. Lin, J.-X.; Chen, G.; Pan, H.-S.; Wang, Y.-C.; Guo, Y.-C.; Jiang, Z.-X. Analysis of stress-strain behavior in engineered geopolymer composites reinforced with hybrid PE-PP fibers: A focus on cracking characteristics. *Compos. Struct.* **2023**, *323*, 117437. [CrossRef]
53. Wang, Y.; Chan, C.L.; Leong, S.H.; Zhang, M. Engineering properties of strain hardening geopolymer composites with hybrid polyvinyl alcohol and recycled steel fibres. *Constr. Build. Mater.* **2020**, *261*, 120585. [CrossRef]
54. Alrefaei, Y.; Dai, J.-G. Tensile behavior and microstructure of hybrid fiber ambient cured one-part engineered geopolymer composites. *Constr. Build. Mater.* **2018**, *184*, 419–431. [CrossRef]
55. Keck, S.; Fulland, M. Effect of fibre volume fraction and fibre direction on crack paths in flax fibre-reinforced composites. *Eng. Fract. Mech.* **2016**, *167*, 201–209. [CrossRef]
56. Heald, P.T.; Spink, G.M.; Worthington, P.J. Post yield fracture mechanics. *Mater. Sci. Eng.* **1972**, *10*, 129–138. [CrossRef]
57. Carpinteri, A.; Fortese, G.; Ronchei, C.; Scorza, D.; Vantadori, S. Mode I fracture toughness of fibre reinforced concrete. *Theor. Appl. Fract. Mech.* **2017**, *91*, 66–75. [CrossRef]
58. Hou, M.; Xu, K.; Monteiro, P.J.M.; Li, V.C. Rubber particle bridging effect on crack width control of low carbon Engineered Cementitious Composites (ECC). *Cem. Concr. Compos.* **2023**, *140*, 105106. [CrossRef]

materials

MDPI

Article

Seismic Performance of Beam–Column Joints in Seawater Sand Concrete Reinforced with Steel-FRP Composite Bars

Ruiqing Liang [1], Botao Zhang [1], Zhensheng Liang [1], Xiemi Li [2] and Shuhua Xiao [2,3,*]

[1] Guangzhou Electric Power Design Institute Co., Ltd., Guangzhou 510060, China; rq_liang0303@163.com (R.L.); botaoz2024@163.com (B.Z.); zhensheng_l@163.com (Z.L.)
[2] School of Civil and Transportation Engineering, Guangdong University of Technology, Guangzhou 510006, China; 18819487074@163.com
[3] School of Environment and Civil Engineering, Dongguan University of Technology, Dongguan 523080, China
* Correspondence: xiaoshuhua@dgut.edu.cn

Abstract: Steel fiber-reinforced polymer (FRP) composite bars (SFCBs) combine the ductility of steel reinforcement with the corrosion resistance and high strength of FRP, providing stable secondary stiffness that enhances the seismic resistance and safety of seawater sea–sand concrete structures. However, the seismic performance of SFCB-reinforced seawater sea–sand concrete beam–column joints remains underexplored. This study presents pseudo-static tests on SFCB-reinforced beam–column joints with varying column SFCB longitudinal reinforcement fiber volume ratios (64%, 75%, and 84%), beam reinforcement fiber volume ratios (60.9%, 75%, and 86%), and axial compression ratios (0.1 and 0.2). The results indicate that increasing the axial compression ratio enhances nodal shear capacity and bond strength, limits slip, and reduces crack propagation, but also accelerates bearing capacity degradation. Higher column reinforcement fiber volumes improve crack distribution and ductility, while beam reinforcement volume significantly affects energy dissipation and crack distribution, with moderate volumes (e.g., 75%) yielding optimal seismic performance. These findings provide insights for the seismic design of SFCB-composite-reinforced concrete structures in marine environments.

Keywords: Steel-FRP composite bar (SFCB); beam–column joints; seawater sand concrete; seismic performance; energy consumption

1. Introduction

With economic growth and the rising demand for construction, concrete usage has expanded significantly. However, extensive use of river sand negatively impacts river ecosystems, navigation, and flood control [1–3]. Sea sand, preferred for its low mud content, good grading, and excellent workability [4–6], shares a similar mineral composition with river sand, and seawater sea–sand concrete generally performs comparably to ordinary concrete, meeting engineering requirements [7]. Nevertheless, the chloride ions in seawater sand accelerate corrosion in steel reinforcement, which can compromise concrete structures, reduce their service life, and even cause the phenomenon of "seawater sand house" [8], limiting seawater sand concrete applications. Various corrosion mitigation methods, such as galvanizing, stainless steel, and epoxy coatings, have been tested but remain suboptimal [9]. Fiber-reinforced polymer (FRP) reinforcement, known for its strength and corrosion resistance, has gained attention as an alternative to steel in concrete structures [10]. However, pure FRP reinforcement increases brittleness and reduces deformation capacity, resulting in structural responses unlike conventional reinforced concrete (RC) [11]. To address these

 https://doi.org/10.3390/ma18102282

limitations, researchers have proposed steel-FRP composite bars (SFCB), which combine the strengths of steel and FRP, offering excellent mechanical properties with stable post-yield stiffness and high ductility [12–15].

SFCB combines the corrosion resistance of FRP reinforcement with the ductility and stable post-yield stiffness of steel [16], while also exhibiting excellent short- and long-term bonding with concrete [12,17,18], thereby enhancing overall structural performance. Dong et al. [19] found that the maximum bond strength between SFCB and sea–sand concrete increased by 14.9%, 17.4%, and 7.1% after 30, 60, and 90 days of seawater dry–wet cycling, respectively, supporting its potential for seawater sea–sand concrete applications [20]. In additional studies, Xiao et al. [21] reported that SFCB beams demonstrate strong ductility, with load-deflection curves that harden post-yield, and increase in ultimate load, bending moment, ductility, and energy absorption with thicker GFRP protective layers. Ge et al. [22] found that as the area ratio of SFCB inner core reinforcement increased, the deflection and crack width of RC columns decreased, while the ultimate load rose due to higher core reinforcement yield strength. Han et al. [23,24] observed that SFCB beams designed for equal stiffness showed similar damage modes to ordinary reinforced beams, with comparable deflections and controlled crack widths. However, they demonstrated a 14.3–54.5% higher load capacity and 47.0–189.1% increase in ductility coefficient. Ding et al. [25] highlighted the superior stiffness, load capacity, and crack control of SFCB-RC members compared to standard RC. Similarly, quasi-static tests by Sun et al. [26] on SFCB-reinforced precast concrete columns showed smaller residual drifts at equivalent lateral displacements. Su et al. [27] reported that, under low circumferential cyclic loading, SFCB beams had comparable cumulative energy consumption to conventional RC beams but showed enhanced seismic performance due to concrete stabilization restraints, achieving a 27.0% increase in load capacity and 36.5% in energy consumption. Additionally, Su et al. [28] developed an analytical model for long-term seismic performance based on bond strength degradation, confirming its accuracy experimentally. This model showed that the seismic performance of SFCB-RC beams in marine environments significantly degrades within the first year, stabilizing after three years of corrosion. Consequently, SFCB is regarded as a promising alternative to steel reinforcement in marine and related structural applications.

Special attention must be given to seismic effects when applying SFCBs, as maintaining the integrity of beam–column joints during earthquakes is crucial. Ensuring that joints exhibit ductile failure modes enhances structural seismic performance [29–31]. Extensive research on reinforced concrete frame joints over recent decades has led to theoretical models, computational methods, and design codes worldwide, promoting safety and reliability in structural design [32,33]. Various shear load capacity models have been developed, including shear friction, empirical equations, constraint, combined block, diagonal compression bar, and truss models [34]. A variety of studies have investigated the use of FRP for strengthening existing beam–column joints to enhance seismic performance. Typically, carbon fiber, glass fiber, or basalt fiber reinforced polymer sheets are applied circumferentially around the joint core to confine concrete deformation, thereby significantly improving the shear capacity and ductility of the joint [35–37]. However, the inherently brittle nature of FRP materials often leads to sudden interfacial debonding failures [38]. In comparison, the seismic performance of SFCB-reinforced joints is less studied. Xu [39] found that SFCB concrete joints with basalt or glass fiber wraps demonstrated superior seismic performance, such as improved energy consumption and ductility, maintaining high load capacity post-yield and exhibiting good secondary stiffness. Conversely, SFCB joints wrapped with carbon fibers had reduced ductility, which is less favorable for seismic resistance. Currently, research on the seismic behavior of SFCB-reinforced concrete joints remains limited, especially in fully replacing steel reinforcement. Key aspects such as force-transfer mechanisms, damage modes, energy consumption, ductility, stiffness, and stress–strain behavior of SFCB-reinforced seawater sea–sand concrete joints are not well understood, restricting the practical application of SFCBs. Thus, further study on the seismic performance of SFCB frame joints is essential.

In this paper, a test program involving six SFCB-reinforced seawater sea–sand concrete beam–column joint specimens was designed and implemented to assess the seismic performance of these new joints. The test parameters included axial compression ratio, column SFCB longitudinal reinforcement fiber volume rate, and beam SFCB longitudinal reinforcement fiber volume rate. The results are presented and discussed concerning damage modes, hysteresis curves, skeleton curves, energy consumption analysis, and beam–column end angles. This study aims to enhance the seismic performance data of SFCB-reinforced seawater sea–sand concrete structures and establish a solid foundation for the future application of SFCBs. This study systematically investigates the effects of SFCB parameters (e.g., fiber volume ratio, axial compression ratio) on the seismic performance of beam–column joints, aiming to address these gaps and provide design recommendations for marine applications.

2. Experimental Programs

2.1. Seawater Sea–Sand Concrete

In this study, C30 seawater sea–sand concrete was formulated with the following proportions: cement: sand: stone: fly ash: mineral powder: water: water reducer = 1:3.08: 4.09:0.215:0.215:0.67:0.031, resulting in a water–binder ratio of 0.47 and a sand rate of 43%. The specific dosages are detailed in Table 1. Ordinary silicate cement with a strength grade of 42.5 R was used. The sea sand, sourced from Zhanjiang, falls within grading area II, with a fineness modulus of 2.8, bulk density of 1520 kg/m^3, and apparent density of 2630 kg/m^3. Crushed stone was obtained from Qingyuan, and its performance indexes are shown in Table 2. The fly ash used was Grade II, with a mixing amount of 15%, a loss on ignition of 2.16%, a moisture content of 0.1%, and a water demand ratio of 101%. The mineral powder was S95 grade, with a dosage of 15%. A polycarboxylic acid water-reducing agent was employed at 2.2%, achieving a water reduction rate of 25.4%, a water ratio of 58%, a solids content of 10.88%, a density of 1.025 kg/cm^3, and a chloride ion content of 0.02%. Additionally, artificial seawater, formulated according to ASTM D1141-98 [40], was used, with the specific configuration ratio detailed in Table 3.

Table 1. Mix proportion of seawater sand concrete.

Cement (kg/m^3)	Sand (kg/m^3)	Stone (kg/m^3)	Fly Ash (kg/m^3)	Mineral Powder (kg/m^3)	Seawater (kg/m^3)	Water Reducer (kg/m^3)
252	779	1034	54	54	170	7.96

Table 2. Physical properties of crushed stone.

Particle Size (mm)	Flakiness and Elongation Index (%)	Apparent Density (kg/m^3)	Bulk Density (kg/m^3)	Fines Content (%)
5~25	4	2620	1510	0.40

The mechanical properties of the concrete were tested according to GB/T 50081-2019 [41] and ASTM C469 [42]. Three 150 mm cube samples and three cylinders were prepared for curing, followed by compressive testing using an MATEST-500t universal testing machine. The cubic and cylindrical compressive strengths of the concrete were found to be 32.59 MPa and 29.42 MPa, respectively, with a modulus of elasticity of 27.73 GPa.

Table 3. Artificial seawater composition mix ratio.

Component	Concentration (g/L)
NaCl	24.530
$MgCl_2$	5.200
Na_2SO_4	4.090
$CaCl_2$	1.160
KCl	0.695
$NaHCO_3$	0.201
KBr	0.101
H_3BO_3	0.027
$SrCl_2$	0.025
NaF	0.003

2.2. Mechanical Properties of SFCBs

The SFCB consists of an HRB400 steel core and an outer layer of GFRP. The HRB400 steel bars are available in diameters of 8 mm, 10 mm, and 12 mm, with the outer fiber layers having thicknesses of 6 mm, 5 mm, and 4 mm, respectively, as shown in Figure 1. All bars were produced or supplied by Shenzhen Haichuan New Materials Technology Co., Ltd. (Shenzhen, China). The fibers for the composite bars and GFRP bars are made from glass fibers produced by Chongqing International Composite Materials Co., Ltd. (Chongqing, China). The resin matrix used is unsaturated polyester resin. The composite bars are manufactured using pultrusion technology and reinforced with nylon fibers. Following the GB/T 30022-2013 [43], tensile tests were conducted on SFCBs. Seamless steel tubes served as anchors at both ends of the tendons, with expanded cement used as the filler material. Five identical specimens for each reinforcement material were prepared, and the sampling lengths are presented in Table 4. The tensile tests were performed using a material testing machine with a loading speed of 2 mm/min. The basic mechanical properties of the SFCB core reinforcement were evaluated in accordance with the GB/T 228.1-2010 [44]. Additionally, the load was measured by the force sensor of the testing machine, while the strain was measured using an extensometer and strain gauges at the center of the specimen. The extensometer is integrated with the tensile testing machine, and the strain gauges are supplied by Chengdu Siweide Technology Co., Ltd. (Chengdu, China). The load and strain data were collected using a TDS-530 static data acquisition system supplied by Tokyo Measuring Instruments Lab. (Tokyo, Japan). The mechanical properties of the bars are summarized in Table 5 and illustrated in Figure 2. When the inner-core steel bars yield, the yield strength of the SFCB is calculated by dividing the tensile force by the area of the composite bar, and therefore it is lower than the yield strength of the steel bars. The stress–strain curves of the SFCB exhibited distinct bipartite stiffness, and the damage mode was characterized by typical necking of the inner-core reinforcement following the failure of the outer fiber layer (Figure 3).

Table 4. Reinforcement tensile specimen size.

Specimens	Diameter d (mm)	Anchor Section Length l_1 (mm)	Working Section Length l_0 (mm)	Overall Length l (mm)
SFCB	16	200	350	750
SFCB	20	300	400	1000
GFRP	10	150	200	500

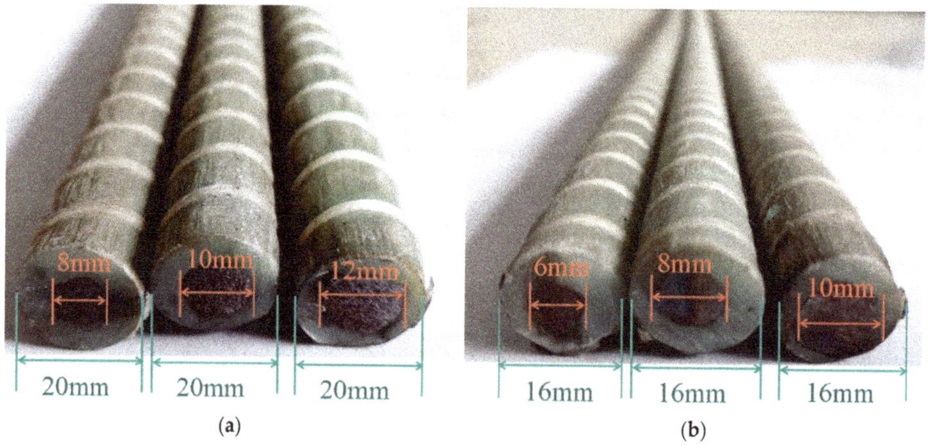

Figure 1. Longitudinal bars: (**a**) column longitudinal bars; (**b**) beam longitudinal bars.

Table 5. Mechanical properties of reinforcement.

Specimens [1]	Diameter d (mm)	Modulus of Elasticity E_s (GPa)	Secondary Rigidity E_s' (GPa)	Yielding Strength f_y (MPa)	Ultimate Strength f_u (MPa)
S6	6	206.21	/	559.28	759.31
S8	8	186.86	/	525.36	714.63
S10	10	202.33	/	523.73	649.38
S12	12	205.72	/	518.17	622.66
G10	10	44.75	/	/	770.82
S6G16	16	71.35	35.81	177.48	771.28
S8G16	16	84.22	32.19	211.76	746.75
S10G16	16	104.69	31.01	233.73	722.28
S8G20	20	64.92	34.71	189.06	811.66
S10G20	20	72.87	32.97	181.05	793.27
S12G20	20	90.15	22.39	235.23	785.68

[1] "S" represents steel, "G" denotes GFRP, and "S6G16" refers to an SFCB with an inner-core steel bar of 6 mm and an outer diameter of 16 mm. Other specimens are named similarly.

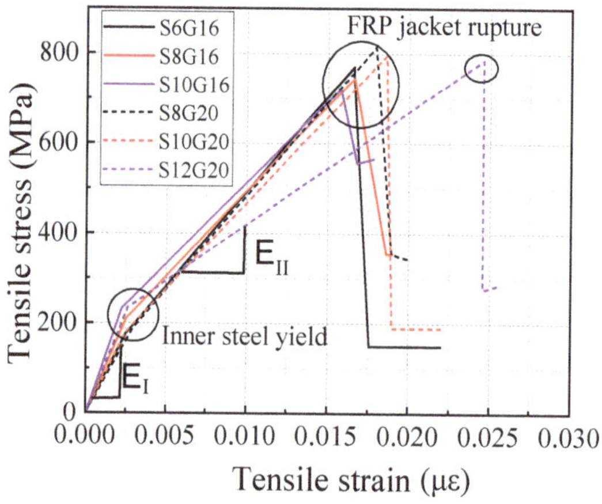

Figure 2. SFCB stress–strain curves.

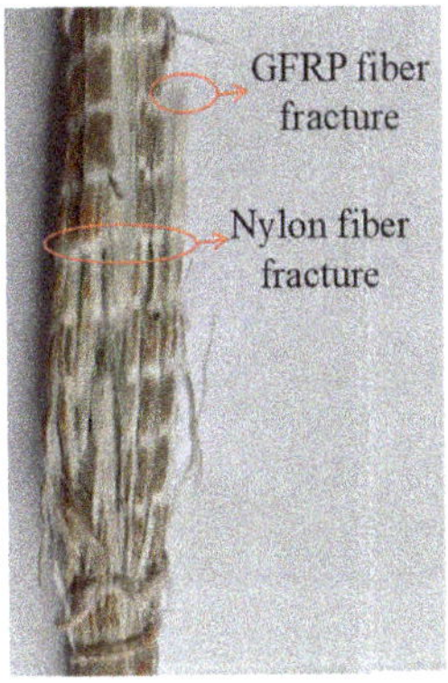

Figure 3. SFCB failure mode.

2.3. SFCB Reinforced Concrete Beam–Column Joints Tested

In this study, pseudo-static tests were conducted on column joints in SFCB-reinforced seawater sea–sand concrete structures. In frame structures, there is no bending moment at the counterbending points of beams and columns, and no vertical uniform load is applied to the beams during testing. Therefore, the section between the counterbending points of the beams and columns was selected as the test object to accurately reflect the structural response [45]. Following the GB 50011-2010 [46], the joints were designed according to the principle of "strong joints and weak members, strong columns and weak beams", resulting in three groups of six joint specimens. The longitudinal reinforcement of the beams consists of steel-FRP composite reinforcement with a diameter of 16 mm, while the column longitudinal reinforcement uses steel-FRP composite reinforcement with a diameter of 20 mm. The hoop reinforcement comprises glass fiber reinforcing bars with a diameter of 10 mm, spaced at 100 mm. The variables in this study include the axial compression ratio and the fiber volumetric rates of the longitudinal reinforcement for both the column and beam SFCB. The design parameters of the specimens are presented in Table 6. The concrete cover of the beam is 25 mm, and that of the column is 30 mm.

Table 6. Specimen design parameters.

Specimens	Column Longitudinal Bar			Beam Longitudinal Bar			Axial Compression Ratio
	Configuration	RR [1] (%)	FVR [1] (%)	Configuration	RR (%)	FVR (%)	
C8B10	6S8G20	2.09	84	4S10G16	1.58	60.9	0.1
C8B10-0.2	6S8G20	2.09	84	4S10G16	1.58	60.9	0.2
C10B10	6S10G20	2.09	75	4S10G16	1.58	60.9	0.1
C12B10	6S12G20	2.09	64	4S10G16	1.58	60.9	0.1
C8B6	6S8G20	2.09	84	4S6G16	1.58	86	0.1
C8B8	6S8G20	2.09	84	4S8G16	1.58	75	0.1
C8B10	6S8G20	2.09	84	4S10G16	1.58	60.9	0.1
C8B10-0.2	6S8G20	2.09	84	4S10G16	1.58	60.9	0.2

[1] RR-Reinforcement ratio, FVR-Fiber volume ratio, in which fibers and resin are considered together.

As shown in Figure 4, The dimensions of the column cross-section are as follows: the total height is 2040 mm, with the upper column measuring 1050 mm and the lower column 690 mm. The column is reinforced with six 20-mm SFCB longitudinal bars. Additionally, the hoops are configured with FRP reinforcement of Φ10@100 along the full length. The beam cross-section has a span of 1000 mm on both sides. It includes four 16-mm SFCB longitudinal bars at the upper and lower corners, with inner core diameters of 6 mm, 8 mm, and 10 mm made from HRB400 rebar, and outer fiber thicknesses of 5 mm, 4 mm, and 3 mm for the SFCB. The hoop bars are also made of FRP and are arranged continuously along the full span. The dimensions and reinforcement details of the joint are shown in Figure 3. The specimens are named based on the core steel of the SFCB in the longitudinal reinforcement. The letter "C" indicates column longitudinal reinforcement, while "B" indicates beam longitudinal reinforcement, with the number following each letter representing the core steel diameter of the SFCB. For example, C8B10-0.2 refers to a nodal specimen with SFCB core reinforcements of 8 mm in the column and 10 mm in the beam longitudinal bars, respectively, and an axial compression ratio of 0.2. The axial compression ratio refers to the ratio of the applied axial force to the product of the concrete strength and the cross-sectional area.

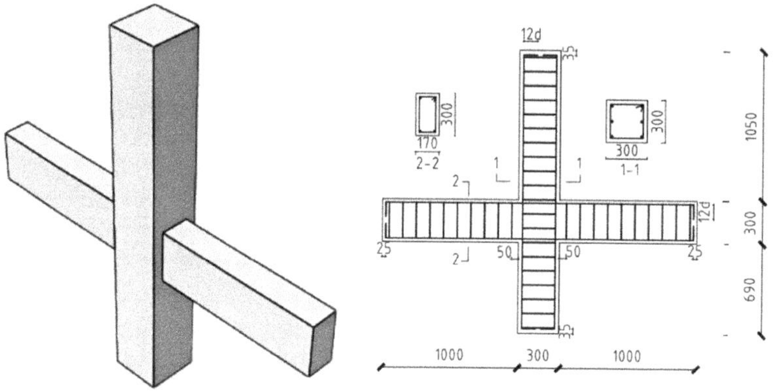

Figure 4. Joint size.

2.4. Loading and Measurement Programs

A 500-ton hydraulic jack was connected to the upper column to apply axial force, while a 300-ton Zhoubang actuator, also connected to the upper column, was used to apply low-cycle horizontal loading. Data acquisition was conducted using a TZT3826E static collector, and the loading equipment is illustrated in Figure 5. The tests adhered to the specifications outlined in JGJ/T101-2015 [47]. The loading procedure included two preloads, each not exceeding 30% of the cracking load, followed by sequential application of 40% and 60% of the cracking loads, gradually increasing to 100% of the cracking load. The load was repeated once for each level of displacement angle before yielding of the specimen, and three times for each level of interstory displacement angle after yielding, until the bearing capacity dropped to 85% of the peak value or severe damage occurred. The test utilized column end loading with displacement control, and the specific loading steps are detailed in Figure 6. The loading process was divided into several stages, with sequential loads applied to 1/750 (0.13%), 1/500 (0.20%), 1/250 (0.40%), and 1/150 (0.67%) of the interlayer displacement angle, with only one application at each of these levels. Subsequent loading involved reaching 1/100 (1%), 3/200 (1.5%), 1/50 (2%), 1/40 (2.5%), 3/100 (3%), 1/25 (4%), 1/20 (5%), and 3/50 (6%) of the interlayer displacement angle, with three repetitions for each layer. The drift angle shown in Figure 6 refers to the interstory drift angle, and the loading sequence indicates the N-th cycle of loading.

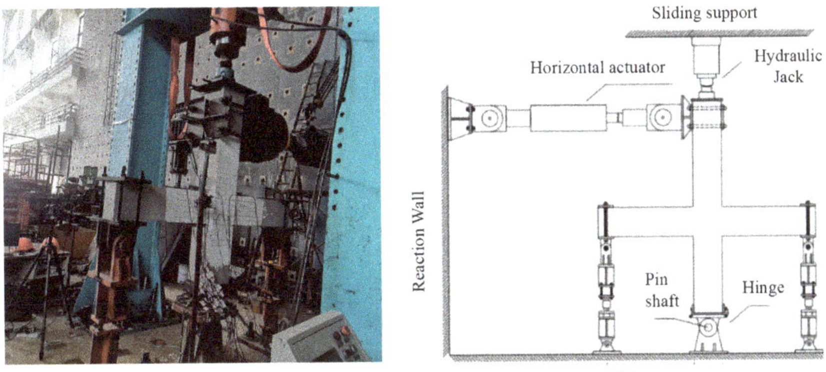

(a) (b)

Figure 5. Test set-up: (**a**) joint specimen; (**b**) schematic drawing.

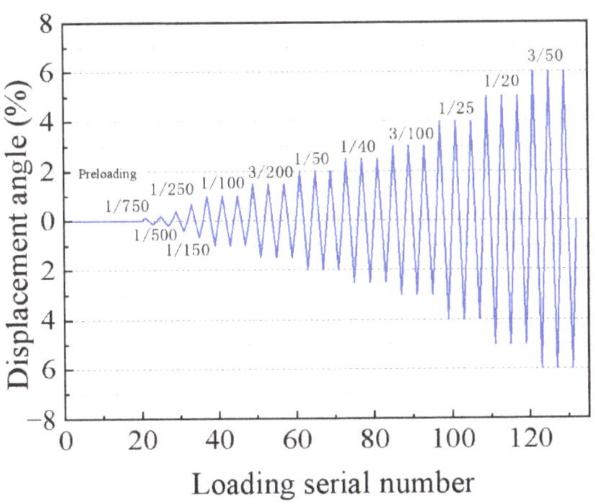

Figure 6. The loading scheme of the specimens.

In this test, the force-displacement data was collected using an electro-hydraulic servo system. To monitor the strain and deformation behavior in critical areas, strain gauges (SGs) and linear variable displacement transducers (LVDTs) were strategically placed on the longitudinal and hoop reinforcement, as well as in the core area of the joint. During the test, displacement and strain data were collected using the TZT3826E static data acquisition system. Specifically, the longitudinal reinforcement of the beam was equipped with 20 mm long strain gauges, each 300 mm apart. For the longitudinal reinforcement of the column, one strain gauge was positioned 150 mm above and below the core area. The hoop reinforcement gauges were placed near the core area to capture relevant data. To measure the corners of the beam and column ends, four displacement gauges were installed on the top and bottom of the left and right beams, each positioned 200 mm from the column face. On the left and right columns, two displacement gauges were arranged on each side at a distance of 200 mm from the beam face to measure the column end corners. Additionally, four strain gauges were affixed at the corners of the concrete surface in the core area of the joint to monitor deformation during the test. The arrangements of these gauges are illustrated in Figures 7 and 8.

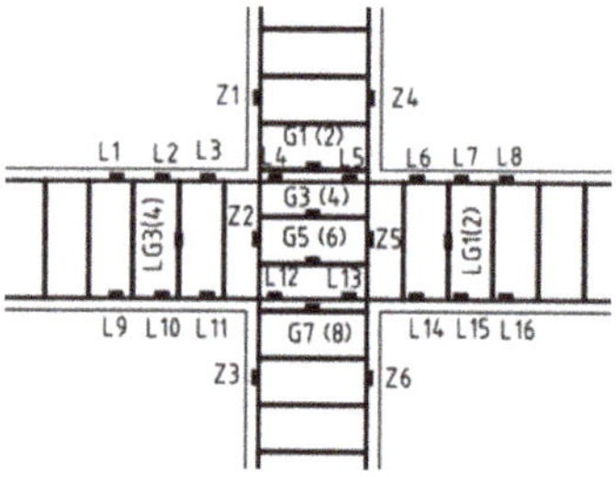

Figure 7. Reinforcement strain gauge layout.

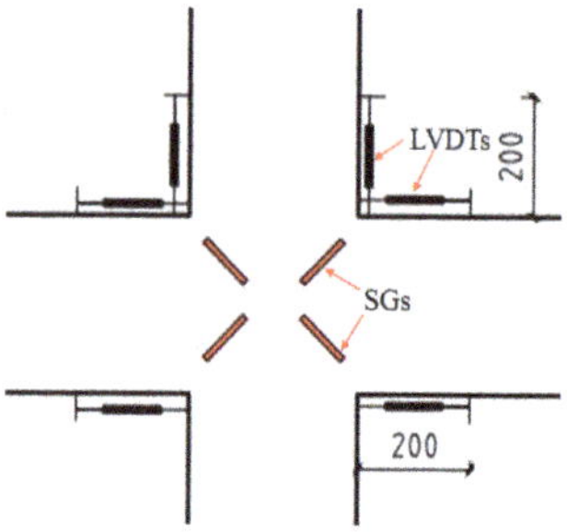

Figure 8. LVDTs and joint core area strain gauge arrangement.

3. Results and Discussion

3.1. Failure Modes

As seen in Figure 9, in specimen C8B10, initial cracks appeared in the right beam–column contact section at an interstory displacement angle of 1/750. By the time the displacement angle reached 1/500, cracks had also formed in the left beam–column contact section. As the displacement angle increased to 1/100 and 3/200, crack propagation intensified, resulting in additional cracks appearing in the beams. At an interstory displacement angle of 1/40, diagonal cracks emerged in the core region of the joint, forming cross-cracks. Ultimately, when the displacement angle reached 1/20, a horizontal crack in the lower part of the left beam extended 200 mm, leading to a penetrating crack in the contact section between the beam and column, concrete spalling, and the termination of the test. Throughout the loading process, the FRP reinforcement in the specimen exhibited no slip damage, indicating a strong bond.

The damage patterns of the other specimens were similar to that of C8B10, all displaying bending failure at the beam ends. Unlike FRP-strengthened joints where debonding dominates [35], no interfacial slip occurred in SFCB specimens, validating their superior bond durability in seawater environments [18]. This behavior aligns with Dong et al.'s [19] findings on SFCB-concrete bond enhancement under cyclic corrosion. However, there are differences in the crack development and damage progression of the specimens. For instance, specimen C8B10-0.2 had fewer cracks, which were primarily concentrated at the beam end, with large cracks extending from initial fractures. The cracks in the joint core were finer, and the concrete spalling was significant. This behavior can be attributed to the increased axial compression ratio, which enhances the nodal shear bearing capacity and the bond strength of the beam's longitudinal tendons in the core region, thereby reducing slip and effectively limiting crack propagation in the joint core [48]. When comparing specimens C8B10, C10B10, and C12B10, C8B10 exhibited the most extensive and widely distributed cracks at the beam end, along with more cracks in the lower column section.

Specimen C10B10 followed with a moderate number of cracks, while C12B10 showed the least amount of cracking. This variation may be linked to the stiffness of the SFCB in the columns: the lower stiffness in column C8B10 allowed the beam to better absorb horizontal loads, leading to more fully developed cracks. Conversely, the greater stiffness of the SFCB longitudinal reinforcement in the columns of C12B10 increased the joint stiffness and correspondingly reduced beam cracking. In another comparison involving C8B6, C8B8, and C8B10, C8B6 demonstrated the least number of cracks, which were wider and accompanied by severe concrete spalling. C8B8 had a greater number of finer cracks and exhibited better concrete integrity, while C8B10 had both a larger number of cracks and a wide distribution. The stabilizing secondary stiffness of the SFCB contributed to crack extension while effectively confining the concrete. However, the lower stiffness of the longitudinal bars in the SFCB beams of C8B6 resulted in excessive concrete cracking under similar displacement conditions, leading to more severe concrete damage and overall specimen deterioration.

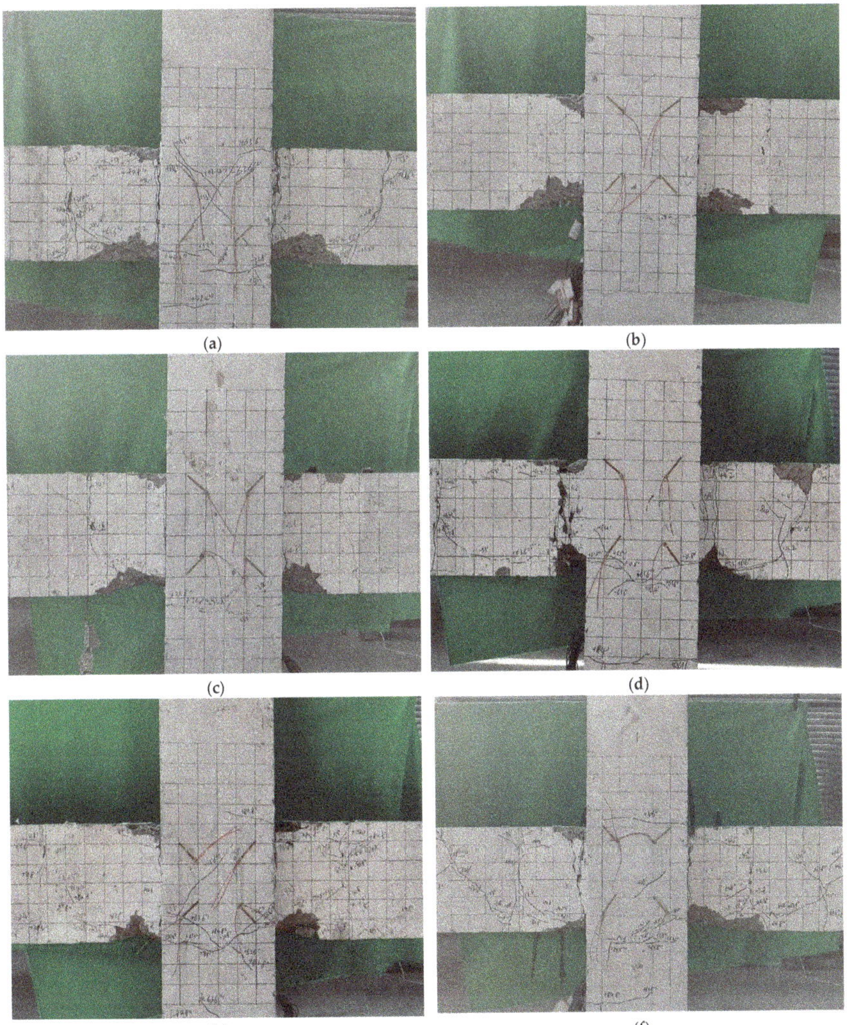
(a) (b) (c) (d) (e) (f)

Figure 9. Failure modes: (**a**) C10B10; (**b**) C8B10-0.2; (**c**) C12B10; (**d**) C8B6; (**e**) C8B8; (**f**) C8B10.

3.2. Hysteresis Curves

Figure 10 illustrates the hysteresis curves of lateral load versus column drift ratio at the column loading point for each specimen. Under cyclic loading, the horizontal load–displacement curve at the top of the column forms multiple loops, known as hysteresis loops. These loops reflect the deformation characteristics, energy dissipation capacity, and stiffness degradation of the structure during cyclic loading, and are important indicators of seismic performance [28,45]. In order to accurately determine the yield displacement of the joint, the energy equivalence method is used, and the specific procedure is shown in Figure 10. First, a point A is identified on the skeleton curve. Then, a polyline OA-BC is drawn such that the area enclosed by this polyline and the X-axis is equal to the area enclosed by the skeleton curve OAC and the X-axis. In other words, the two shaded areas in Figure 10 are equal. When these areas are equal, the displacement value corresponding to point B on the polyline OA-BC, along the horizontal axis, is taken as the yield displacement of the joint. The ductility factor of the specimen is shown in Table 7. Table 7 summarizes the key results, including yield load (P_y), yield displacement (Δ_y), peak load (P_m), peak displacement (Δ_m), ultimate load (Pu), ultimate displacement (Δ_u), and displacement coefficient (DC). The ductile properties of the specimens are characterized using the displacement ductility coefficient, defined as the ratio of the ultimate displacement to the yield displacement of the specimen (Δ_u/Δ_y).

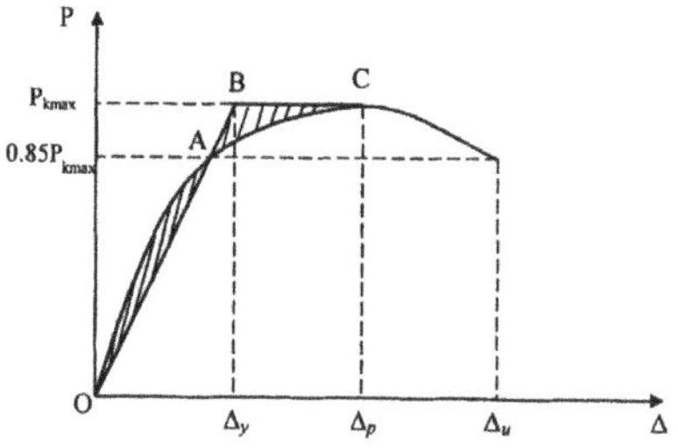

Figure 10. Yield point determination method.

Table 7. Specimen ductility ratio.

Specimens	P_y (kN)	Δ_y (mm)	P_m (kN)	Δ_m (mm)	P_u (kN)	Δ_u (mm)	DC	Average Value
C8B10-0.2	45.80 −44.13	35.80 −37.92	53.94 −54.70	84.52 −85.09	45.80 −49.82	105.95 −106.2	2.96 2.80	2.88
C10B10	37.19 −42.36	28.80 −29.90	43.23 −49.23	45.30 −63.70	36.46 −46.21	88.28 −106.10	3.07 3.55	3.31
C12B10	45.36 −41.00	35.61 −40.64	50.46 −47.81	67.05 −67.25	47.66 −44.68	105.95 −106.05	2.98 2.61	2.79
C8B6	39.13 −44.13	40.07 −40.14	44.24 −53.48	72.62 −63.88	37.60 −45.71	103.64 −106.28	2.59 2.65	2.62
C8B8	39.10 −49.81	33.02 −39.49	46.38 −56.06	65.44 −87.26	40.28 −49.51	127.14 −127.26	3.85 3.22	3.54
C8B10	41.53 −50.11	33.05 −35.17	46.50 −53.78	65.97 −88.64	41.58 −49.94	108.77 −113.54	3.29 3.23	3.26

The hysteresis curves of specimens C8B10-0.2 and C8B10 (Figure 11a,b) exhibit an inverse S-shape, demonstrating a "pinching" effect. In the initial elastic stage, the hysteresis curves of both specimens largely overlap, with no significant residual deformation. As the interlayer displacement angle increases, the specimens enter the yielding stage, leading to residual deformation upon unloading. Stiffness gradually decreases, and the growth rate of the hysteresis curve area slows, indicating reduced energy-dissipating capacity. Eventually, the bearing capacity of the specimens diminishes, and the pinch effect becomes pronounced. Notably, C8B10-0.2 shows a higher load-carrying capacity and a fuller hysteresis loop compared to C8B10, consistent with Su et al.'s [27] observation that higher axial compression ratios enhance energy dissipation in SFCB-reinforced beams. However, this improvement is accompanied by accelerated stiffness degradation, suggesting a trade-off between strength and ductility in high-axial-load scenarios.

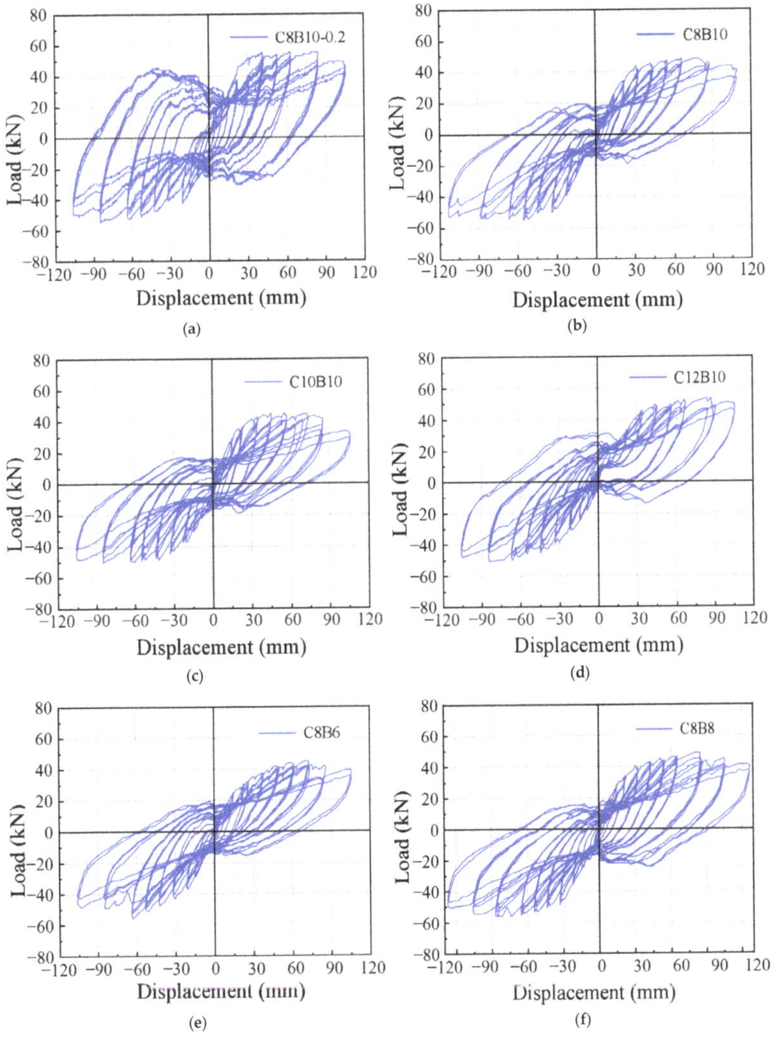

Figure 11. Hysteresis curves of specimens: (**a**) C8B10-0.2; (**b**) C8B10; (**c**) C10B10; (**d**) C12B10; (**e**) C8B6; (**f**) C8B8.

The hysteresis curves of specimens C8B10, C10B10, and C12B10 (Figure 11b–d) also follow an inverse S-shape, with minimal differences in the area of the single-cycle hysteresis loops. Their residual deformations are comparable, indicating that the fiber volume ratio of the longitudinal reinforcement in the column SFCB has a limited impact on the hysteresis curves. However, the hysteresis area and load capacity of C12B10 are slightly higher, showing increases of 8% and 16.7% compared to C8B10 and C10B10, respectively. This improvement may be attributed to the greater stiffness of the column longitudinal reinforcement in C12B10, which enhances lateral stiffness. In contrast, the hysteresis curves of specimens C8B6, C8B8, and C8B10 (Figure 11b,e,f) demonstrate a significant influence of the fiber volume ratio of the longitudinal reinforcement in the beam SFCB on seismic performance. In this group, C8B8 and C8B10 exhibit similar maximum load-carrying capacities and single-cycle hysteresis loop areas, whereas C8B6 shows lower energy consumption capacity and load-carrying capacity. The 4.8% increase in load-carrying capacity of C8B8 over C8B6 may be attributed to the higher fiber volumetric rate, which enhances secondary stiffness and improves the seismic energy consumption capacity of the joints. Conversely, the smaller diameter of the inner core reinforcement in C8B6 resulted in insufficient stiffness of the SFCB, hindering effective crack control and leading to severe concrete damage and relatively poor seismic performance.

3.3. Skeleton Curves

The skeleton curve represents the envelope created by connecting the maximum horizontal loads of the specimens under identical loading displacements, reflecting the various stages and characteristics of force and deformation in the members. As shown in Figure 12a, the skeleton curves of specimens C8B10-0.2 and C8B10 initially coincide, presenting a linear form. This indicates that the axial compression ratio does not significantly affect the bearing capacity and stiffness of the specimens during the early loading stage, where the joint remains in the elastic phase. As the interlayer displacement increases, the joint experiences plastic deformation, causing the curve's slope to gradually decrease, which reflects a reduction in joint stiffness. In contrast, the bearing capacity of C8B10-0.2 increased by 16%, but the rate of decline was more rapid. This suggests that while a higher axial compression ratio enhances stiffness and bearing capacity, it also accelerates the decay of load-bearing capacity. The reduction in deformation capacity can be attributed to the fact that greater axial pressure imposes more constraints on the specimen, leading to increased bond friction.

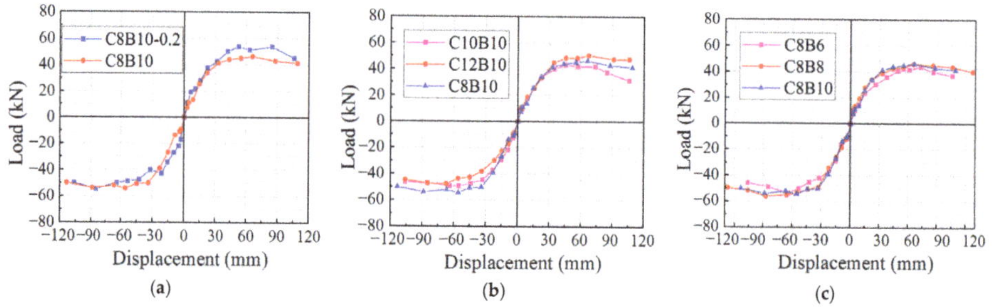

Figure 12. Skeleton curves of specimens: (**a**) C8B10-0.2 and C8B10; (**b**) C12B10, Z10L10, and C8B10; (**c**) C8B6, C8B8, and C8B10.

Figure 12b presents the skeleton curves for specimens C12B10, C10B10, and C8B10, which exhibit similar development trends. In the elastic stage, the initial stiffness of these three specimens is comparable, as it is primarily governed by the inner-core ordinary steel reinforcement when subjected to equal axial compression ratios. As the load increases, the skeleton curves decline relatively gradually after reaching the peak load. This behavior may be due to the secondary stiffness of the SFCB, which enables the structure to sustain a high load-carrying capacity even as damage occurs. Among these specimens, C12B10 exhibited the highest load capacity (50 kN), while C10B10 had the lowest, indicating that increased stiffness in the column longitudinal reinforcement effectively enhances the load capacity of the specimens.

Figure 12c illustrates the skeleton curves for specimens C8B6, C8B8, and C8B10. While the three curves are close during the elastic stage, noticeable differences emerge in the plastic stage, indicating that the fiber volume ratio of the beam longitudinal reinforcement significantly influences the seismic performance of the specimens. The load-carrying capacities of C8B8 and C8B10 are similar, whereas C8B6 demonstrates a lower load capacity. This reduction can be attributed to the smaller diameter of the inner core reinforcement in the SFCB, which hampers effective crack control, resulting in coarser cracks and more severe damage, ultimately reducing the specimen's bearing capacity.

3.4. Energy Consumption

The energy consumption capacity of the specimens is analyzed through unicircular dissipation, cumulative dissipation, and equivalent viscous damping coefficient. The circumferential energy consumption of the specimen, represented by the area enclosed within a single hysteresis curve, is illustrated in Figure 13. The data indicates that increasing the axial compression ratio significantly enhances the unicircular energy consumption of the specimen, particularly in the later loading stages. Specifically, at the 23rd cycle (corresponding to an interlaminar displacement angle of 1/20), the axial compression ratio of 0.2 resulted in a 52% increase in energy consumption compared to a ratio of 0.1. Furthermore, among the specimens in the column SFCB longitudinal fiber volume rate group, the differences in unicircular energy consumption were minimal, with only slight variations observed in the number of loading cycles before damage occurred. At the 23rd cycle, specimen C12B10 exhibited a 24% increase in unicircular energy consumption relative to specimen C10B10, while specimen C8B10 showed a 16% increase. In the beam SFCB longitudinal bars fiber volume rate group, the unicircular energy consumption for the three specimens was comparable during the first 16 cycles. This can be attributed to the larger fiber volume rate in the SFCB longitudinal bars of specimen C8B6, as well as the elasticity of the outsourced FRP material, which contributes to strong deformation capacity and enhances energy consumption. However, in the later stages of loading, excessive concrete deformation led to a reduction in both the load-carrying capacity and energy consumption capacity of specimen C8B6. In contrast, specimen C8B8 demonstrated superior deformation and energy consumption capacities compared to both C8B6 and C8B10, owing to its relatively higher secondary and initial stiffness.

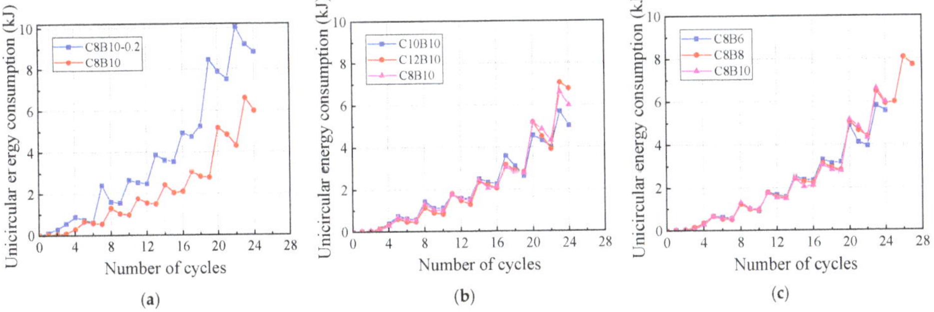

(a) (b) (c)

Figure 13. Unicircular energy consumption of specimens: (**a**) axial compression ratio group specimens; (**b**) column longitudinal reinforcement fiber volume rate group specimens; (**c**) beam longitudinal reinforcement fiber volume rate group specimens.

Cumulative energy consumption reflects the total energy absorbed by the specimen throughout the static test and is calculated by summing the energy dissipated during each loading cycle. The cumulative energy consumption of the specimens is illustrated in Figure 14. Comparative analysis reveals that the axial compression ratio significantly influences cumulative energy consumption; specimens with an axial compression ratio of 0.2 exhibited much higher cumulative energy consumption than those with a ratio of 0.1. Notably, at the point of damage, the cumulative energy consumption of specimen C8B10-0.2 was 80% greater than that of specimen C8B10. Additionally, the cumulative energy consumption among specimens in the column SFCB longitudinal fiber volume rate group and the beam SFCB longitudinal fiber volume rate group was similar. However, specimen C12B10 demonstrated slightly higher cumulative energy consumption compared to C10B10, while specimen C8B8 significantly outperformed specimens C8B6 and C8B10, with cumulative energy consumption values that were 43% and 41.6% higher, respectively, due to its ability to sustain larger displacements.

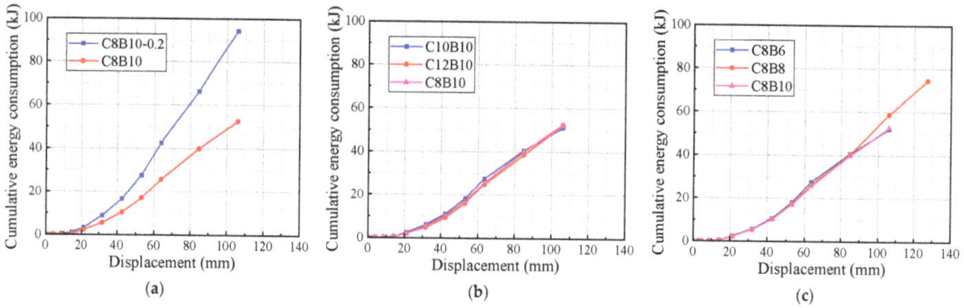

Figure 14. Cumulative energy consumption of specimens: (**a**) axial compression ratio group specimens; (**b**) column longitudinal reinforcement fiber volume rate group specimens; (**c**) beam longitudinal reinforcement fiber volume rate group specimens.

According to the JGJ/T101-2015 [43], the calculation schematic for the equivalent viscous damping coefficient and energy consumption coefficient is illustrated in Figure 15. The formulas for calculating the energy consumption coefficient and the equivalent viscous damping coefficient are provided in Equations (1) and (2).

$$E = \frac{S_{ABCD}}{S_{OBE} + S_{ODF}} \quad (1)$$

$$\xi_{eq} = \frac{1}{2\pi}\frac{S_{ABCD}}{S_{OBE} + S_{ODF}} \quad (2)$$

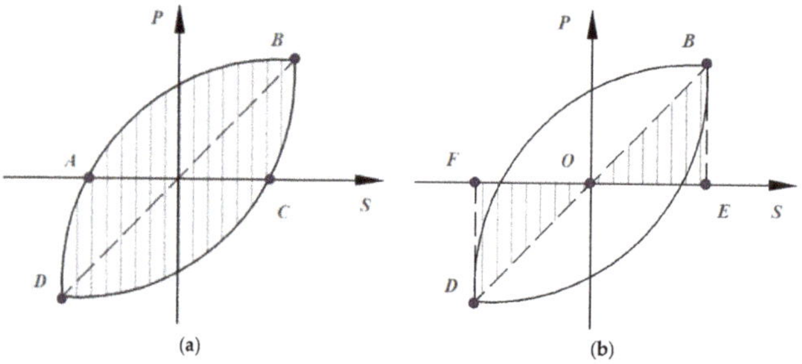

Figure 15. Diagram of equivalent viscous coefficient and energy consumption coefficient: (**a**) S_{ABCD}; (**b**) S_{OBE} and S_{ODF}.

As shown in Figure 16, the specimens continued to exhibit a certain growth trend even after damage. Notably, the growth slopes of specimens C8B8 and C12B10 were steeper at the time of damage, suggesting that the SFCB specimens retained good energy consumption capacity even after damage. This behavior highlights the contribution of the secondary stiffness of the SFCB, which enhances the energy consumption capacity of the specimens. Similarly, the axial compression ratio significantly affected the equivalent viscous damping coefficient of the joints. Increasing the axial compression ratio substantially improved the equivalent viscous damping coefficient. On the other hand, the fiber volumetric ratio of the column SFCB longitudinal reinforcement and the beam SFCB longitudinal reinforcement had little impact on the equivalent viscous damping coefficient. Specimens C10B10 and C8B6 exhibited slightly higher equivalent viscous damping coefficients in the intermediate section, which may be attributed to the larger cracks that developed in the concrete during the intermediate loading stage, including diagonal cracks in the core region. However, as the loading displacement increased, the concrete in these specimens experienced excessive deformation and damage, which slowed the growth of their viscous damping coefficients. Eventually, the equivalent viscous damping coefficients of these specimens became equal to or slightly lower than those of the other specimens in the same group.

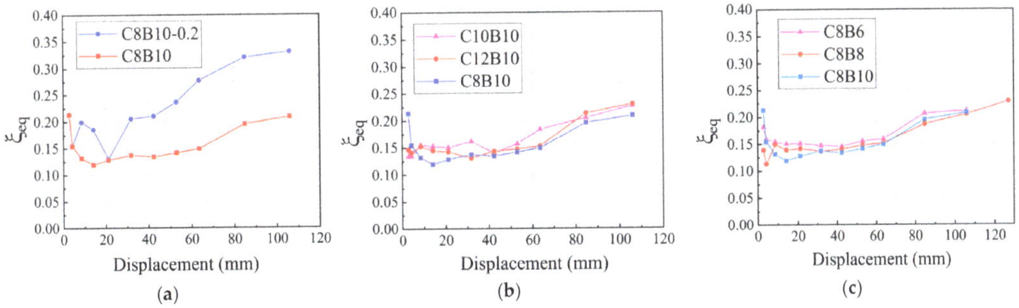

Figure 16. Equivalent viscous damping coefficient of the specimen: (**a**) axial compression ratio group specimens; (**b**) column longitudinal reinforcement fiber volume rate group specimens; (**c**) B = beam longitudinal reinforcement fiber volume rate group specimens.

3.5. Fixed-End Rotations of Column and Beam

The effects of the column longitudinal reinforcement fiber volume rates on the beam and column end angles of the specimens are illustrated in Figure 17. From the figure, it is evident that the three longitudinal fiber volume rates have minimal influence on the beam end angle of the specimens. The relationship between the beam end angle and the displacement for all three specimens appears to be approximately linear. In contrast, the column end angle is affected by the "off-axis" phenomenon during the loading process, resulting in asymmetrical column end angles at the same positive and negative loading displacements. Additionally, as the fiber volume rate of the longitudinal bars increases—leading to a decrease in longitudinal bar stiffness—the column end angle of the specimen at the same loading displacement tends to increase.

Figure 18 depicts the effect of the beam longitudinal reinforcement fiber volume rate on the beam and column end turn angles of the specimens. In the final loading stage (3/50), the beam end turn angle of the specimens decreases as the beam longitudinal reinforcement fiber volume rate increases. This phenomenon may be attributed to two main factors: first, an increase in the longitudinal reinforcement fiber volume rate enhances the stiffness of the tendons in the beam, allowing it to resist a larger bending moment, which typically would result in an increase in the beam end turn angle. However, second, when the longitudinal bar fiber volume rate is increased, the exterior fibers experience greater forces after the reinforcement yields, which in turn reduces the beam end turn angle at the joint. The column end turn angle of the specimens exhibited a nonlinear relationship with displacement, with specimen C8B6 demonstrating a smaller column end turn angle compared to the others.

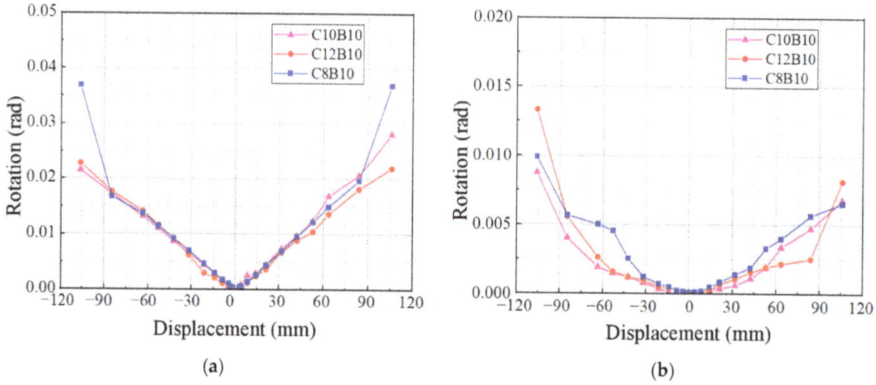

Figure 17. Effect of longitudinal reinforcement fiber volume rate in columns: (**a**) beam end corners; (**b**) column end corners.

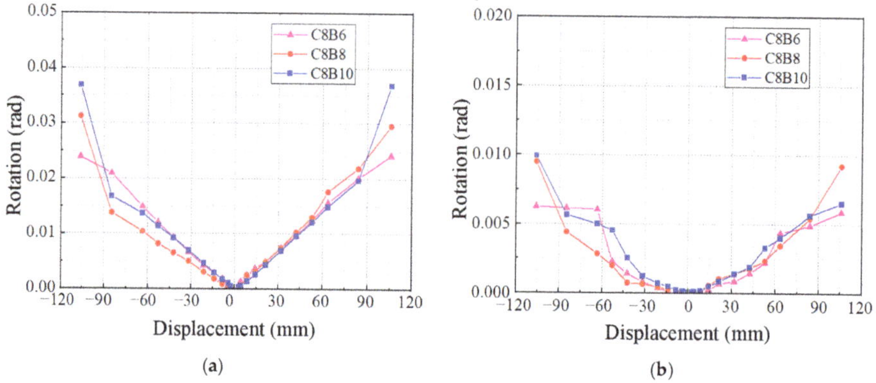

Figure 18. Effect of longitudinal reinforcement fiber volume rate in beams: (**a**) beam end corners; (**b**) column end corners.

3.6. SFCB Strains

Due to the presence of numerous strain gauge measurement points and the premature failure of some strain gauges during the loading process, this study selects the most significant locations for strain measurements, prioritizing those where the strain gauges remained intact throughout loading. The longitudinal tendon strain curves of the beam are presented in Figure 19a,b. The data indicate that the longitudinal bar strains of all specimens exceeded 2200, signifying that the core reinforcement of the SFCB bars reached yielding, with the reinforcement primarily in a tensile state during the late loading stages. Notably, the beam longitudinal bar strain of specimen C12B10 is relatively small, while specimen C8B6 exhibits a larger strain. This discrepancy suggests that concrete cracks in specimen C8B6 developed more extensively, causing upward growth of the tensile cracks and resulting in increased tensile height, which in turn elevated the longitudinal tendon strains in this specimen. Additionally, the strain at the position of column longitudinal bar Z2 is depicted in Figure 19c. The results show that the strains of the column longitudinal tendons did not reach the yield strain of the reinforcement, indicating that the column cross-section remained relatively intact during the loading process. This observation aligns with the design specification's requirement for "strong columns and weak beams". During the push–pull loading, the longitudinal bars of all specimens were in tension. The strain curves of the hoop bars are illustrated in Figure 20. The strains of all the hoops are relatively low, with the strains of the beam hoops, column hoops, and joint core area hoops remaining below 800, significantly smaller than those of the beam longitudinal tendons. This finding corresponds with the observation that all specimens experienced damage at the beam end, characterized by small and fine diagonal cracks.

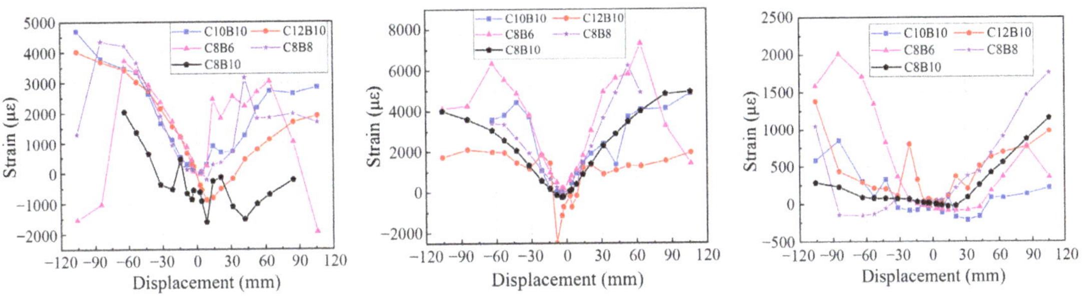

(a) (b) (c)

Figure 19. Strain–displacement curves of longitudinal bars: (**a**) beam longitudinal bar L3 position; (**b**) beam longitudinal bar L5 position; (**c**) column longitudinal bar Z2 position.

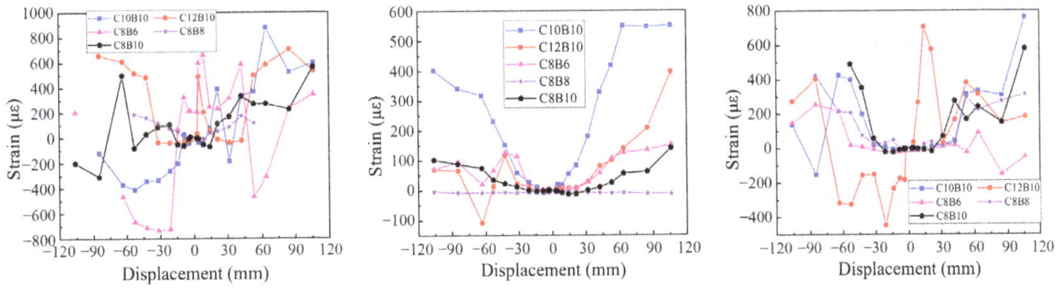

(a) (b) (c)

Figure 20. Strain–displacement curves of stirrups: (**a**) beam hoop LG1 position; (**b**) column hoop G1 position; (**c**) column hoop G5 position.

4. Conclusions

This study addresses the critical gap in understanding the seismic behavior of SFCB-reinforced seawater sea–sand concrete beam—column joints, which has hindered their marine applications. Key findings include the following:

(1) All specimens showed beam end bending failure, in line with the design principle of 'strong column and weak beam'. The longitudinal reinforcement of the column section did not reach yield strain, and the stirrup strain in the core area was low (<800 με), indicating no significant shear failure occurred in the core area of the joint. Additionally, the combination of SFCBs and seawater sand concrete exhibits excellent

bonding properties and crack resistance in a corrosive environment, with no sliding failure of the FRP bars.

(2) Increasing the axial compression ratio (e.g., from 0.1 to 0.2) significantly enhances the shear and bond strength of the joints, effectively inhibiting crack propagation and slip, but also accelerating bearing capacity degradation. At the same displacement angle, the energy consumption of the specimen with an axial compression ratio of 0.2 (e.g., C8B10-0.2) is 52% higher than that of the specimen with 0.1, and the cumulative energy consumption is 80% higher.

(3) An increase in the longitudinal SFCB fiber volume ratio (e.g., from 64% to 84%) improves joint stiffness, fracture distribution uniformity, and ductility. The C12B10 specimen, with the highest fiber volume ratio, exhibited the highest bearing capacity (50 kN), 16.7% and 8% higher than C8B10 and C10B10, respectively.

(4) The longitudinal SFCB fiber volume ratio significantly affects the seismic performance of the beam. The specimen with a medium fiber volume ratio (75%) (e.g., C8B8) demonstrated the best performance, with cumulative energy consumption 43% and 41.6% higher than C8B6 and C8B10, respectively. Its cracks were finer, and the concrete integrity was better. A low volume ratio (e.g., 60.9%) resulted in significant cracking of the concrete and a decrease in bearing capacity, while a high volume ratio (e.g., 86%) may reduce energy dissipation efficiency due to stiffness imbalance.

(5) The stable secondary stiffness of SFCBs provides continuous bearing capacity during the joint plastic stage, delays stiffness degradation, and improves the equivalent viscous damping coefficient of the specimen. For instance, C8B8 still maintains high energy dissipation capacity after failure.

Author Contributions: Conceptualization, S.X. and X.L.; methodology, R.L.; investigation, R.L.; data curation, S.X. and X.L.; writing—original draft preparation, S.X.; writing—review and editing, R.L., Z.L. and B.Z.; supervision, R.L.; funding acquisition, R.L. All authors have read and agreed to the published version of the manuscript.

Funding: This research is supported by the Guangzhou Electric Power Design Institute 2024 Substation Engineering Mechanization Research Topic Technical Services (20240946).

Institutional Review Board Statement: Not applicable.

Informed Consent Statement: Not applicable.

Data Availability Statement: The original contributions presented in this study are included in the article. Further inquiries can be directed to the corresponding author.

Conflicts of Interest: Authors Ruiqing Liang, Botao Zhang, Zhensheng Liang were employed by the company Guangzhou Electric Power Design Institute Co., Ltd. The remaining authors declare that the research was conducted in the absence of any commercial or financial relationships that could be construed as a potential conflict of interest.

References

1. Rentier, E.S.; Cammeraat, L.H. The environmental impacts of river sand mining. *Sci. Total Environ.* **2022**, *838*, 155877. [CrossRef] [PubMed]
2. Guo, Y.C.; Cai, Y.J.; Xie, Z.H.; Xiao, S.H.; Zhuo, K.X.; Cai, P.D.; Lin, J.X. Experimental investigation of GFRP bar bonding in geopolymer concrete using hinged beam tests. *Eng. Struct.* **2025**, *322*, 119036. [CrossRef]
3. Su, J.Y.; Chen, G.; Pan, H.S.; Lin, J.X.; Zhang, J.; Zhuo, K.X.; Chen, Z.B.; Guo, Y.C. Rubber modified high strength-high ductility concrete: Effect of rubber replacement ratio and fiber length. *Constr. Build. Mater.* **2023**, *404*, 133243. [CrossRef]
4. Liu, X.H.; Chen, Z.X.; Yu, Z.C.; Chen, P.; Zhang, Y.Z. A review on ultra-high performance seawater sea sand concrete: Hydration, microstructure and properties. *Constr. Build. Mater.* **2024**, *438*, 136945. [CrossRef]

5. Zhang, W.; Ding, D.W.; Sun, J.X.; Guo, F.; Feng, H.B.; Ma, H.Y.; Hong, S.X.; Dong, B.Q.; Hou, D.S. Novel insights on the setting process, hardened properties, and durability of sustainable ultra-high-performance seawater sea sand concrete. *Constr. Build. Mater.* **2024**, *447*, 138092. [CrossRef]
6. Hu, X.; Bai, T.; Zhao, Y.; Ren, Q.; Chen, Y.; Li, H.; Shi, C. Interface mechanical properties of CFRP-seawater sea sand concrete under simulated seawater immersion. *J. Build. Eng.* **2024**, *97*, 110768. [CrossRef]
7. Mashayekhi, A.; Hassanli, R.; Zhuge, Y.; Ma, X.; Chow, C.W.K.; Bazli, M.; Manalo, A. Structural behaviour of small GFRP-reinforced seawater sea-sand fiber reinforced concrete culverts. *Structures* **2024**, *69*, 107492. [CrossRef]
8. News, S.L.C. Corrosion News. *Mater. Corros.* **2020**, *71*, 1407–1414. [CrossRef]
9. Cristoforetti, A.; Rossi, S.; Deflorian, F.; Fedel, M. Recent progress in understanding filiform corrosion on organic coated steel: A comprehensive review. *Prog. Org. Coat.* **2024**, *192*, 108469. [CrossRef]
10. Dong, S.C.; Zhou, P.; Ning, Z.; Wu, X.; Li, C.G.; Xian, G.J. Durability of carbon- and glass-fiber reinforced thermoplastic polymer composites: A literature review. *J. Build. Eng.* **2024**, *98*, 111055. [CrossRef]
11. Karim, H.; Sheikh, M.N.; Hadi, M.N.S. Longitudinal Reinforcement Limits for Fiber-Reinforced Polymer Reinforced Concrete Members. *ACI Struct. J.* **2017**, *114*, 687–696. [CrossRef]
12. Xiao, S.H.; Cai, Y.C.; Xie, Z.H.; Guo, Y.C.; Zheng, Y.; Lin, J.X. Bond durability between steel-FRP composite bars and concrete under seawater corrosion environments. *Constr. Build. Mater.* **2024**, *419*, 135456. [CrossRef]
13. Cai, Y.J.; Xiao, S.H.; Chen, Y.W.; Huang, Z.R.; Lin, J.X.; Guo, Y.C.; Peng, Y.Q.; Xie, Z.H. Tensile behavior and durability prediction of GFRP-steel composite bars under chloride environments. *J. Mater. Res. Technol.* **2023**, *23*, 5746–5759. [CrossRef]
14. Guo, Y.C.; Xiao, S.H.; Zeng, J.J.; Su, J.Y.; Li, T.Z.; Xie, Z.H. Behavior of concrete-filled FRP tube columns internally reinforced with FRP-steel composite bars under axial compression. *Constr. Build. Mater.* **2022**, *315*, 125714. [CrossRef]
15. Cai, Y.C.; Xie, Z.H.; Xiao, S.H.; Huang, Z.R.; Lin, J.X.; Guo, Y.C.; Zhuo, K.X.; Huang, P.Y. An investigation of fatigue behavior and residual strength model of steel-GFRP composite bar. *Compos. Struct.* **2024**, *327*, 117685. [CrossRef]
16. Xiao, S.H.; Cai, Y.J.; Xie, Z.H.; Zheng, Y.; Lin, J.X.; Guo, Y.C. Effects of marine environment and fatigue pre-damage on the residual tensile properties of SFCBs. *Constr. Build. Mater.* **2024**, *449*, 138273. [CrossRef]
17. Zhao, D.B.; Pan, J.; Zhou, Y.W.; Sui, L.L.; Ye, Z.H. New types of steel-FRP composite bar with round steel bar inner core: Mechanical properties and bonding performances in concrete. *Constr. Build. Mater.* **2020**, *242*, 118062. [CrossRef]
18. Dong, Z.Q.; Wu, G.; Zhao, X.L.; Zhu, H.; Lian, J.L. Bond durability of steel-FRP composite bars embedded in seawater sea-sand concrete under constant bending and shearing stress. *Constr. Build. Mater.* **2018**, *192*, 808–817. [CrossRef]
19. Dong, Z.Q.; Wu, G.; Xu, Y.Q. Experimental study on the bond durability between steel-FRP composite bars (SFCBs) and sea sand concrete in ocean environment. *Constr. Build. Mater.* **2016**, *115*, 277–284. [CrossRef]
20. Han, S.W.; Zhou, A.; Fan, C.; Xiao, G.; Ou, J.P. Seawater sea-sand concrete columns using durable and ductile steel-FRP composite bars and FRP ties: A promising hybrid scheme for marine engineering. *Eng. Struct.* **2023**, *294*, 116731. [CrossRef]
21. Xiao, S.H.; Lin, J.X.; Li, L.J.; Guo, Y.C.; Zeng, J.J.; Xie, Z.H.; Wei, F.F.; Li, M. Experimental study on flexural behavior of concrete beam reinforced with GFRP and steel-fiber composite bars. *J. Build. Eng.* **2021**, *43*, 103087. [CrossRef]
22. Ge, W.J.; Chen, K.B.; Guan, Z.W.; Ashour, A.; Lu, W.G.; Cao, D.F. Eccentric compression behaviour of concrete columns reinforced with steel-FRP composite bars. *Eng. Struct.* **2021**, *238*, 112240. [CrossRef]
23. Han, S.W.; Fan, C.; Zhou, A.; Ou, J.P. Shear behavior of concrete beams reinforced with corrosion-resistant and ductile longitudinal steel-FRP composite bars and FRP stirrups. *Eng. Struct.* **2023**, *278*, 115520. [CrossRef]
24. Han, S.W.; Fan, C.; Zhou, A.; Ou, J.P. Simplified implementation of equivalent and ductile performance for steel-FRP composite bars reinforced seawater sea-sand concrete beams: Equal-stiffness design method. *Eng. Struct.* **2022**, *266*, 114590. [CrossRef]
25. Ding, L.N.; Lei, M.K.; Wang, X.; Shi, Y.W.; Zhu, Z.G.; Yu, P.C.; Wu, Z.S. Durability of concrete members reinforced with steel-FRP composite bars under dry-wet cycles of seawater. *Structures* **2021**, *33*, 2273–2283. [CrossRef]
26. Sun, Z.Y.; Sun, Y.L.; Zheng, Y.; Iwashita, K.; Zhang, J.; Wu, G. Experimental study on precast concrete columns reinforced with bundled SFCBs under horizontal cyclic loading. *J. Build. Eng.* **2023**, *74*, 106882. [CrossRef]
27. Su, C.; Wang, X.; Ding, L.N.; Chen, Z.Y.; Liu, S.; Wu, Z.S. Experimental study on the seismic behavior of seawater sea sand concrete beams reinforced with steel-FRP composite bars. *Eng. Struct.* **2021**, *248*, 113269. [CrossRef]
28. Su, C.; Wang, X.; Ding, L.N.; Wu, Z.S. Seismic degradation behavior of steel-FRP composite bar reinforced concrete beams in marine environment based on experimental and numerical investigation. *Compos. Struct.* **2023**, *313*, 116949. [CrossRef]
29. Niroomandi, A.; Maheri, A.; Maheri, M.R.; Mahini, S.S. Seismic performance of ordinary RC frames retrofitted at joints by FRP sheets. *Eng. Struct.* **2010**, *32*, 2326–2336. [CrossRef]
30. Sadeghi-Movahhed, A.; De Domenico, D.; Majdi, A. Structural flexibility impact on pounding severity and seismic performance of adjacent isolated buildings. *Soil Dyn. Earthq. Eng.* **2024**, *181*, 108667. [CrossRef]
31. Huang, Z.; Gong, L.; Jiang, L.; Chen, Y.F.; Qi, J.; Huang, Y. Study on seismic performance and restoring force model of framed CFST column-composite box beam joints. *Structures* **2024**, *69*, 107458. [CrossRef]

32. Yuen, Y.P.; Kuang, J.S. Nonlinear seismic responses and lateral force transfer mechanisms of RC frames with different infill configurations. *Eng. Struct.* **2015**, *91*, 125–140. [CrossRef]
33. Favvata, M.J.; Izzuddin, B.A.; Karayannis, C.G. Modelling exterior beam–column joints for seismic analysis of RC frame structures. *Earthq. Eng. Struct. Dyn.* **2008**, *37*, 1527–1548. [CrossRef]
34. Mousavi, S.S.; Dehestani, M. Influence of mixture composition on the structural behaviour of reinforced concrete beam-column joints: A review. *Structures* **2022**, *42*, 29–52. [CrossRef]
35. del Rey Castillo, E.; Niroomandi, A.; Triantafillou, T. Seismic Retrofitting of Realistic Beam–Column Joints with Shear Failure Using FRP Sheets and FRP Anchors. *J. Compos. Const.* **2024**, *28*, 1–16. [CrossRef]
36. Attari, N.; Youcef, Y.S.; Amziane, S. Seismic performance of reinforced concrete beam–column joint strengthening by frp sheets. *Structures* **2019**, *20*, 353–364. [CrossRef]
37. Yu, F.; Xu, B.; Wang, J.M.; Fang, Y.; Yang, Y.Z.; Chen, Z.P.; Zhang, Y.F. Influence of joint forms on the seismic behavior of concrete-filled PVC-CFRP tubular column-RC beam joints connected with core steel tubes. *Eng. Struct.* **2024**, *303*, 1. [CrossRef]
38. Shokrzadeh, M.R.; Aziminejad, A.; Moghaddam, A.S. Evaluation of various FRP strengthening configurations for RC beam-column joints. *Bridg. Struct.* **2024**, *20*, 71–91. [CrossRef]
39. Xu, W. Experimental Study on Seismic Performance of Steel-Continuous Fiber Composite Reinforced Concrete Joints. Master's Thesis, Southeast University, Nanjing, China, 2010. (In Chinese).
40. *ASTM D1141-98*; Standard Practice for the Preparation of Substitute Ocean Water. ASTM International: West Conshohocken, PA, USA, 2013.
41. *GB/T 50081-2019*; Standard for Test Methods of Concrete Physical and Mechanical Properties. Ministry of Housing and Urban-Rural Development of the PRC; State Administration for Market Regulation: Beijing, China, 2019.
42. *ASTM C469*; Standard Test Method for Static Modulus of Elasticity and Poisson's Ratio of Concrete in Compression. ASTM International: West Conshohocken, PA, USA, 2010.
43. *GB/T 30022-2013*; Test Method for Basic Mechanical Properties of Fiber Reinforced Polymer Bar. Standardization Administration of China: Beijing, China, 2013.
44. *GB/T 228.1-2010*; Metallic Materials-Tensile Testing at Ambient Temperature. State Administration for Market Regulation; Standardization Administration of the People's Republic of China: Beijing, China, 2010.
45. Lin, G.; Zeng, J.J.; Liang, S.D.; Liao, J.J.; Zhuge, Y. Seismic behavior of novel GFRP bar reinforced concrete beam-column joints internally reinforced with an FRP tube. *Eng. Struct.* **2022**, *273*, 115100. [CrossRef]
46. *GB 50011-2010*; Code for Seismic Design of Buildings. Ministry of Housing and Urban-rural Construction of the People's Republic of China and the General Administration of Quality Supervision; Inspection and Quarantine of the People's Republic of China: Beijing, China, 2016.
47. *JGJ/T101-2015*; Specifications for Seismic Test of Buildings. China Architecture & Building Press: Beijing, China, 2015.
48. Fu, J.P.; Chen, T.; Wang, Z.; Bai, S.L. Effect of axial load ratio on seismic behaviour of interior beam–column joints. In Proceedings of the 2th World Conference on Earthquake Engineering, Auckland, New Zealand, 30 January–4 February 2000.

materials

MDPI

Article

Hydration and Mechanical Properties of Low-Carbon Binders Using CFBC Ash

Young-Cheol Choi

Department of Civil and Environmental Engineering, Gachon University, Seongnam 13120, Gyeonggi-do, Republic of Korea; zerofe@gachon.ac.kr; Tel.: +82-31-750-5721

Abstract: Circulating fluidized bed combustion (CFBC) ash, a byproduct typically generated from coal-fired CFBC power plant boilers, contains high content of free lime and anhydrite. Due to its chemical composition, CFBC ash exhibits self-cementing properties; however, its performance is limited. One approach to enhancing the self-cementing properties of CFBC ash is through the incorporation of mineral admixtures such as gypsum. This study investigated the influence of desulfurization gypsum (DG) on the self-cementing behavior of CFBC ash. To this end, paste and mortar specimens were prepared and evaluated for their hydration and mechanical characteristics. The hydration behavior was analyzed using isothermal calorimetry, thermogravimetric analysis (TGA), setting time measurements, and X-ray diffraction (XRD) analysis. Mechanical properties were assessed by measuring the compressive strength at various curing ages. Additionally, changes in microstructure were examined by evaluating the pore size distribution through mercury intrusion porosimetry (MIP). The experimental results indicate that the appropriate incorporation of DG enhances the hydraulic reactivity of CFBC ash and significantly improves the compressive strength.

Keywords: fluidized bed combustion ash; desulfurization gypsum; self-hardening property; hydration; compressive strength

1. Introduction

Globally, fossil fuels continue to dominate as primary energy sources due to their current reserves, existing infrastructure, and relatively low fuel costs. However, traditional coal combustion technologies are known to emit sulfur dioxides (SO_x) and nitrogen oxides (NO_x), which contribute directly to air pollution [1]. In recent years, the adoption of circulating fluidized bed combustion (CFBC) technology has been gaining attention for its high combustion efficiency and its significant reduction in SO_x and NO_x emissions, thereby contributing to improved air quality [2,3]. With the increasing deployment of CFBC technology, the amount of CFBC ash generated as a byproduct is also growing substantially, raising concerns regarding its environmental impact and the need for effective utilization strategies [4]. Unlike conventional pulverized coal combustion fly ash (FA), CFBC ash is typically produced at a relatively low combustion temperature of around 900 °C. As a result, CFBC ash tends to exhibit an irregular particle morphology rather than spherical, and it generally contains a higher level of unburned carbon [5,6].

Furthermore, due to the addition of limestone during the desulfurization process, CFBC ash contains elevated levels of CaO and SO_3, resulting in higher free CaO content compared to FA. This composition imparts excellent self-cementing properties to the CFBC ash, enabling it to function as a binder material with performance similar to that of cement [7–9]. Additionally, its chemical similarity to fly ash suggests potential pozzolanic reactivity, indicating its applicability as a supplementary cementitious material (SCM) [10,11]. However, the high free CaO content can lead to excessive heat evolution, and the presence of SO_3 may induce undesirable expansion, both of which can adversely affect the performance of cement and concrete products [12].

Numerous researchers have investigated the hydration reactions of CFBC ash [13–15]. CFBC ash, which exhibits self-cementing properties, typically forms calcium silicate hydrate (C-S-H), ettringite, and $Ca(OH)_2$ as its primary hydration products, which are similar to those observed in ordinary Portland cement (OPC) systems [6,14,16]. In CFBC ash with high SO_3 content, the rapid formation of ettringite in the early stages of hydration has been reported to cause significant expansion. Sheng et al. [16] investigated the effects of the particle size on the self-cementitious properties of CFBC ash. According to their study, a lower 45 μm sieve residue of CFBC ash led to increased self-cementitious strength and contributed to the mitigation of volumetric instability. Sheng et al. [17] investigated the influence of lime content in CFBC ash on its self-cementing behavior and concluded that the formation of $Ca(OH)_2$ through the slaking of free lime plays a dominant role in the development of self-cementing properties. Several studies have explored the use of CFBC ash in various construction materials, aiming to utilize it as a value-added component [18–26]. Due to its production characteristics, CFBC ash is expected to exhibit pozzolanic reactivity when blended with cementitious materials, which enhances its potential as a mineral admixture. The applicability of CFBC ash as an SCM has been validated by many researchers [27–30]. Shen et al. [31] further demonstrated that, due to the high SO_3 content in CFBC ash, it could serve as a substitute for conventional gypsum in cement formulations, acting as a set-retarding agent in hydration processes.

In recent years, numerous studies have been conducted to significantly reduce environmental impacts by utilizing CFBC ash in the development of low-cement or zero-cement binders and related products, either by drastically reducing or entirely eliminating the use of OPC [32–34]. These studies primarily focus on blending CFBC ash with ground granulated blast furnace slag—a common precursor in alkali-activated cement systems—or activating CFBC ash directly using alkaline agents such as sodium carbonate or calcium hydroxide to formulate cementless binders. Moreover, owing to its inherent mineralogical composition and morphological advantages, CFBC ash has also been investigated for use in environmental applications such as heavy metal removal from contaminated media [35]. However, its practical application in the construction industry remains limited. This is largely due to issues such as excessive heat release during early hydration and expansion-induced cracking, which stem from the high content of free CaO and SO_3 in CFBC ash [36]. Wang and Song [5] reported that the volumetric stability of CFBC-based cementitious systems can be maintained if the SO_3 content is controlled below 3.5%. Gao et al. [37] conducted a study on SCMs containing CFBC ash in concrete. According to their research, f-CaO and SO_3 in CFBC ash synergistically enhanced early hydration when combined with FA and ground granulated blast furnace slag (GGBS), with the optimal mix being 30% CFBC ash, 60% GGBS, and 10% FA. Rust et al. [38] conducted a study on the development of low-energy cement using CFBC ash and flue gas desulfurization (FGD) gypsum. They reported that FGD gypsum converted to hemihydrate effectively promoted early strength development in the CFBC ash-based binder. CFBC ash-based binders offer significant potential as low-carbon alternatives for sustainable construction by reducing the consumption of cement. However, to fully realize this potential, it is essential to enhance the self-cementing properties of CFBC ash.

In this study, the effect of DG on the self-cementing characteristics of CFBC ash was investigated. The hydration characteristics were evaluated through isothermal calorimetry, thermogravimetric analysis (TGA), setting time tests, and X-ray diffraction (XRD) analysis. Furthermore, mechanical performance was assessed by measuring the compressive strength at various curing ages. Microstructural evolution was also examined by analyzing the pore size distribution of specimens using mercury intrusion porosimetry (MIP) at different ages.

2. Experimental Details

2.1. Materials

In this study, CFBC ash was obtained from a CFBC boiler using bituminous coal as fuel, operated at a combustion temperature range of 800–950 °C. The ash was collected from the electrostatic precipitator after combustion. As shown in Figure 1, the CFBC ash exhibited a light brown color. The effects of DG on the hydration behavior of CFBC ash were investigated. The DG used in this study was obtained from Company H in South Korea.

Figure 1. Optical image of CFBC ash.

Figure 2 presents the SEM images of CFBC ash and DG. While some spherical particles are observed in CFBC ash, the majority of the particles exhibit angular, fractured shapes. The DG particles exhibit irregular shapes with a wide range of sizes and relatively rough surface textures.

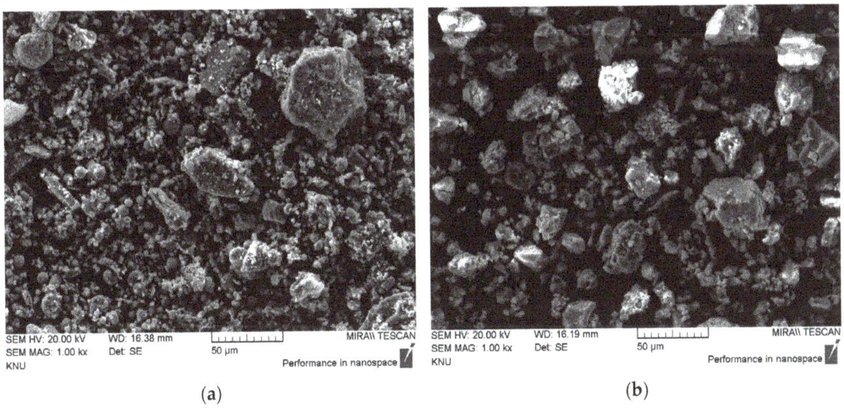

Figure 2. SEM images of raw materials: (**a**) CFBC ash; (**b**) DG.

The chemical compositions of CFBC ash and DG were analyzed using X-ray fluorescence (XRF) spectroscopy, and the results are summarized in Table 1. Here, LOI stands for loss on ignition. The free CaO content was also determined using the ethylene glycol extraction method [39]. CFBC ash was found to primarily consist of CaO, SiO_2, Al_2O_3, and Fe_2O_3, with SO_3 content of 5.57%. The free CaO content was measured to be 11.1%. The free CaO and SO_3 content in CFBC ash is attributed to the in-furnace desulfurization process, in which limestone is directly injected into the combustor, resulting in the formation of

CaO and $CaSO_4$. For DG, CaO and SO_3 were the predominant components, accounting for 62.4% and 22.7%, respectively, and the free CaO content was significantly high at 44.5%. The specific gravities of CFBC ash and DG were 2.85 g/cm^3 and 2.61 g/cm^3, respectively.

Table 1. Chemical oxide compositions of CFBC ash and DG.

	Chemical Oxide Composition (wt%)								
	SiO_2	Al_2O_3	Fe_2O_3	CaO	MgO	K_2O	Na_2O	SO_3	LOI
CFBC ash	27.1	11.3	11.1	35.5	4.6	0.8	1.2	5.6	1.9
DG	4.9	0.4	0.3	62.5	1.7	0.2	-	22.8	7.0

Figure 3 shows the XRD patterns of CFBC ash and DG. Based on the XRD patterns in Figure 3, a quantitative phase analysis was conducted using the Rietveld refinement method. Corundum was used as the internal standard for the analysis. The phase composition of CFBC ash was determined to be 22.7% SiO_2, 27.2% CaO, and 19.1.0% $CaSO_4$. In the case of DG, the main phases were 41.3% CaO and 30.7% $CaSO_4$. Additionally, $Ca(OH)_2$ and $CaCO_3$ were identified, which was attributed to partial hydration and carbonation due to exposure to atmospheric moisture.

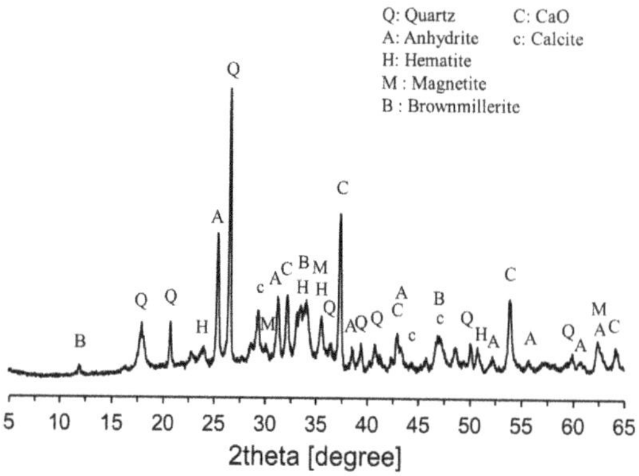

(a)

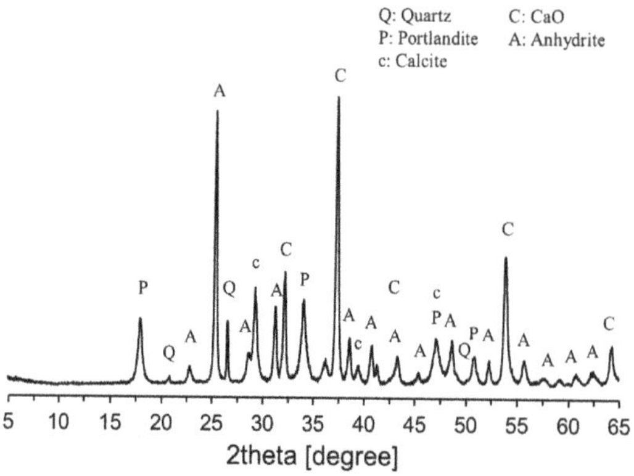

(b)

Figure 3. XRD patterns of raw materials: (**a**) CFBC ash; (**b**) DG.

Figure 4 presents the particle size distributions of CFBC ash and DG. The average particle sizes of CFBC ash and DG were measured to be 17.3 μm and 19.6 μm, respectively.

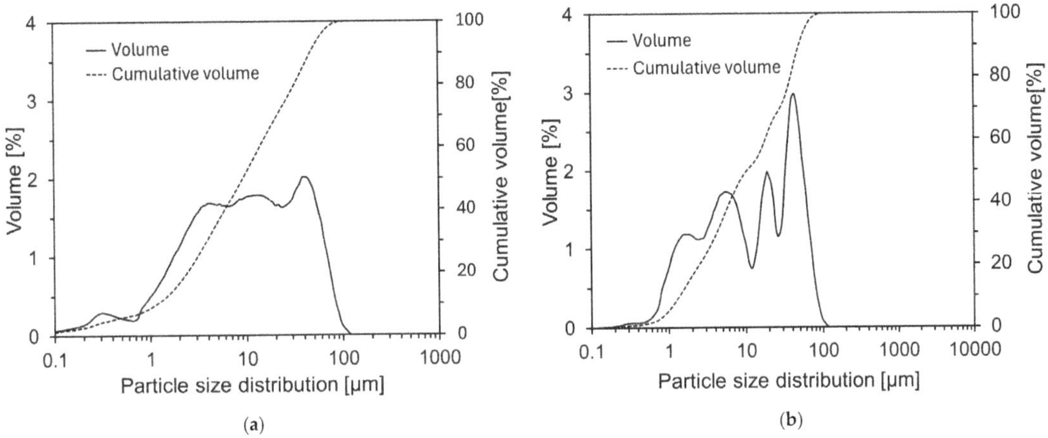

Figure 4. Particle size distributions of raw materials: (**a**) CFBC ash; (**b**) DG.

2.2. Mixture Proportions and Test Methods

In this study, paste and mortar specimens were prepared based on the mixture proportions shown in Table 2 to investigate CFBC ash-based binders as a substitute for cement. The primary variable was the replacement ratio of DG, aimed at evaluating its effects on the self-cementing behavior of CFBC ash. To examine the influence of the DG content, binder systems were prepared with DG replacing cement at 5%, 10%, and 15% by weight. To ensure the homogeneous mixing of CFBC ash and DG, the dry powders were thoroughly blended using a V-type mixer prior to specimen fabrication. For all specimens, the water-to-binder ratio was fixed at 0.5 by weight. ISO-standard sand was used for mortar specimens at a sand-to-binder ratio of 3:1 by weight.

Table 2. Mixture proportions of specimens containing CFBC ash and DG.

Specimen	W/B (–)	CFBC Ash (g)	DG (g)	Sand (g)
Plain	0.5	100	-	300
CFDG05		95	5	
CFDG10		90	10	
CFDG15		85	15	
CFDG20		80	20	

The fresh mortar was prepared according to Table 2; then, it was placed into 40 mm × 40 mm × 160 mm prismatic molds and compacted using a vibration table to eliminate entrapped air; it was then finished to a smooth surface. After casting, the specimens were cured at 20 ± 1 °C and relative humidity greater than 90% for 24 h. Following demolding, the specimens were submerged in water for continued curing. For compressive strength testing, the specimens were cut in half to obtain 40 mm × 40 mm × 80 mm prisms, and a compressive load was applied uniformly to the surfaces. The final compressive strength values for each variable were determined by averaging six replicate measurements. The setting time (initial and final) was measured according to ISO 9597 [40].

The heat of hydration was measured under isothermal conditions using a TAM Air isothermal calorimeter (TA Instruments, New Castle, DE, USA). Approximately 4 g of paste, prepared according to the mixture proportions in Table 2, was sealed in an ampoule and placed in the calorimeter. The heat flow due to hydration was continuously recorded for 24 h at a constant temperature of 23 ± 0.1 °C. X-ray diffraction (XRD) patterns were obtained using a PANalytical X'Pert Pro MPD diffractometer (Malvern PANalytical, Almelo, The Netherlands) to analyze the crystalline phases. Measurements were conducted with Cu-Kα radiation over a 2θ range of 5° to 65°, using a step size of 0.02° and a scan rate of 0.5°/min. Thermogravimetric analysis (TGA) was performed using a Thermo plus EVO2 thermal analyzer (Rigaku, Tokyo, Japan) under a nitrogen atmosphere. Approximately 10 mg of a powder sample was heated from room temperature to 1000 °C at a heating rate of 10 °C/min. The mass loss of the sample was continuously recorded as a function of the temperature. The pore structure analysis of the paste specimens was conducted using a Micromeritics Autopore IV 9500 mercury intrusion porosimeter (Micromeritics Instrument Corporation, Norcross, GA, USA). Before testing, the samples were thoroughly dried and then mounted in the sample cell of the instrument. The pore size distribution was measured over a pressure range of 0.003 psia to 30,000 psia.

3. Results and Discussion

3.1. Hydration of CFBC Ash Containing DG

Figure 5 presents the normalized heat flow and cumulative heat release profiles of CFBC ash paste samples incorporating various amounts of DG. In typical OPC systems, a sharp exothermic peak—referred to as the initial hydrolysis peak—appears immediately upon contact with water. This is followed by a dormant period, after which the hydration of tricalcium silicate (C_3S) begins in earnest, leading to the formation of C-S-H gel and $Ca(OH)_2$ during the acceleration period. As the reaction rate gradually decreases, a distinct second peak (the main hydration peak) typically appears. However, here, CFBC ash exhibited different hydration behavior compared to OPC. As shown in Figure 5a, the CFBC ash system exhibited an initial peak approximately 5 min after water contact—similar in timing to OPC—but no distinct second peak was observed. This suggests a deviation from the typical C_3S-driven hydration process found in OPC. As the DG content increased, the magnitude of the initial peak also slightly increased, accompanied by an overall increase in the peak area. This phenomenon was likely attributed to the presence of free CaO in the DG, which reacts exothermically with water.

Figure 5b illustrates the normalized cumulative heat released over time. The plain sample composed solely of CFBC ash showed a 24 h cumulative heat release rate of approximately 38.5 J/g. In contrast, samples containing DG exhibited higher cumulative heat release. Notably, the sample in which 20 wt% of CFBC ash was replaced with DG (CFDG20) exhibited a cumulative heat release rate of 45.6 J/g at 24 h—an 18.4% increase compared to the plain sample.

Figure 6 shows the XRD patterns of the plain sample cured for 7 and 28 days. As shown in Figure 6, the main hydration products identified in the CFBC ash system were portlandite and ettringite. In addition, unreacted components inherent in CFBC ash, such as CaO, quartz, anhydrite, and gypsum, were partially detected. In OPC systems, the ettringite formed during early hydration typically converts to monosulfate as the sulfate content diminishes and tricalcium aluminate (C_3A) reacts further. However, in the CFBC ash system, monosulfate was not observed, likely due to the relatively high sulfate content and the low C_3A concentration in the CFBC ash compared to OPC. Generally, the free CaO present in CFBC ash reacts with water to produce $Ca(OH)_2$, which subsequently reacts with alumina or silica phases to form ettringite and C-S-H gel. These hydration reactions confer self-cementing properties to CFBC ash [7–9].

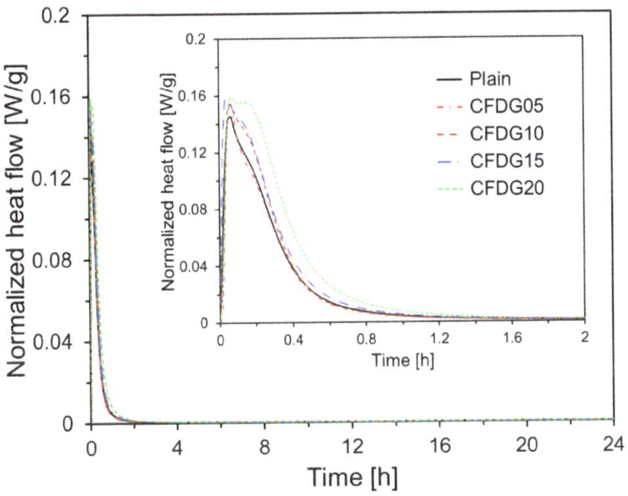

(**a**)

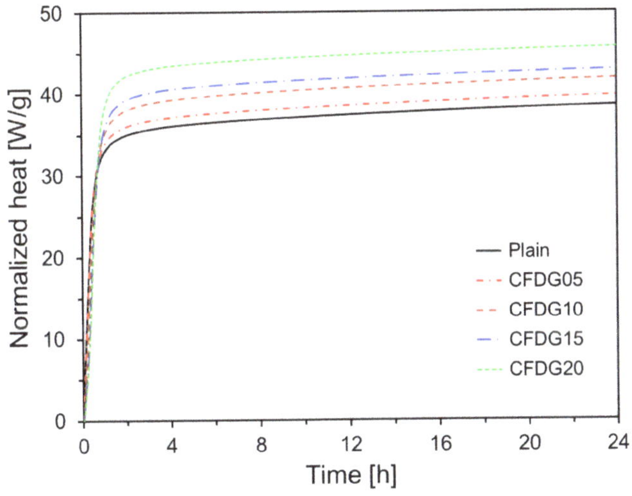

(**b**)

Figure 5. Isothermal calorimetry results: (**a**) normalized heat flow; (**b**) normalized heat.

Figure 7 presents the XRD analysis results of CFBC ash pastes with varying DG content. The overall hydration products were similar to those observed in the plain sample (Figure 6). As shown in Figure 7a, at 7 days of curing, both the anhydrite peak near 25.5° and the portlandite peak around 34° increased with increasing DG content. This trend is attributed to the hydration of free CaO contained in DG. At 28 days, as illustrated in Figure 7b, the intensity of the ettringite peak increased compared to that at 7 days, while the intensities of the anhydrite and portlandite peaks decreased. These changes suggest that anhydrite may have undergone phase transformation through hemihydrate to dihydrate gypsum, which then reacted with portlandite, Al_2O_3, and water to form ettringite.

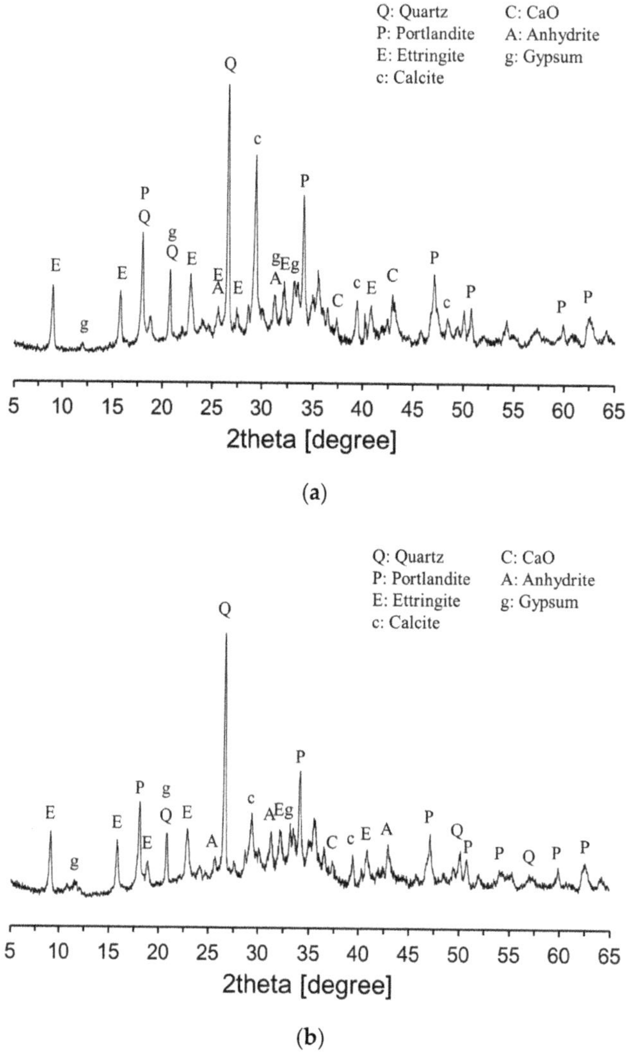

Figure 6. XRD patterns of plain sample: (**a**) 7 days; (**b**) 28 days.

Figure 8 shows the TG curves of CFBC ash pastes with varying DG content at different curing ages. Weight losses were observed near 100 °C and 450 °C. The weight loss around 100 °C is generally attributed to the evaporation of free water and the decomposition of hydration products such as ettringite and C-S-H gel that had not reacted with the CFBC ash [41,42]. The weight loss near 450 °C corresponds to the thermal decomposition of $Ca(OH)_2$, a key hydration product [41,42]. As the curing age increased, a greater degree of hydration occurred, leading to more significant weight loss near 100 °C. Additionally, as shown in Figure 8, the amount of $Ca(OH)_2$ formed from the hydration of free CaO decreased over time, as indicated by the reduced weight loss near 450 °C. This suggests that $Ca(OH)_2$ gradually transformed into ettringite and C-S-H gel over time.

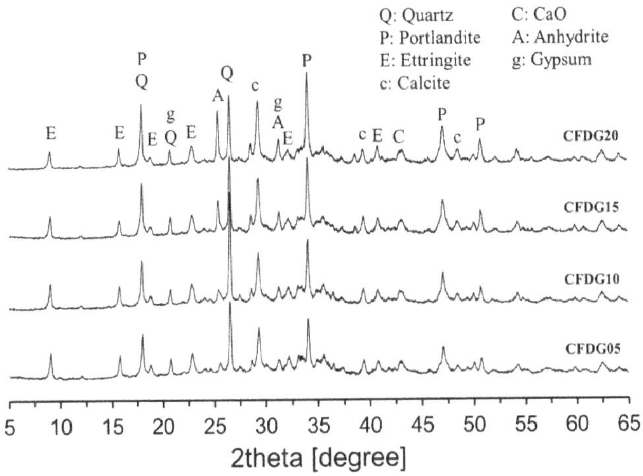

(**a**)

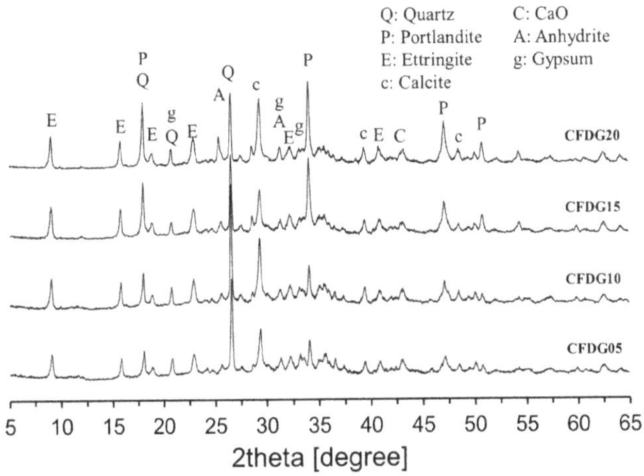

(**b**)

Figure 7. XRD patterns of specimens containing DG: (**a**) 7 days; (**b**) 28 days.

Figure 9 presents the quantitative results regarding the $Ca(OH)_2$ content derived from the thermogravimetric analysis shown in Figure 8. As illustrated in the graph, the $Ca(OH)_2$ content increased with higher DG replacement levels, regardless of the curing age (excluding the 28-day value for the plain sample). At 7 days, the $Ca(OH)_2$ content in the plain sample was 4.63%, while the sample with 20 wt% DG (CFDG20) exhibited $Ca(OH)_2$ content that was approximately 2.73 times greater than that of the plain sample. With increasing curing ages, the $Ca(OH)_2$ content generally decreased. In the plain sample, the $Ca(OH)_2$ content decreased by 0.62% from day 7 to day 14, and it further decreased by 0.74% from day 14 to day 28. In contrast, samples containing DG exhibited a slightly different trend. Although the $Ca(OH)_2$ content also decreased over time, the degree of reduction was not significantly affected by the DG replacement level. On average, the $Ca(OH)_2$ content in DG-blended samples decreased by 0.74% from day 7 to day 14 and by 2.8% from day 14 to day 28.

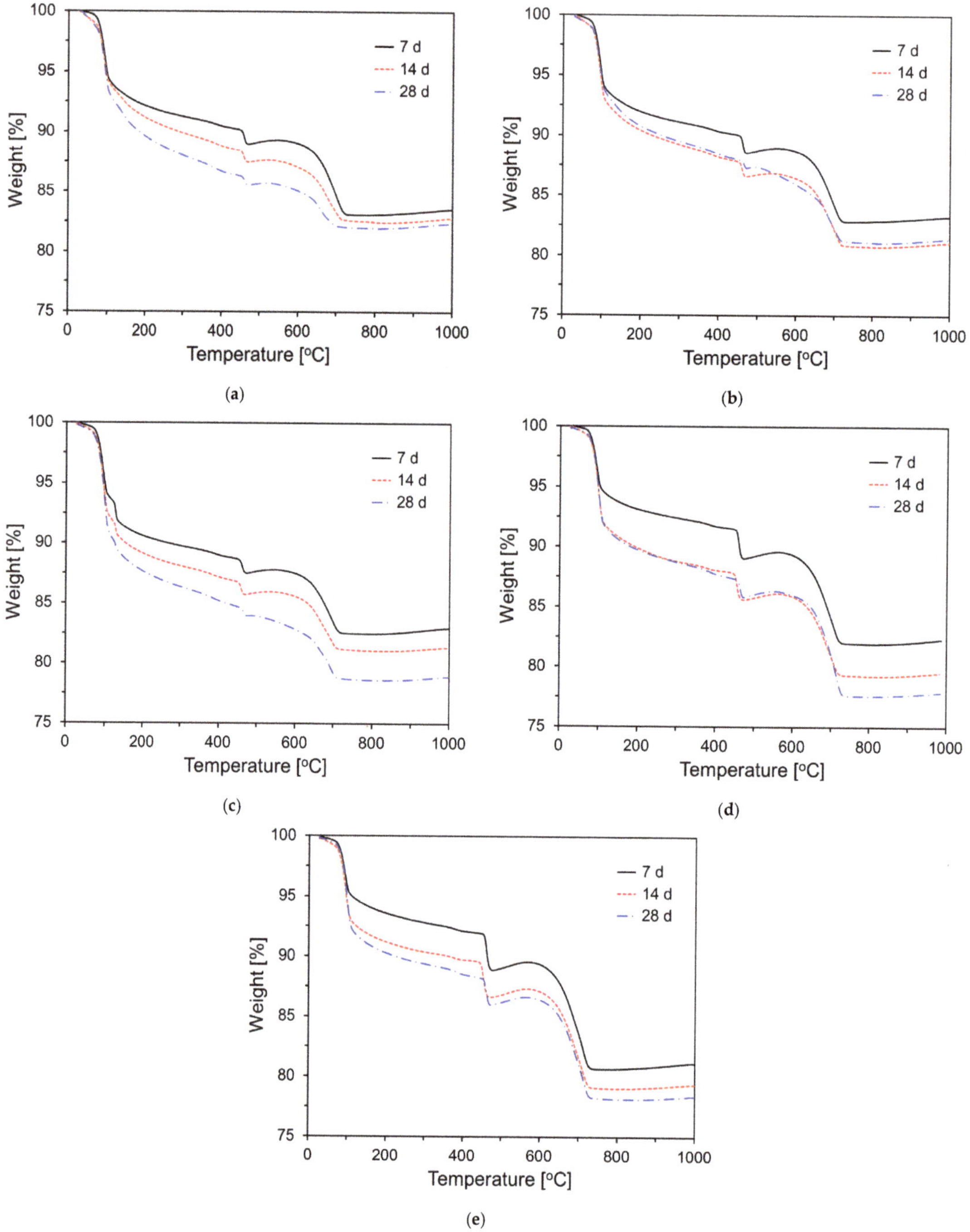

Figure 8. TG curves of specimens with varying DG content at different curing ages: (**a**) plain; (**b**) CFDG05; (**c**) CFDG10; (**d**) CFDG15; (**e**) CFDG20.

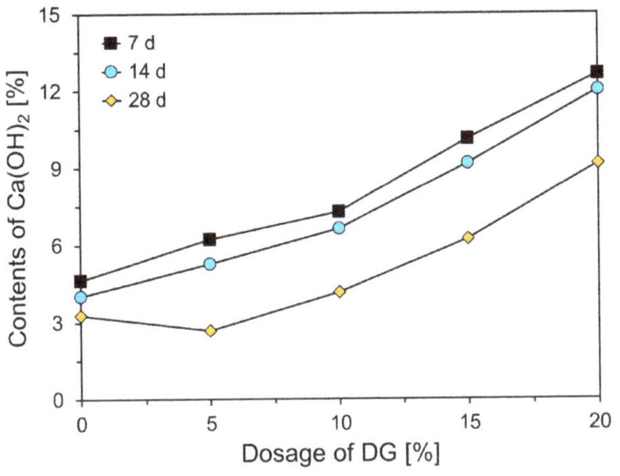

Figure 9. Quantitative analysis of $Ca(OH)_2$ content from thermogravimetric data.

3.2. Setting Time and Compressive Strength

Figure 10 illustrates the changes in the initial and final setting times of CFBC ash paste as a function of the DG content. Both the initial and final setting times increased with higher DG replacement levels. Specifically, as the DG content increased from 0% to 20%, the initial setting time increased from 48 min to 70 min, while the final setting time increased from 94 min to 117 min. This trend indicates that DG exhibits a retarding effect on the setting behavior of CFBC ash paste.

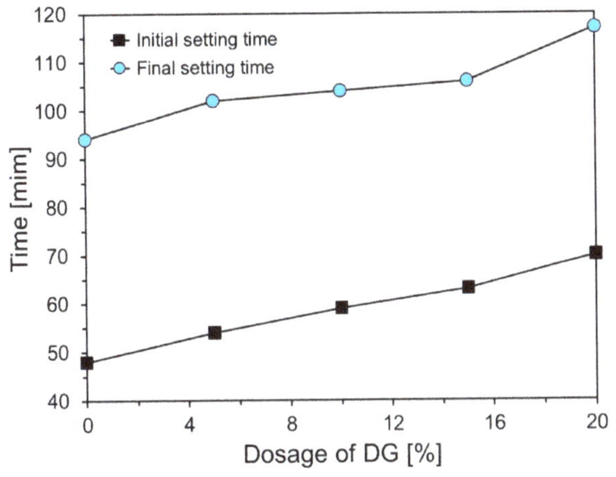

Figure 10. Setting times of CFBC ash pastes with varying DG content.

To evaluate the effects of the DG content on the mechanical performance of the CFBC ash mortar, compressive strength tests were conducted at curing ages of 3, 7, 14, and 28 days. The results are presented in Figure 11. DG was incorporated at varying levels of 0%, 5%, 10%, 15%, and 20% by weight. As shown in Figure 11, the development of the compressive strength was significantly influenced by the DG content. At early ages (3 and 7 days), the DG-blended specimens exhibited lower compressive strength compared to the plain sample. This reduction is attributed to the retardation of early hydration reactions caused by DG. The plain sample showed continuous strength development over time, reaching 26.4 MPa at 28 days. Notably, the specimen with 5% DG (CFDG05) achieved the highest 28-day strength of 29.2 MPa. While the incorporation of 5% DG clearly improves the 28-day compressive strength, the observed reduction in early-age strength and the sharp decline in strength at higher DG levels point to a very narrow optimal range for DG addition. This behavior reflects a delicate balance between the beneficial and detrimental effects of SO_3 and free CaO introduced through DG. At a moderate level (5%), DG likely provides a controlled supply of sulfate ions, facilitating the formation of stable ettringite and enhancing the densification of the microstructure over time. However, the initial retardation of hydration, evidenced by the lower early-age strength, suggests that DG delays the formation of strength-contributing hydration products in the first week of curing.

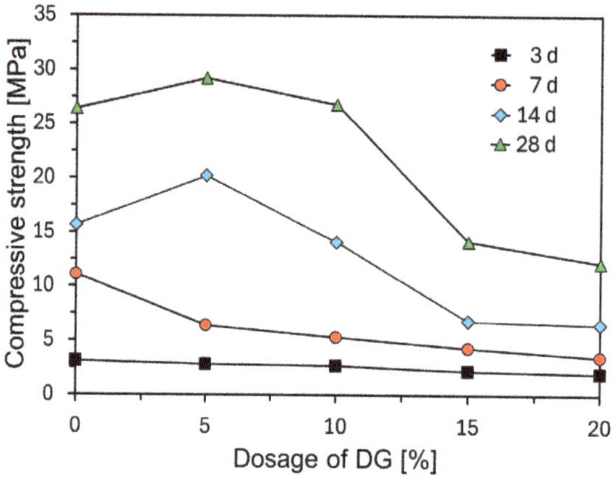

Figure 11. Compressive strength of CFBC ash mortars with varying DG content over curing time.

However, when the DG content exceeded 5%, a substantial reduction in compressive strength was observed at all curing ages. The 28-day strength of the specimen with 10% DG was slightly reduced to 26.8 MPa, whereas specimens with 15% and 20% DG (CFDG15 and CFDG20) showed significantly lower strength of 14.2 MPa and 12.2 MPa, respectively. More critically, beyond 5% DG, the dramatic reduction in compressive strength across all ages reveals that the system becomes sulfate-rich, which shifts the hydration equilibrium unfavorably. Excessive ettringite formation may not only increase the porosity due to volume expansion but may also block the ongoing hydration of calcium silicate phases, leading to an underdeveloped C–S–H matrix. This is consistent with the significant strength losses observed at 15% and 20% DG, where the 28-day strength was less than half that of the optimum. Furthermore, the narrow window between beneficial and detrimental DG content highlights a critical challenge in practical applications—even small deviations in the DG dosage can result in large variations in mechanical performance.

These findings indicate that achieving a durable and high-strength CFBC ash-based binder using DG requires the precise optimization and stringent quality control of the DG content. Future studies should focus on systematically mapping this narrow optimal range across different curing regimes and mix designs, as well as investigating the synergistic use of DG with other mineral admixtures to broaden the performance window.

3.3. Microstructure of CFBC Ash Paste Containing DG

To assess the influence of DG incorporation on the pore structure development of CFBC ash pastes, MIP tests were conducted at curing ages of 7, 14, and 28 days. Two types of measurements were performed: incremental intrusion (mL/g), which reflects the pore size distribution, and cumulative intrusion (mL/g), which represents the total accessible porosity. These were measured for CFBC ash paste specimens with DG content of 0%, 5%, 10%, 15%, and 20%. Figure 12 displays the incremental intrusion profiles over time. In general, all specimens exhibited a reduction in larger pores with increasing curing ages. Specifically, as shown in Figure 12, specimens such as the plain sample, CFDG05, and CFDG10 showed a clear trend in which larger pores decreased while finer pores increased over time. This effect was most pronounced in CFDG05. Specimens with higher DG content, namely CFDG15 and CFDG20, also showed similar trends but to a lesser extent. The overall reduction in pore size with the curing age can be attributed to the continued hydration of the CFBC ash, during which hydration products such as C–S–H and C–A–H are formed and progressively fill the capillary pores, thereby reducing both the porosity and average pore diameter.

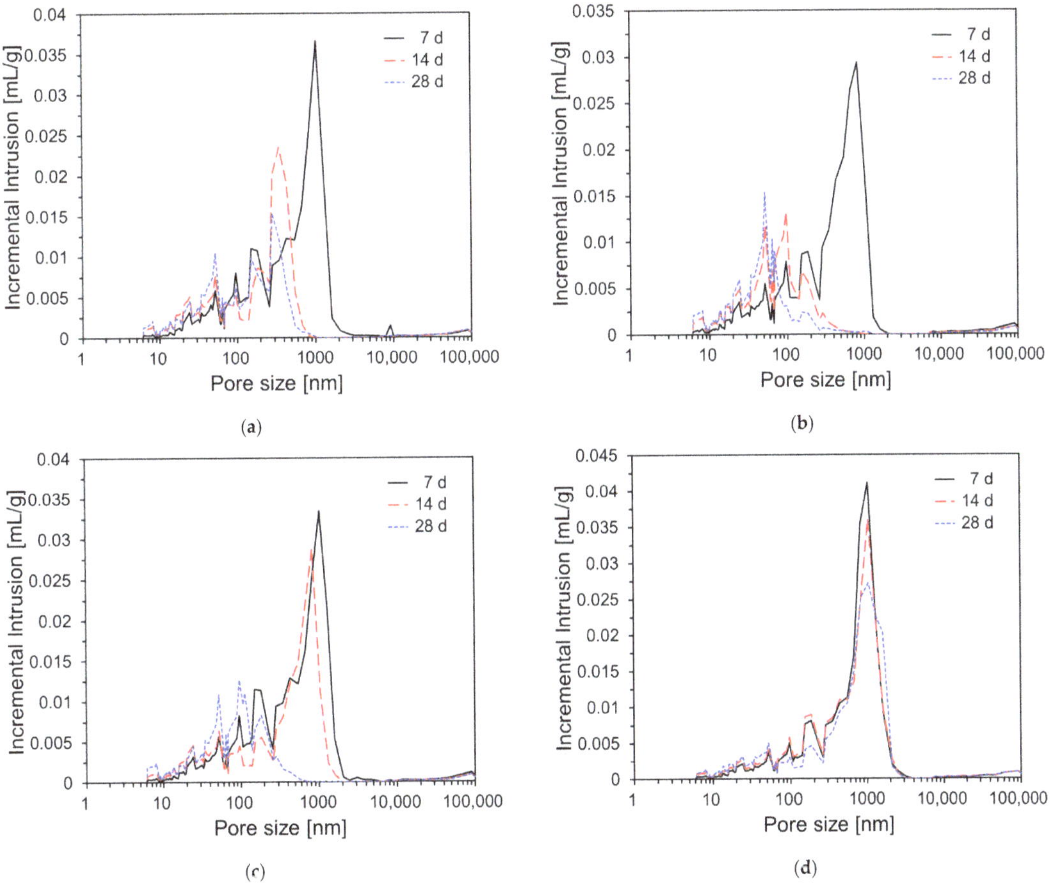

Figure 12. *Cont.*

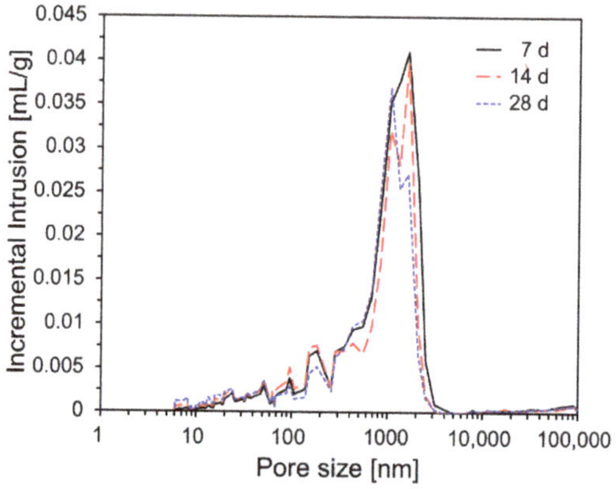

(e)

Figure 12. Incremental intrusions of CFBC ash pastes with DG over curing time: (**a**) plain; (**b**) CFDG05; (**c**) CFDG10; (**d**) CFDG15; (**e**) CFDG20.

Figure 13 presents the cumulative intrusion curves of each specimen over time. As illustrated, all samples showed a decreasing trend in cumulative intrusion with the curing age, regardless of the DG content. This corresponds to a reduction in total porosity and is consistent with the compressive strength results shown in Figure 13, indicating progressive microstructural densification due to the continued hydration of the CFBC ash. For the plain sample, the cumulative intrusion decreased from 0.307 mL/g at 7 days to 0.282 mL/g at 28 days. However, the extent of this reduction varied with the DG content. In specimens with 5–10% DG, the cumulative intrusion values at 28 days were similar to or slightly higher than those of the plain sample, suggesting a marginal increase in total porosity. In contrast, specimens with higher DG content (15–20%) exhibited consistently higher cumulative intrusion values throughout the curing period. Notably, CFDG20 recorded the highest porosity at 28 days with 0.299 mL/g. This trend is attributable to the free CaO present in DG. A moderate amount of free CaO can promote early hydration reactions and enhance the microstructure through $Ca(OH)_2$ formation [43]. However, excessive DG incorporation may impair the microstructure and reduce the compressive strength, as shown in Figure 11. This degradation is due to the reduction in CFBC ash content and the excessive presence of free CaO [44]. These findings demonstrate that, while moderate DG addition has a limited influence on the pore structure, excessive DG use significantly increases the total porosity, thereby compromising the mechanical strength and durability of the material.

Based on the results of this study, further research on the long-term durability of CFBC ash-based binders incorporating DG is deemed necessary. In particular, performance verification under harsh environmental conditions—such as sulfate attack, carbonation, and freeze–thaw cycles—should be prioritized. Through such studies, a deeper understanding of the role of DG in regulating the hydration kinetics and enhancing the durability of CFBC ash-based systems can be achieved, which is expected to contribute to the development of high-performance, low-carbon binders for sustainable construction.

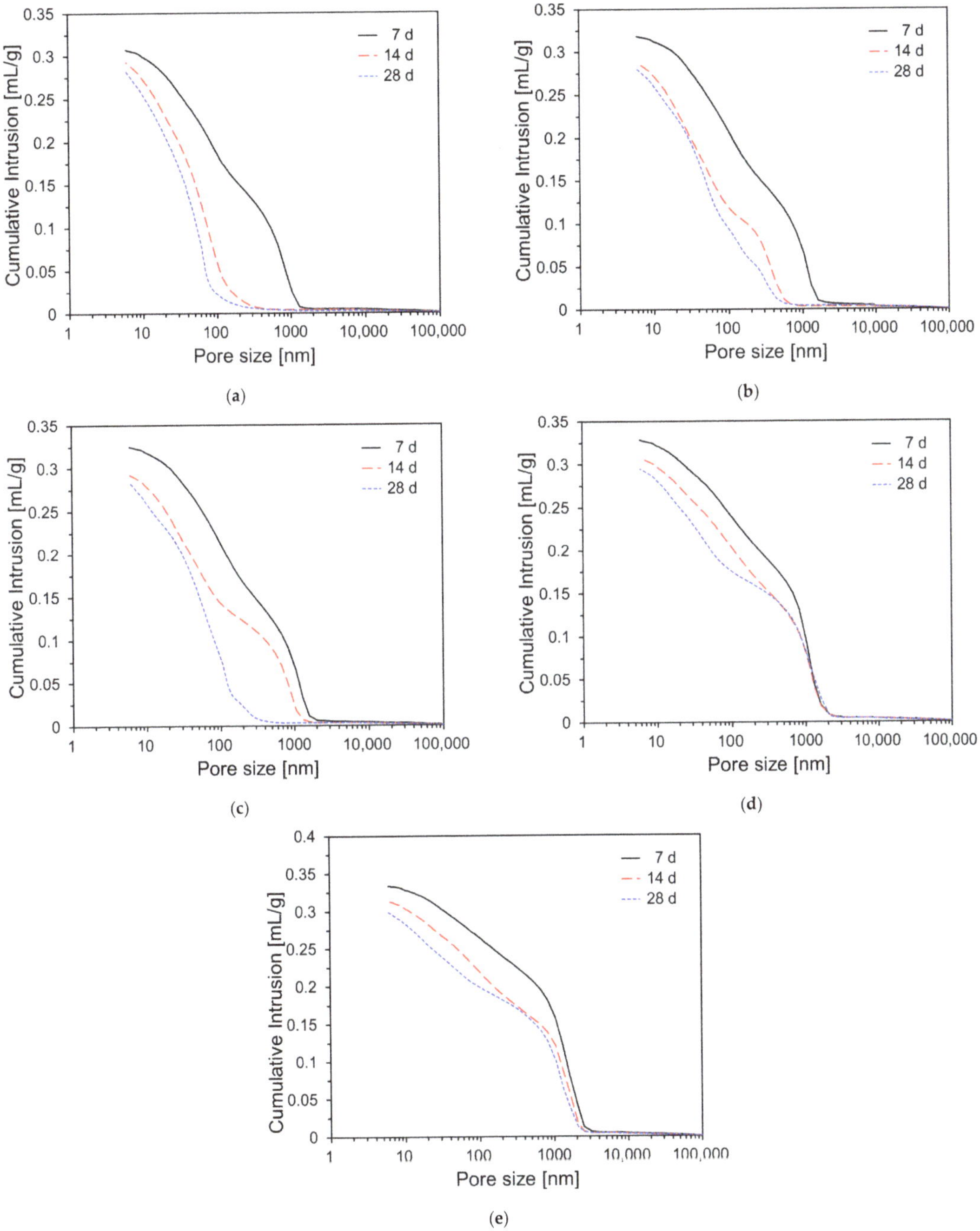

Figure 13. Cumulative intrusions of CFBC ash pastes with DG over curing time: (**a**) plain; (**b**) CFDG05; (**c**) CFDG10; (**d**) CFDG15; (**e**) CFDG20.

4. Conclusions

This study experimentally investigated the intrinsic hydraulic properties of CFBC ash and the influence of DG on its hydration behavior. A comprehensive analysis was conducted to evaluate the heat evolution, hydration products, setting behavior, compressive strength development, and microstructural changes of CFBC ash-based binders.

The hydration reaction of CFBC ash exhibited distinct characteristics compared to OPC. Isothermal calorimetry revealed that, unlike OPC, CFBC ash displayed a pronounced first exothermic peak upon contact with water but no distinct second peak. This suggests a deviation from conventional C_3S-dominated hydration. The magnitude of the first peak slightly increased with higher DG content, likely due to the presence of free CaO in DG. CFBC ash contains inherent free CaO, which reacts with water to form $Ca(OH)_2$. This compound subsequently reacts with alumina and silica components to form ettringite and C–S–H gel, contributing to the self-cementing behavior of the material. The XRD analysis at an early age (7 days) showed that increasing DG content resulted in the greater formation of anhydrite and portlandite, attributable to the hydration of DG's free CaO. At 28 days, anhydrite was observed to transform into dihydrate gypsum via hemihydrate intermediates. The dihydrate phase then reacted with portlandite, Al_2O_3, and H_2O, leading to the increased formation of ettringite. This was confirmed by a higher ettringite peak and reduced anhydrite and portlandite peaks at 28 days. The setting time measurements indicated that both the initial and final setting times increased with greater DG content, confirming that DG has a retarding effect on the setting behavior of CFBC ash paste.

Compressive strength testing revealed that the DG content significantly affected the mechanical properties of the CFBC ash mortar. At early curing ages, the suppressive effect of DG on hydration led to reduced strength compared to the plain sample. At 28 days, however, the sample with 5% DG (CFDG05) achieved the highest strength of 29.2 MPa, suggesting that moderate DG incorporation can effectively regulate the hydration kinetics and enhance the strength. In contrast, DG content above 5% led to marked reductions in compressive strength across all curing ages, indicating that excessive DG negatively impacts binder performance by hindering the formation and densification of hydration products. Mercury intrusion porosimetry (MIP) demonstrated that large pores decreased and fine pores increased with the curing age, particularly in the CFDG05 specimen. This trend is attributed to the gradual filling of pores by hydration products such as C–S–H and C–A–H, resulting in microstructural densification.

Moderate amounts of free CaO promote early hydration and improve the microstructure through $Ca(OH)_2$ formation. However, excessive free CaO, resulting from high DG content, leads to increased total porosity and microstructural degradation, which in turn reduce the compressive strength. These findings clearly indicate that, while moderate DG addition can enhance the performance of CFBC ash-based binders, excessive DG incorporation significantly increases the porosity and compromises both the mechanical strength and durability. Therefore, careful optimization of the DG dosage is essential for the effective utilization of CFBC ash in sustainable binder systems. To promote the use of CFBC ash as a low-carbon binder for sustainable construction, future research is essential on blended systems incorporating CFBC ash, as well as conducting field tests and performing durability tests under aggressive environments.

Funding: This work was supported by a National Research Foundation of Korea (NRF) grant funded by the Korean government (MSIT) (RS-2025-00520142). This work was also supported by the Industrial Technology Innovation Program (No. RS-2024-00438194, Development of technology of raw material substitute of using recycling resource to reduce sintering temperature by 100 degrees Celsius) funded by the Ministry of Trade, Industry & Energy (MOTIE, KOREA).

Institutional Review Board Statement: Not applicable.

Informed Consent Statement: Not applicable.

Data Availability Statement: The original contributions presented in this study are included in the article. Further inquiries can be directed to the corresponding author.

Conflicts of Interest: The author declares no conflicts of interest.

References

1. Havlica, J.; Brandstetr, J.; Odler, I. Possibilities of Utilizing Solid Residues from Pressured Fluidized Bed Coal Combustion (PSBC) for the Production of Blended Cements. *Cem. Concr. Res.* **1998**, *28*, 299–307. [CrossRef]
2. Anthony, E.J. Fluidized Bed Combustion of Alternative Solid Fuels; Status, Successes and Problems of the Technology. *Prog. Energy Combust. Sci.* **1995**, *21*, 239–268. [CrossRef]
3. Anthony, E.J.; Granatstein, D.L. Sulfation Phenomena in Fluidized Bed Combustion Systems. *Prog. Energy Combust. Sci.* **2001**, *27*, 215–236. [CrossRef]
4. Anthony, E.J.; Jia, L.; Wu, Y. CFBC Ash Hydration Studies. *Fuel* **2005**, *84*, 1393–1397. [CrossRef]
5. Wang, B.; Song, Y. Methods for the Control of Volume Stability of Sulfur Rich CFBC Ash Cementitious System. *Mag. Concr. Res.* **2013**, *65*, 1168–1172. [CrossRef]
6. Chi, M.; Huang, R. Effect of Circulating Fluidized Bed Combustion Ash on the Properties of Roller Compacted Concrete. *Cem. Concr. Compos.* **2014**, *45*, 148–156. [CrossRef]
7. Kang, Y.H.; Choi, Y.C. Development of Non-Sintered Zero-OPC Binders Using Circulating Fluidized Bed Combustion Ash. *Constr. Build. Mater.* **2018**, *178*, 562–573. [CrossRef]
8. Song, Y.; Qian, J.; Wang, Z.; Wang, Z. Self-Cementing Mechanism of CFBC Coal Ashes at Early Ages. *J. Wuhan Univ. Technol. Mater. Sci. Ed.* **2008**, *23*, 338–341. [CrossRef]
9. Illikainen, M.; Tanskanen, P.; Kinnunen, P.; Körkkö, M.; Peltosaari, O.; Wigren, V.; Österbacka, J.; Talling, B.; Niinimäki, J. Reactivity and Self-Hardening of Fly Ash from the Fluidized Bed Combustion of Wood and Peat. *Fuel* **2014**, *135*, 69–75. [CrossRef]
10. Wu, T.; Chi, M.; Huang, R. Characteristics of CFBC Fly Ash and Properties of Cement-Based Composites with CFBC Fly Ash and Coal Fired Fly Ash. *Constr. Build. Mater.* **2014**, *66*, 172–180. [CrossRef]
11. Chi, M.C.; Huang, R.; Wu, T.H.; Fou, T.C. Utilization of Circulating Fluidized Bed Combustion (CFBC) Fly Ash and Coal-Fired Fly Ash in Portland Cement. *Key Eng. Mater.* **2014**, *629–630*, 306–313. [CrossRef]
12. Zhang, Y.; Sun, W.; Shang, L. Mechanical Properties of High Performance Concrete Made with High Calcium High Sulfate Fly Ash. *Cem. Concr. Res.* **1997**, *27*, 1093–1098. [CrossRef]
13. Li, X.; Chen, Q.; Huang, K.; Ma, B.; Wu, B. Cementitious Properties and Hydration Mechanism of Circulating Fluidized Bed Combustion (CFBC) Desulfurization Ashes. *Constr. Build. Mater.* **2012**, *36*, 182–187. [CrossRef]
14. Sheng, G.; Li, Q.; Zhai, J. Investigation on the Hydration of CFBC Fly Ash. *Fuel* **2012**, *98*, 61–66. [CrossRef]
15. Zhou, M.; Chen, P.; Chen, X.; Ge, X.; Wang, Y. Study on Hydration Characteristics of Circulating Fluidized Bed Combustion Fly Ash (CFBCA). *Constr. Build. Mater.* **2020**, *251*, 118993. [CrossRef]
16. Sheng, G.; Li, Q.; Zhai, J.; Li, F. Self-Cementitious Properties of Fly Ashes from CFBC Boilers Co-Firing Coal and High Sulphur Petroleum Coke. *Cem. Concr. Res.* **2007**, *37*, 871–876. [CrossRef]
17. Sheng, G.; Zhai, J.; Li, Q.; Li, F. Utilization of Fly Ash Coming from a CFBC Boiler Co-Firing Coal and Petroleum Coke in Portland Cement. *Fuel* **2007**, *86*, 2625–2631. [CrossRef]
18. Chen, G.-Y.; Huang, W.-H. Activation of Blast Furnace Slag with CFB Fly Ash as a Supplementary Binder Material: Hydration Products and Effects of Sulfate Attack. *Crystals* **2022**, *12*, 41. [CrossRef]
19. Ren, K.; Ma, S.; Feng, Y.; Xu, N.; Bai, S. Study on the Composite Gravel Preparation and the Synergistic Absorption of CO_2 by Fly Ash of CFB Boiler. *Fuel* **2023**, *342*, 127843. [CrossRef]
20. Zhang, W.; Liu, X.; Zhang, Z.; Wang, Y.; Xue, Y.; Hao, X.; Lu, Y. Circulating Fluidized Bed Fly Ash Mixed Functional Cementitious Materials: Shrinkage Compensation of F-CaO, Autoclaved Hydration Characteristics and Environmental Performance. *Materials* **2021**, *14*, 6004. [CrossRef]
21. Gazdič, D.; Fridrichová, M.; Kulísek, K.; Vehovská, L. The Potential Use of the FBC Ash for the Preparation of Blended Cements. *Procedia Eng.* **2017**, *180*, 1298–1305. [CrossRef]
22. Jia, G.; Wang, Y.; Yang, F. A Review on the Application of Circulating Fluidized Bed Fly Ash in Building Materials. *Adv. Mater. Sci. Eng.* **2022**, *2022*, 7099430. [CrossRef]
23. Ohenoja, K.; Pesonen, J.; Yliniemi, J.; Illikainen, M. Utilization of Fly Ashes from Fluidized Bed Combustion: A Review. *Sustainability* **2020**, *12*, 2988. [CrossRef]
24. Alemu, A.S.; Lee, B.Y.; Park, S.; Kim, H.K. Self-Healing of Portland and Slag Cement Binder Systems Incorporating Circulating Fluidized Bed Combustion Bottom Ash. *Constr. Build. Mater.* **2022**, *314*, 125571. [CrossRef]
25. Ma, X.; He, T.; Xu, Y.; Da, Y.; Wang, H.; Yang, R. Properties of Composite Sintered Modified Fluidized Bed Incineration Fly Ash as Cement Admixture. *Constr. Build. Mater.* **2023**, *378*, 131210. [CrossRef]

26. Lee, J.; Lee, T. Effects of High CaO Fly Ash and Sulfate Activator as a Finer Binder for Cementless Grouting Material. *Materials* **2019**, *12*, 3664. [CrossRef]
27. Chi, M. Synthesis and Characterization of Mortars with Circulating Fluidized Bed Combustion Fly Ash and Ground Granulated Blast Furnace Slag. *Constr. Build. Mater.* **2016**, *123*, 565–573. [CrossRef]
28. Lin, W.-T.; Weng, T.-L.; Cheng, A.; Chao, S.-J.; Hsu, H.-M. Properties of Controlled Low Strength Material with Circulating Fluidized Bed Combustion Ash and Recycled Aggregates. *Materials* **2018**, *11*, 715. [CrossRef]
29. Nguyen, H.A.; Chang, T.P.; Shih, J.Y.; Chen, C.T.; Nguyen, T.D. Influence of Circulating Fluidized Bed Combustion (CFBC) Fly Ash on Properties of Modified High Volume Low Calcium Fly Ash (HVFA) Cement Paste. *Constr. Build. Mater.* **2015**, *91*, 208–215. [CrossRef]
30. Zheng, D.; Wang, D.; Li, D.; Ren, C.; Tang, W. Study of High Volume Circulating Fluidized Bed Fly Ash on Rheological Properties of the Resulting Cement Paste. *Constr. Build. Mater.* **2017**, *135*, 86–93. [CrossRef]
31. Shen, Y.; Qian, J.; Zhang, Z. Investigations of Anhydrite in CFBC Fly Ash as Cement Retarders. *Constr. Build. Mater.* **2013**, *40*, 672–678. [CrossRef]
32. Xu, H.; Li, Q.; Shen, L.; Zhang, M.; Zhai, J. Low Reactive Circulating Fluidized Bed Combustion (CFBC) Fly Ashes as Source Material for Geopolymer Synthesis. *Waste Manag.* **2010**, *30*, 57–62. [CrossRef] [PubMed]
33. Chindaprasirt, P.; Thaiwitcharoen, S.; Kaewpirom, S.; Rattanasak, U. Controlling Ettringite Formation in FBC Fly Ash Geopolymer Concrete. *Cem. Concr. Compos.* **2013**, *41*, 24–28. [CrossRef]
34. Liu, Z.; Li, L.; Shao, N.; Hu, T.; Han, L.; Wang, D. Geopolymerization Enhanced Hydrothermal Synthesis of Analcime from Steel Slag and CFBC Fly Ash and Heavy Metal Adsorption on Analcime. *Environ. Technol.* **2018**, *41*, 1753–1765. [CrossRef]
35. González, A.; Moreno, N.; Navia, R.; Querol, X. Study of a Chilean Petroleum Coke Fluidized Bed Combustion Fly Ash and Its Potential Application in Copper, Lead and Hexavalent Chromium Removal. *Fuel* **2010**, *89*, 3012–3021. [CrossRef]
36. Anthony, E.J.; Berry, E.E.; Blondin, J.; Bulewicz, E.M.; Burwell, S. Advanced Ash Management Technologies for CFBC Ash. *Waste Manag.* **2003**, *23*, 503–516. [CrossRef]
37. Gao, Y.; Li, J.; Lu, Z.; Zhang, M.; Duan, L.; Wang, J. Utilization of Composite Supplementary Cementitious Material Containing Circulating Fluidized Bed Ash in Concrete. *Constr. Build. Mater.* **2025**, *479*, 141486. [CrossRef]
38. Rust, D.; Morrison, M.; Lawrence, M.; Lynsdale, C. Development of Low-Energy Cement Products Using Circulating Fluidized Bed Combustion Ash and FGD Gypsum. *J. Mater. Civ. Eng.* **2012**, *24*, 1125–1131. [CrossRef]
39. Javellana, M.P.; Jawed, I. Extraction of Free Lime in Portland Cement and Clinker by Ethylene Glycol. *Cem. Concr. Res.* **1982**, *12*, 399–403. [CrossRef]
40. *ISO 9597:2020*; Cement—Test Methods—Determination of Setting Time and Soundness. International Organization for Standardization: Geneva, Switzerland, 2020.
41. Pane, I.; Hansen, W. Investigation of Blended Cement Hydration by Isothermal Calorimetry and Thermal Analysis. *Cem. Concr. Res.* **2005**, *35*, 1155–1164. [CrossRef]
42. Monteagudo, S.M.; Moragues, A.; Gálvez, J.C.; Casati, M.J.; Reyes, E. The Degree of Hydration Assessment of Blended Cement Pastes by Differential Thermal and Thermogravimetric Analysis: Morphological Evolution of the Solid Phases. *Thermochim. Acta* **2014**, *592*, 37–51. [CrossRef]
43. Nawaz, A.; Julnipitawong, P.; Krammart, P.; Tangtermsirikul, S. Effect and Limitation of Free Lime Content in Cement–Fly-Ash Mixtures. *Constr. Build. Mater.* **2016**, *102*, 515–530. [CrossRef]
44. Mtarfi, N.H.; Rais, Z.; Taleb, M. Effect of Clinker Free Lime and Cement Fineness on the Cement Physicochemical Properties. *J. Mater. Environ. Sci.* **2017**, *8*, 2541–2548.

materials

MDPI

Article

Study on Mechanical Properties of Alkali-Activated Coal Gasification Slag Concrete

Rongjian Shen [1], Xiaojun Li [2,*] and Shen Li [2]

1 Shanxi Anjian Investment and Construction Co., Ltd., Xi'an 710003, China; 18391867275@163.com
2 College of Safety science and Engineering, Xian University of Science and Technology, Xi'an 710054, China; 15991609639@163.com
* Correspondence: lxjun@xust.edu.cn

Abstract

Coal gasification slag (CGS) is a solid byproduct generated during coal gasification. Stacking and land-filling of CGS wastes substantial land resources and has significant environmental risks. In this paper, based on the Ca/Si and Si/Al ratios of the raw materials, the mix design of alkali-activated CGS concrete was optimized using a pure center-of-gravity design method. The compressive and flexural strengths of geopolymer concrete with varying mix proportions were measured to investigate the effects of sodium silicate modulus, material content, and dry density on its mechanical properties. Specimens of different sizes were prepared to analyze the influence of testing methods on the compressive, flexural, and tensile properties. The results indicate that the mechanical properties of geopolymer concrete are significantly influenced by the raw material composition and the modulus of the activator. With increasing curing age, both compressive and flexural strengths exhibit varying degrees of improvement. The stress-strain behavior of alkali-activated CGS concrete aligns closely with that of ordinary concrete. A comparative analysis of 100 mm length and 20 mm length cubic specimens revealed a compressive strength size conversion coefficient of approximately 0.456, while the flexural specimen exhibited a coefficient of 0.599. For tensile strength evaluation, both the Brazilian splitting method and the double punch test method yielded consistent and reliable results, demonstrating their suitability for assessing CGS-based concrete.

Keywords: coal gasification slag; alkali-activated; mechanical properties; testing method

1. Introduction

China is the largest producer and consumer of the coal in the world [1]. The advancement of coal conversion technologies toward higher efficiency, lower energy consumption, and minimized environmental impact has become an inevitable trend in energy development [2]. Researchers have focused on finding new low-carbon cementitious materials that can replace Portland cement, and promoting the green transformation of coal [3]. Among them, coal gasification is the main method for clean utilization of coal, but a large amount of CGS is produced in the process of gasification. CGS is a primary solid byproduct of coal chemical processes and poses significant environmental and health risks when improperly managed [4]. The research on comprehensive development and utilization of CGS has great ecological and economic significance, aiming to avoid environmental pollution, and promote green development.

Regarding the utilization of CGS, scholars have conducted extensive research on its applications in construction and highway engineering fields, and have also explored its potential uses in the ceramic industry [5]. Based on the potential resource value of CGS, scholars have conducted much research on the physical properties and chemical components of CGS. The mineral composition of CGS is similar to that of FA, and it is rich in Si, Al and Ca, which has potential pozzolanic activity, resulting in the potential to replace cement in the preparation of composite cementitious materials [6]. Furthermore, the coarse slag contains more aluminate, silicate and other mineral phases. In the process of hydration with cement or hardening with lime, hydrated calcium silicate and calcium aluminate gel with dense structures are formed, which improves the strength and durability of cement paste [7]. Li et al. studied the characteristics of hydration products of CGS and cement and showed that a high content of residual carbon in fine slag inhibits the hydration reaction between CGS and cement. The residual carbon in the coarse slag is lower, and the active oxides such as SiO_2 promote the hydration reaction, produce more gel, and improve the compressive strength of the cement mortar [8]. Li et al. [9] prepared thermally conductive UHPC with CGS. Lian et al. [10] investigated the reaction mechanism of CGS in relation to cement; in contrast to CGFS, the presence of abundant active mineral phases in CGCS promoted the cementitious reaction and improved the strength of the mortar. Jiao et al. [11] found that substituting quartz sand and river sand with CGS (possessing a higher crushing value) in UHPC mixtures led to increased compressive strength under standard curing. The maximum strength occurred when 75% of the conventional fine aggregates were replaced. Yoshitaka [12] added CGS as an aggregate to concrete and tested the aggregate performance and mechanical properties of the concrete, reporting that the 28 d compressive strength of CGS concrete and natural sand concrete showed no difference. ZHAO et al. [13] found that ceramsite prepared by cement, quartz and CGS according to a certain ratio can be used as concrete aggregate and admixture. Yang et al. [14] discussed the feasibility of using gasification slag as auxiliary cementitious material (SCM) and systematically studied the working performance, mechanical properties and microstructure characteristics of cement paste after adding gasification slag. The results show that the compressive strength of cement paste with gasification slag is about 3.7–9.3% lower than that of ordinary Portland cement (PC), but it can still meet the design requirements of 42.5 composite Portland cement. Fu et al. [15] demonstrated that when the CGS content is below 10%, it can effectively function as both nucleation sites and pozzolanic material in cement paste. Compared to ordinary Portland cement (PC) without CGS addition, the G1P specimen exhibited 7.1%, 6.9%, and 5.4% increases in compressive strength at 1 d, 7 d, and 28 d, respectively. The unreacted gasification slag primarily exists in aggregated form. Luo et al. [16] systematically investigated the effects of coal gasification coarse slag (CGA) and fine slag (CGS) on cement paste hydration and microstructure development, determining their strength activity indices to be 100.9% and 82.7%, respectively. Current research on utilizing CGS as supplementary cementitious materials (SCMs) remains limited, and large-scale application as SCMs proves difficult without prior carbon-ash separation treatment. According to Wei et al. [17], CGS functions effectively as a raw material in nonburnt brick production, with elevated dosages yielding products compliant with the MU20 standard [18]. Separately, Zhang et al. [19] determined that CGS pozzolanic mineral composition boosts mortar strength in alkaline settings, qualifying it as a sustainable alternative to conventional fine aggregates and partial cement replacement in roadway applications.

In summary, scholars have carried out research on different engineering applications of CGS, but as yet, there is no unified understanding for the rational and effective use of CGS. In this study, using differently sized specimens, the compressive strength, flexural strength and tensile strength of concrete were studied. The influence of different mix ratios and specimen sizes on the mechanical properties of concrete was studied, and the evaluation system of the mechanical properties of concrete was further improved, which has reference value for the utilization of CGS.

2. Materials and Methods

2.1. Raw Materials

2.1.1. Cementitious Materials

CGS: CGS is a solid waste byproduct of coal gasification methanol production. The CGS used in this study was sourced from a chemical company in Shanxi Province, China. The sieve analysis results and physical properties of the material are presented in Tables 1 and 2 [10], respectively.

Table 1. Particle size distribution of coal gasification slag from sieve analysis.

Size of Screen Mesh (mm)	9.5	4.75	2.36	1.18	0.6	0.3	0.15	0.075
Percent of Pass (%)	100	94.09	73.38	55.89	40.65	16.63	6.37	1.27

Table 2. Physical properties of coal gasification slag.

Surface Area (m^2/g)	pH	Burning Loss (%)	Density (g/cm^3)	Water Absorption (%)
84.50	9.33	3.3	1.63	4.0

According to the preliminary screening statistics, the particles with a particle size of less than 2.36 mm in the CGS account for about 73%, particles of 2.36–4.75 mm account for about 21%, and the rest are particles with a particle size of more than 4.75 mm 20.

Supplementary cementitious materials: FA, SF, and limestone flour (LF) were used as supplementary cementitious materials to produce geopolymer concrete. The chemical composition of the binders is listed in Table 3 [10].

Table 3. Chemical constituents of cementitious materials used in experiment (Mass%).

Material	SiO_2	Al_2O_3	Fe_2O_3	CaO	MgO	Na_2O	K_2O
CGS	41.19	15.46	17.70	12.61	1.68	1.71	1.36
FA	42.43	21.38	12.81	1.12	2.12	2.04	1.02
SF	96.78	0.78	0.56	0.64	0.73	0.78	0.67

Coal gasification slag: CGS, fly ash: FA, silica fume: SF.

2.1.2. Alkali Activator

In this experiment, sodium silicate is used as an activator to configure the cementitious materials; a sodium silicate solution produced by a chemical plant in Wuxi was used as the activator. The chemical composition and basic properties of the sodium silicate are shown in Table 4.

Table 4. Chemical composition and basic properties of sodium silicate solution.

Chemical Composition (%)			Density(g/cm^3)	Modulus	pH
Na_2O	SiO_2	H_2O			
7.96	26.10	65.94	1.47	3.30	10–13

Chemical analysis grade, pure granular sodium hydroxide with a purity of 99% was used to adjust the modulus of the sodium silicate solution. The sodium silicate modulus ($Na_2O{\cdot}nSiO_2$) represents the molar ratio of SiO_2 to Na_2O in the sodium silicate solution, where n is the modulus of sodium silicate. By adding NaOH particles to the sodium silicate solution, the sodium silicate modulus is adjusted to that required for each mix ratio. The principle is shown in Formula (1).

$$Na_2SiO_3 + 2NaOH = Na_2SiO_3 + Na_2O + H_2O \quad (1)$$

The amount of NaOH added in 100 g sodium silicate solution was determined according to Formula (2).

$$n = \frac{26.1/60}{7.96/62 + x/80} = \frac{0.435}{0.128 + x/80} \quad (2)$$

where 26.1 is the content of SiO_2 in the solution, and 60 is its molecular weight; 7.96 is the content of Na_2O in the solution, and 62 is its molecular weight; and 80 is the molecular weight of 2 mol NaOH.

2.2. *Specimen Preparation*

The preparation procedure was conducted as follows: Step 1: The sodium silicate solution was prepared one day in advance according to the required modulus and cooled for subsequent use. Step 2: The binders were mixed at low speed for 2 min. Step 3: After thorough mixing, the aggregates were added and stirred for 1 min. Step 4: The sodium silicate solution and water were slowly poured into the mixer, followed by slow stirring for 7 min and then rapid stirring for 1 min. Step 5: The mixture was poured into molds and vibrated for 1 min on a vibration table, after which the surface was sealed with plastic film for moisture retention; the test blocks in their molds were placed in a standard curing room maintained at (20 ± 2) °C with 95% relative humidity for 48 h, then demolded and continuously cured until reaching the predetermined age.

2.3. *Test Methods*

2.3.1. Compressive Strength Testing Method

In this experiment, cubic test samples with side lengths of 20 mm and 100 mm were used to test the compressive strength of alkali-activated CGS concrete. After 48 h of pouring, the test block was demolded, and the conventional curing of 7 d, 14 d and 28 d was carried out to test the cube compressive strength of concrete at three ages [20].

$$f_c = \frac{F_{c,\max}}{A} \quad (3)$$

where f_c represents the compressive tensile strength of the concrete cube specimens; $F_{c,max}$ is the failure load of the specimen; and A is the bearing area of the specimen.

2.3.2. Flexural Strength Testing Method

The test samples of 20 × 20 × 80 mm and 40 × 40 × 160 mm were used to test the flexural strength. After 48 h of pouring, the test blocks were demolded and cured at room temperature for 7 d, 14 d and 28 d. The flexural strength of concrete at three ages was tested. The flexural strength was calculated according to Formula (4):

$$f_f = \frac{1.5PL}{b^3} \quad (4)$$

where f_f represents the flexural strength of the concrete specimen; P is the failure load of the specimen; L is the distance of two fulcrums; and b is the side length of the prism square section.

2.3.3. Tensile Strength Testing Method

(1) Brazilian splitting method

Tensile properties represent crucial mechanical characteristics of geotechnical materials. The tensile strength of concrete represents its capacity to withstand tensile fracture under tensile stress. In this study, the Brazilian splitting method and double punch test were employed to investigate the tensile strength of alkali-activated CGS concrete with various mix proportions.

The Brazilian splitting [21] method involves placing a cylindrical specimen horizontally at the center of the testing machine's bearing plate. Two loading baselines are drawn along the axial direction on the specimen's side, and two spacers are aligned along these baselines. When loaded in compression, the specimen experiences tensile failure along its radial direction, enabling the determination of its tensile strength. The calculation formula is as follows:

$$\sigma_t = \frac{2p}{\pi DH} \tag{5}$$

where is the maximum tensile stress at the specimen's center, MPa; p represents the failure load, N; D is the specimen diameter, mm; and H is the specimen height, mm.

The test specimen, with a diameter (D) of 50 mm and height (H) of 25 mm (height-to-diameter ratio of 0.5), complies with the specification requirements for cylindrical specimens (diameter range: 48–54 mm; height-to-diameter ratio: 0.5–1.0).

(2) Double punch testing method

The double punch test [22] can indirectly determine the tensile strength of geotechnical materials. Fang and Chen proposed a new tensile test method based on the radial fracturing mechanism, which is called the double punch test. The layout is shown in Figure 1 [22].

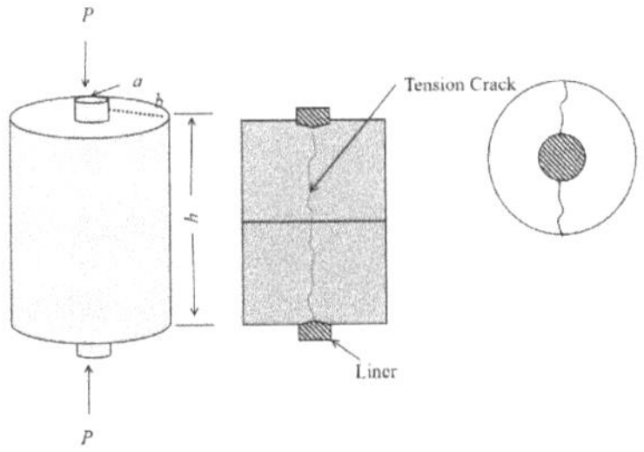

Figure 1. Double punch test layout and failure diagram.

The test method to determine the tensile strength involves vertically positioning the cylindrical specimen between the press's bearing plates with cylindrical steel pads serving as liners at both ends, then applying an axial load until tensile failure occurs. Based on limit analysis and experimental data, the calculation formula is as follows:

$$\sigma_t = \frac{P}{\pi(kbh - a^2)} \tag{6}$$

where σ_t represents the specimen's tensile strength (MPa), P denotes the ultimate pressure at failure (kN), b is the specimen radius (m), H stands for the specimen height (m), a indicates the liner diameter (m), and k is the material constant (approximately 1.0 according to experimental measurements on concrete, mortar, asphalt cement, and rock).

The cylindrical specimens were designed with a diameter-to-height ratio of 1:1.15 and a short-cylinder liner diameter-to-specimen diameter ratio of 1:4. For this test, the actual dimensions were as follows: specimen radius b = 30 mm, height H = 70 mm, and liner diameter a = 15 mm. These all meet the specified dimensional requirements.

2.3.4. Particle Size Distribution

The particle size distribution of the materials was tested according to the sieve analysis method specified in "Test Methods of Aggregate for Highway Engineering" (JTG 3432-2024) [23], specifically following the T0327-2005 fine aggregate sieve analysis procedure. The testing was conducted using the square-hole sieves prescribed in the standard

2.3.5. SEM Tests

Scanning electron microscopy (SEM, VEGA I-XMU, TESCEN, Prague, Czech Republic) was used to examine the effect of the mix ratio on the mechanical properties of specimens along with the modulus of activator.

2.4. Mix Ratio Design

The pure centroid design method is an effective approach for optimizing the composition of concrete materials. In this study, this method was employed to optimize the relative components of the concrete mixture. Based on a review of the relevant literature and preliminary experiments, the principles for configuring CGS concrete using sodium silicate as an alkali activator were established. The process is as follows:

(1) Selection of applicable supplementary cementitious materials: In the process of preparing geopolymer concrete and determining the mix ratio, the selection of raw materials, especially cementitious materials, will have an important impact. Cement, FA, SF, lime, slag and red mud are commonly used in the production of cementitious materials; the main components and alkali-binder ratio will affect the performance of the concrete. Based on the design principle of alkali-activated cementitious material composition, FA, SF and LF were selected as supplementary cementitious materials. FA is a kind of low-calcium volcanic ash material, which can provide Si and Al elements in the reaction to promote the hydration reaction and generate C-A-S-H as the main reaction product. The addition of LF increases the content of CaO in the reaction system, and then increases the contents of Ca/Si in C-S-H, which makes it easier for the hydration and alkali excitation reactions to occur. The silicon element in SF can promote a more complete polymerization reaction, thereby enhancing the strength of geopolymer concrete. Additionally, due to the small particle size of SF, the unreacted particles can fill the voids in the concrete and increase its density. This improvement in density not only further enhances the strength of the concrete but also significantly improves its resistance to permeability.

(2) Selection of appropriate alkali activator: The selection of activators significantly influences both the early-strength development and long-term durability of materials. Current commonly used activators are primarily categorized into three types: alkali activators, salt activators, and acid activators. Research indicates that sodium silicate as an alkali activator offers distinct advantages: silicon-aluminum ratio modulation, and reaction optimization. Sodium silicate provides soluble silicates that effectively increase the Si/Al ratio in cementitious systems, facilitating formation of more stable geopolymer network structures. The active SiO_2 and Al_2O_3 in FA and SF dissolve in alkaline environments and combine with $[SiO_4]^{4-}$ from sodium silicate, accelerating the formation of three-dimensional N-A-S-H gel that enhances early strength. Pozzolanic reaction enhancement: The alkaline environment (pH $\approx$ 13) of sodium silicate rapidly activates depolymerization of glassy phases in FA, releasing reactive silicon-aluminum components, while the high specific surface area of SF increases reactive interfaces, enabling synergistic geopolymerization and cement hydration that reduces setting time and improves interfacial transition zone density.

(3) The adequate utilization of CGS: The process of configuring concrete also follows the principle of environmental protection and relevant design indicators. The proportion of CGS in cementitious materials should be between 20% and 50% to maximize the utilization and consumption of CGS.

(4) Determining the mix ratios of Si/Al and Ca/Si: Based on the prior literature, theoretical calculations and analyses indicate that the optimal performance range for cementitious materials is achieved when the Si/Al ratio is maintained in the range of 2.8–3.5, and the optimal range for the Ca/Si ratio is 0.3–0.5 [24]. Building upon this foundation, this study designed a ternary cementitious system primarily composed of CGS, FA, and SF. While the Si/Al ratio of the system met the target range, the Ca/Si ratio consistently remained below the lower design threshold of 0.3 due to insufficient calcium content. To address this limitation and optimize the system's composition, LF was introduced as a supplementary calcium source. The modified cementitious system components, along with their corresponding Si/Al and Ca/Si ratios, are presented in Table 5. This adjustment successfully elevated the Ca/Si ratio to the desired range of 0.3–0.5, thereby enhancing the system's performance.

Table 5. The relative composition of four cementitious materials and their Si/Al and Ca/Si ratios.

NO	Mass Ratio				Molar Ratio	
	CGS	FA	SF	LF	Si/Al	Ca/Si
a	0.18	0.5	0.2	0.12	3.46	0.32
b	0.3	0.4	0.1	0.2	2.90	0.48
c	0.35	0.3	0.15	0.2	3.45	0.43
d	0.45	0.25	0.1	0.2	3.11	0.48
e	0.5	0.2	0.1	0.2	3.18	0.47

Determination of the basic experimental parameters: According to the literature and based on previous experiments [25], the water-binder ratio was set to 0.35, and the cement-sand ratio was set to 1.0. When the modulus of sodium silicate is 1.0–2.0, the performance of concrete is the better. Therefore, activators with moduli of 1.0, 1.5 and 2.0 were used in this experiment. The mix ratio scheme is shown in Table 6.

Table 6. Mix design proportions (kg/m^3).

No.	Modulus	CGS	FA	SF	LF	R.S	Water	S.S	W/B
1	1.0	180	500	200	120	1000	125	300	0.35
2	1.0	300	400	100	200	1000	125	300	0.35
3	1.0	350	300	150	200	1000	125	300	0.35
4	1.0	450	250	100	200	1000	125	300	0.35
5	1.0	500	200	100	200	1000	125	300	0.35
6	1.5	180	500	200	120	1000	95	350	0.35
7	1.5	300	400	100	200	1000	95	350	0.35
8	1.5	350	300	150	200	1000	95	350	0.35
9	1.5	450	250	100	200	1000	95	350	0.35
10	1.5	500	200	100	200	1000	95	350	0.35
11	2.0	180	500	200	120	1000	60	400	0.35
12	2.0	300	400	100	200	1000	60	400	0.35
13	2.0	350	300	150	200	1000	60	400	0.35
14	2.0	450	250	100	200	1000	60	400	0.35
15	2.0	500	200	100	200	1000	60	400	0.35

3. Results and Discussion

The compressive and flexural strengths of the specimens prepared using the 15 mix proportions are presented in Table 7.

Table 7. Mechanical properties (compressive and flexural strength) at 7, 14 and 28 days.

NO.	f_c (MPa)			f_f (MPa)		
Age	7 d	14 d	28 d	7 d	14 d	28 d
1	16.73	22.29	24.08	1.83	3.19	3.28
2	16.86	20.42	23.42	2.90	3.78	5.92
3	17.65	20.41	21.82	2.03	3.52	6.57
4	14.00	15.77	18.99	1.60	3.11	5.76
5	15.21	21.45	22.26	2.13	5.87	7.49
6	12.19	16.15	26.25	1.89	3.39	4.67
7	10.44	15.24	22.69	2.40	3.78	5.33
8	12.50	18.81	23.26	1.86	3.26	6.17
9	6.06	10.06	15.36	1.84	5.04	7.97
10	10.26	17.17	21.39	3.36	6.71	9.31
11	10.79	12.85	16.06	3.44	5.45	6.13
12	11.17	16.89	24.61	3.10	5.49	7.29
13	10.03	17.31	18.27	1.83	3.19	3.28
14	11.90	15.79	21.13	2.90	3.78	5.92
15	12.01	16.42	17.31	2.03	3.52	6.57

3.1. Compressive Strength Performance

3.1.1. Effect of Raw Materials on Compressive Strength

Table 7 shows the effective compressive strength of 15 groups of alkali-activated CGS concrete at 7 d, 14 d and 28 d. It can be seen from the table that when the modulus is 1.0, the ratio of CGS and SF has a significant effect on the compressive strength of concrete. When the ratio of CGS is 0.5–0.6 and the SF is 0.1–0.15, the compressive strength of concrete is the smallest, about 18.98–19.62 MPa. Taking the test results of fourth group of mix proportions as an example, the content of CGS slag is 0.56, the proportion of SF is 0.13, and the compressive strength of the concrete is 18.99 MPa. When the modulus of the activator is 1.5, the effect of SF on the compressive strength of concrete is more significant, and with increases in SF content, the compressive strength also increases. When the ratio of CGS, FA and SF is 0.4, 0.35 and 0.25 respectively, the compressive strength of concrete is the largest. When the modulus is 2.0, the effect of SF on the compressive strength of concrete is still significant. However, as the proportion of SF increases, the strength of concrete decreases gradually. When the proportion of SF increases from 0.13 to 0.19 and 0.23, the compressive strength of concrete decreases from 24.61 MPa to 18.27 MPa and 16.06 MPa.

As shown in Table 7, the compressive strength of concrete in Groups 1, 3, and 5 increases with decreasing sodium silicate modulus. This trend can be attributed to the higher alkali concentration in the system at lower modulus, which enhances the dissolution of quartz in CGS and FA. As a result, the concentrations of Ca, Si, and Al increase, promoting the formation of N-A-S-H and C-(A)-S-H gels. These gels densify the microstructure, thereby improving the mechanical strength. The addition of NaOH reduces the sodium silicate modulus and modifies its electric double-layer structure, expanding the diffusion layer. This increases the availability of Na^+ and OH^- ions in the system, leading to more thorough activation of SiO_2 and Al_2O_3 in the cementitious materials. Consequently, the polymerization reaction becomes more extensive, contributing to higher compressive strength. Furthermore, elevated alkalinity facilitates the complete excitation of Ca, Si, and Al, increasing the precursor content in the system. This results in a more robust skeletal structure from polymerization, further enhancing the concrete's strength. However, excessive modulus reduction intensifies the polymerization reaction, causing C-A-S-H gels to encapsulate unreacted particles and form a dense protective film. While this film improves short-term strength, it inhibits further reaction of internal particles and may lead to pronounced efflorescence (Figure 2). This phenomenon explains why some mixtures exhibit reduced compressive strength at very low moduli.

Figure 2. Concrete with efflorescence phenomenon.

3.1.2. Effect of Curing Age on Compressive Strength

Figure 3 shows the variation in compressive strength with curing age for alkali-activated CGS concrete.

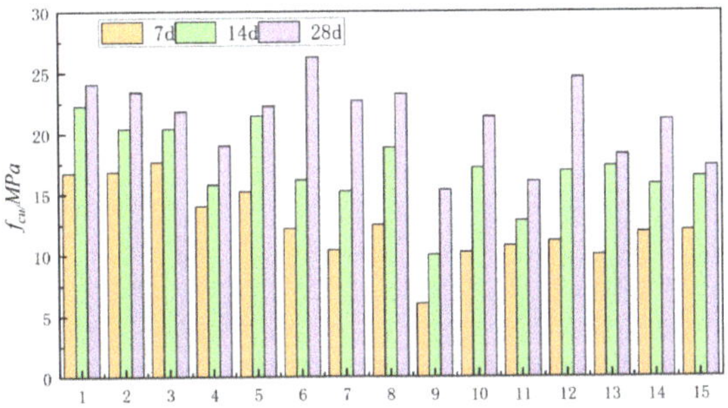

Figure 3. Variation in compressive strength of concrete with curing ages of 7, 14 and 28 days.

The strength development exhibits a distinct two-stage growth pattern: a rapid increase during the early stage (7–14 days), followed by a slower growth rate in the later stage (14–28 days). Group 1 demonstrated a 33% strength increase during the 7–14-day period, approximately 25 percentage points higher than its 14–28-day growth. The most significant difference was observed in Group 13, which showed a 72% strength gain in the first week, 67 percentage points higher than its subsequent growth. In contrast, Groups 4, 11, and 14 exhibited slightly higher strength development during the 14–28-day period, with increases of 8, 6, and 1 percentage points respectively compared to their early-stage growth, indicating relatively stable performance after initial curing.

This behavior can be explained by the reaction mechanisms of alkali-activated materials. In the initial stage, the dissolution of Al and Si oxides from the cementitious material leads to the breaking of Al-O and Si-O bonds in the glassy phase, releasing $[AlO_4]^{5-}$ and $[SiO_4]^{4-}$ tetrahedra. These species rapidly polymerize to form geopolymer networks dominated by Si-O-Al and Si-O-Si linkages, contributing to early strength development. As curing continues, the ongoing decomposition of oxides generates increasing amounts of N-A-S-H and C-(A)-S-H gels. The formation of C-(A)-S-H further promotes the dissolution of Si and Al from raw materials, accelerating the geopolymerization process. Due to its more compact structure compared to N-A-S-H, C-(A)-S-H enhances the continued strength gain, though at a reduced rate in later stages [26].

Additionally, the early rapid strength development is facilitated by the substantial exothermic heat released during the dissolution of cementitious materials and sodium silicate, which drives the geopolymerization reaction. However, as the reaction progresses, the heat release diminishes, and in systems with lower modulus, efflorescence occurs while the alkaline environment weakens. These factors collectively result in the observed slower strength growth rate during the later curing period compared to the initial stage.

3.1.3. Size Effect on Compressive Strength

To investigate the size effect on concrete compressive strength, this study prepared 100 × 100 × 100 mm cube specimens using mix the proportions 1–3, 6–8, and 11–13, with three replicates for each mix design. Following standard curing procedures, all specimens were maintained for 28 days before undergoing uniaxial compressive strength testing. This experimental approach enabled systematic evaluation of how specimen dimensions influence measured compressive strength characteristics.

Figure 4 illustrates the typical failure modes of the specimens. The failure process can be distinctly divided into three characteristic stages: the elastic deformation stage, crack propagation stage, and final failure stage. During the initial elastic stage, the specimens exhibited only elastic deformation under progressively increasing load without visible macroscopic damage. Upon entering the crack development stage, primary penetrating cracks first formed along the principal stress direction, subsequently triggering rapid expansion of multi-directional secondary cracks. The ultimate failure stage manifested as rapid deterioration after peak stress was reached, featuring spalling of the concrete surface layers accompanied by crushing failure in the core region.

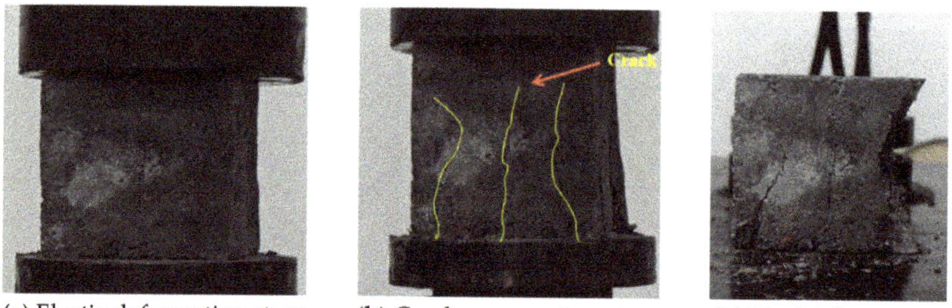

(**a**) Elastic deformation stage (**b**) Crack propagation stage (**c**) Final failure stage

Figure 4. The failure mode of a specimen is three stages.

The unconfined compressive strength results for the nine mix proportions are presented in Table 8. The data reveal that the 100 mm specimens consistently demonstrate lower compressive strength compared to the 20 mm specimens. This size effect primarily stems from the presence of initial microcracks in aggregates that are inevitably introduced during specimen preparation. Larger specimens contain a greater quantity of aggregates per unit volume, consequently incorporating more inherent microcracks which ultimately reduce the measured compressive strength. The calculation shows that when the modulus of sodium silicate is 1.0, f_{cu}, 100/f_{cu}, 20 is about 0.60. When the modulus of sodium silicate is 1.5, the ratios are about 0.31, 0.33 and 0.50, respectively. When the modulus of sodium silicate is 2.0, the ratios are 0.34 and 0.38, indicating that the modulus of sodium silicate will affect the compressive strength of differently sized specimens.

Table 8. Test results for average unconfined compressive strength of concrete at 28 days.

Mix Scheme	1	2	3	6	7	8	11	12	13
f_{cu} (MPa)	14.83	14.14	13.47	8.73	11.80	7.30	5.61	8.39	7.05

To verify the mechanical properties of CGS concrete, a comparative analysis was conducted with ordinary Portland cement concrete (OPC). The experimental results showed that the compressive strength of CGS geopolymer concrete ranged from 5.61 to 14.83 MPa, while OPC concrete with the same curing age and similar mix proportions typically achieved strengths of 20–40 MPa. Further research will be conducted to optimize the CGS mix ratio to enhance its mechanical performance.

Figure 5 presents the correlation analysis between compressive strength measurements of 100 mm and 20 mm cube specimens across all nine mix proportions. The experimental data yield a size conversion coefficient of approximately 0.456, indicating a consistent strength reduction relationship between the two specimen dimensions. This empirically derived coefficient quantitatively characterizes the scale effect observed in the alkali-activated concrete system, where larger specimens exhibit proportionally lower compressive strength values compared to their smaller counterparts.

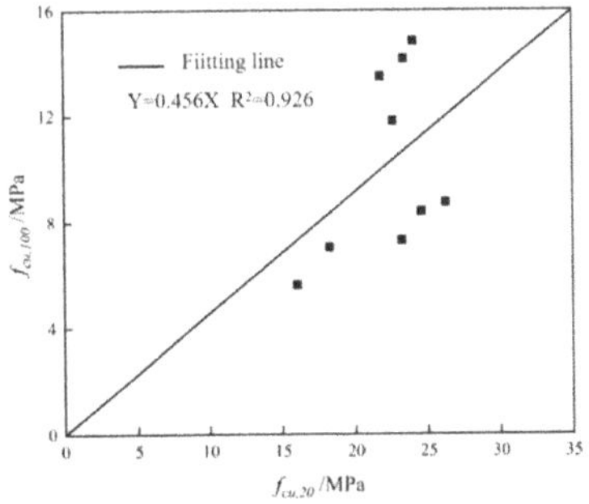

Figure 5. Fitting results of 28-day compressive strength of 20 mm and 100 mm cube specimens.

3.2. *Flexrual Strength Performance*

3.2.1. Flexural Strength

Table 7 presents the flexural strength test results under different mix proportions. When the activator modulus is 1.0, the flexural strength of concrete shows distinct trends based on component ratios. With CGS content between 0.35 and 0.50, increasing the SF proportion enhances flexural strength. For instance, comparing Groups 1 and 2 in which the CGS proportions are 0.20 and 0.38 respectively, and the SF proportions are 0.23 and 0.13, the flexural strength increases from 4.42 MPa to 5.95 MPa with higher SF content. Conversely, when CGS content ranges from 0.50 to 0.65, increasing FA content reduces flexural strength, as demonstrated by Groups 4 and 5 in which FA proportions of 0.25 and 0.31 correspond to strength reductions from 5.92 MPa to 3.28 MPa. The minimum flexural strength (3–4 MPa) occurs when CGS is 0.45–0.55, FA 0.3–0.4, and SF 0–0.15. At an activator modulus of 1.5, FA content significantly influences flexural strength when the CGS proportion is 0.35–0.45, with strength decreasing as FA increases. Groups 7 and 8 exemplify this trend, showing a strength reduction from 7.49 MPa to 5.76 MPa as FA decreases from 0.50 to 0.38. Meanwhile, for CGS proportions of 0.40–0.55, SF content becomes the dominant factor, with Groups 8 and 9 demonstrating a strength reduction from 7.49 MPa to 4.67 MPa as SF increases from 0.13 to 0.19. When the modulus reaches 2.0, FA and SF primarily govern flexural strength behavior. For FA proportions of 0.35–0.55, increasing CGS content decreases strength, as shown by Groups 13 and 14 in which strength drops from 9.31 MPa to 6.13 MPa as CGS increases from 0.44 to 0.56. However, with FA proportions of 0.25–0.35, flexural strength improves with increasing SF content.

The test results demonstrate that the flexural strength of concrete is significantly influenced by the modulus of the activator, showing a clear increasing trend with higher modulus values. For instance, the flexural strengths of Groups 4, 9, and 14 were measured at 3.28 MPa, 4.67 MPa, and 6.13 MPa, respectively, exhibiting consistent enhancement as the activator modulus increased. Notably, concrete specimens prepared with an activator modulus of 1.0 consistently exhibited lower flexural strength compared to those with a modulus of 2.0. This behavior can be attributed to several mechanistic factors. When the alkali activator modulus is too low, the sodium silicate solution releases fewer free Si-based geopolymer groups, which inhibits optimal gel formation. Additionally, excessive alkali content leads to the crystallization of NaOH on the surface of geopolymer concrete during hydration, resulting in efflorescence. This surface deposition hinders further hydration reactions, ultimately compromising strength development. Furthermore, as the modulus decreases, the geopolymerization reaction proceeds more completely, forming either a cross-linked network or a layered structure within the matrix. While this enhances reaction completeness, it also increases the brittleness of the resulting geopolymer concrete, which may contribute to the observed reduction in flexural strength at lower moduli.

3.2.2. Effect of Curing Age on Flexural Strength

Figure 6 illustrates the development of flexural strength of alkali-CGS concrete with curing age. It can be seen in the diagram that the flexural strength growth rate of concrete in the early stage (7–14 d) is faster, while the growth rate in the later stage (14–28 d) is significantly lower than that in the early stage. The strength of the Group 1 concrete increased by 58% during the 7–14 d period, which was about 39 percentage points higher than that during the 14–28 d period. The strength of the Group 12 specimens with 7–14 d of curing increased by 173%, which was about 115 percentage points higher than that of specimens with 14–28 d of curing, and this difference was the largest in Group 15.

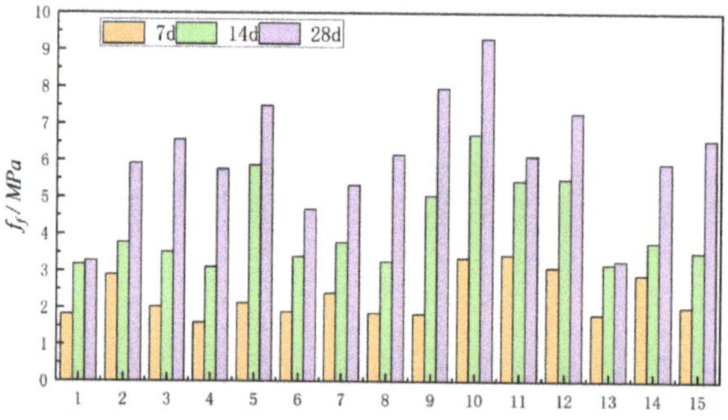

Figure 6. Variation in flexural strength of concrete with curing ages of 7, 14 and 28 days.

3.2.3. Size Effect on Flexural Strength

To investigate the size effect on concrete flexural strength, 40 × 40 × 160 mm prism specimens were prepared using mix proportions 1–3, 6–8, and 11–13, with three replicates for each mix design. All specimens were tested according to standard measurement methods, with flexural strength determined through three-point bending tests.

Figure 7 illustrates the failure mode of the flexural specimens, showing typical brittle fracture characteristics. During loading, no visible cracks appeared in the specimens prior to failure. Upon reaching peak load, the specimens fractured suddenly, demonstrating complete brittle failure behavior.

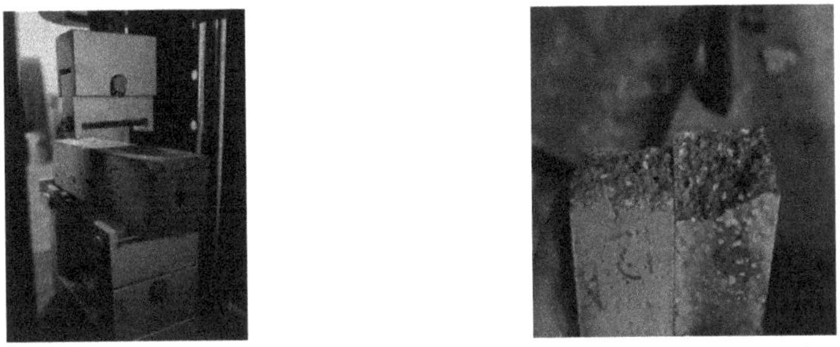

(**a**) Specimen cracks (**b**) Fracture cross section

Figure 7. Specimen failure mode.

The results show that the highest flexural strength was observed for sample 1 (5.11 MPa), while sample 3 showed the lowest flexural strength of 3.29 MPa. Based on previous experimental data analysis, the calculated flexural strength ratios between 20 × 20 × 80 mm and 40 × 40 × 160 mm specimens demonstrate clear modulus-dependent characteristics: For specimens with a sodium silicate modulus of 1.0, the flexural strength ratio ranges between 0.84 and 0.89. At a modulus of 1.5, the ratio decreases to approximately 0.60–0.69. With a modulus of 2.0, the ratio further reduces to 0.4.53. These results revealed that the size effect becomes more pronounced with increasing sodium silicate modulus and that the ratio variation follows a consistent downward trend with modulus elevation

Figure 8 shows the fitting result of the flexural strength for nine groups of 20 × 20 × 80 mm and 40 × 40 × 160 mm cubes, and the size conversion coefficient is about 0.599.

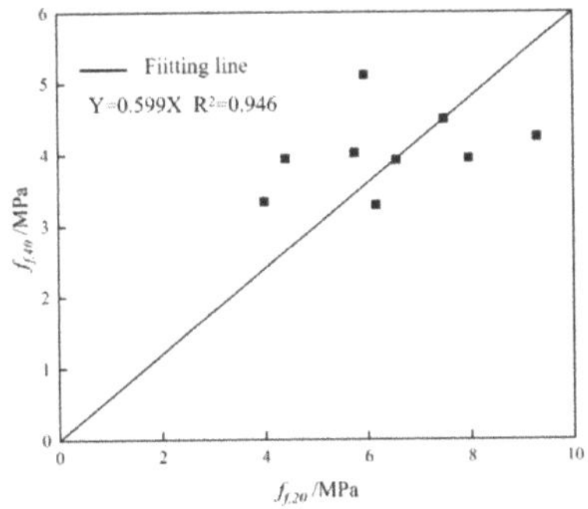

Figure 8. Flexural strength fitting results of 28-day flexural strength for 40 mm × 40 mm × 160 mm and 20 mm × 20 mm × 80 mm specimens.

3.3. *Tensile Strength Performance*

Tensile strength is a crucial mechanical property of geotechnical materials. The tensile strength of concrete refers to its capacity to resist tensile fracture under tensile stress. Based on the fifteen groups of mix proportions determined in the preliminary stage, three moduli were selected, and the three groups of mix proportions were investigated using the Brazilian splitting and double punch test methods to examine the tensile strength of alkali-activated CGS concrete under different mix proportions.

3.3.1. Brazilian Splitting Test

The test specimens were fabricated with dimensions of D = 50 mm and H = 25 mm, meeting the dimensional requirements specified in relevant standards. Figure 9 demonstrates the specimen failure pattern, exhibiting characteristic radial fracture propagation.

Figure 9. Brazilian splitting specimen failure pattern.

The specimens cured to 28 d were placed on the compressive testing machine, and then the radial load was applied to the specimens. When the specimens were damaged, the ultimate pressure was recorded, and the tensile strength of the concrete was calculated according to Formula (5). Three parallel tests were carried out for each mix ratio, and the average value was taken. The Brazilian splitting test was carried out on nine effective specimens, and the measured tensile strength values (MPa) were as follows: 1.80, 1.81, 1.85, 1.72, 1.72, 1.71, 1.64, 1.64, and 1.61. The tensile strength of this group of specimens is in good agreement with the typical variation range of brittle materials.

Figure 10 shows the linear fitting results of Brazilian splitting tensile strength and dry density for all test specimens. It can be seen from the figure that the fitting result is a straight line passing through the origin, and the dry density is proportional to the tensile strength.

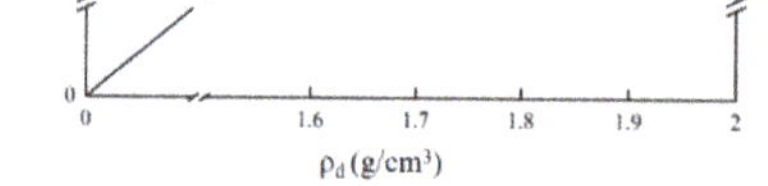

Figure 10. Fitting results of tensile strength and dry density measured by Brazilian splitting method.

3.3.2. Double Punch Test

The double punch test can indirectly measure the tensile strength of geotechnical materials. Fang and Chen proposed this new tensile test method based on the radial fracturing mechanism. The ratio of the diameter to the height of the cylinder sample is 1:1.15, and the ratio of the diameter of the short column liner to the diameter of the specimen is 1:4. The size of the test specimen (b = 30 mm, H = 70 mm, a = 15 mm) meets the required size. The biaxial penetration test diagram and the failure pattern of the specimen are shown in Figure 11.

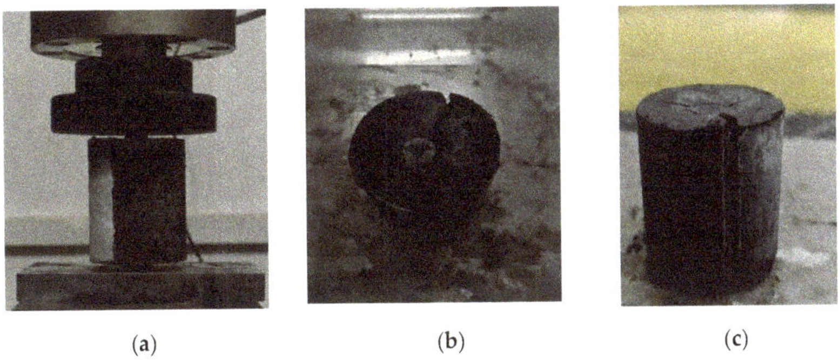

Figure 11. Specimen failure pattern diagrams: (**a**) double punch test diagram, (**b**) axial failure mode of specimen, (**c**) tensile crack.

The specimens with a curing age of 28 days were placed on a compressive testing machine, and then the axial load was applied to the specimens for testing. When the specimens were destroyed, the ultimate pressure was recorded. The tensile strength of the concrete was calculated according to Formula (6). Three parallel tests were carried out for each group of mix ratios, and the average value was taken. The results show that sample 2 had the highest strength with 1.43 MPa, and the lowest flexural strength was observed for sample 13 with 1.20 MPa. The average tensile strength based on the double punch test was 1.33 MPa.

Figure 12 shows the linear fitting results of biaxial penetration tensile strength and dry density of all test specimens. It can be seen from the figure that the dry density is proportional to the tensile strength.

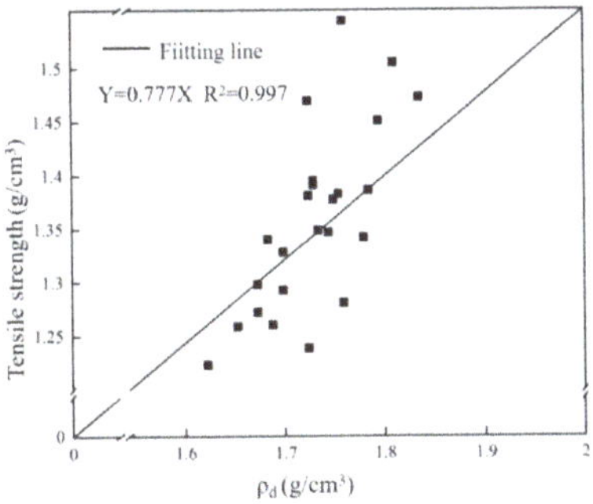

Figure 12. Fitting results of tensile strength and dry density measured by double punch test.

3.3.3. Comparison of the Two Testing Methods

The correlation analysis of the tensile strength measurement results measured by the above two methods was carried out using the analysis software, and the fitting results are shown in Figure 13. It can be seen that the tensile strength measured by the Brazilian splitting and double punch tests has a high degree of correlation, and the tensile strength ratio of the Brazilian splitting and double punch tests is about 0.77. This shows that the results obtained by these two indirect methods for measuring the tensile strength of concrete are basically reliable and can truly reflect the tensile strength characteristics of concrete materials.

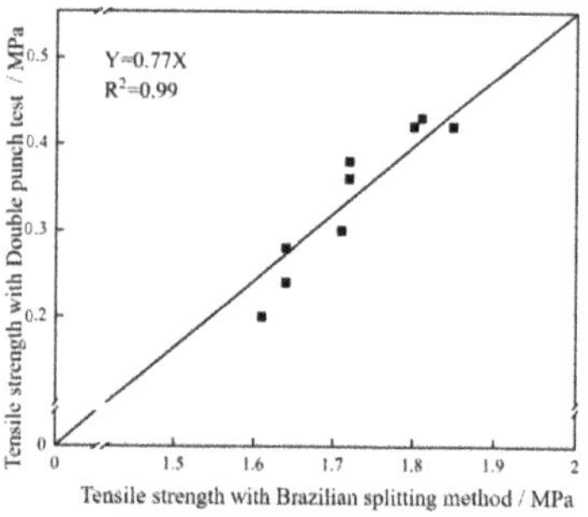

Figure 13. Fitting results of tensile strength measured by Brazilian splitting test and double punch test.

3.3.4. Calculation of Shear Strength by Compression-Tension Test

Fang and Hirst first proposed the method of determining shear strength parameters by a compression-tension test. Then, Li and Dong deduced the calculation formula of c (cohesion) and φ (internal friction angle) based on the Mohr-Coulomb criterion [27].

$$n = \frac{q_u}{\sigma_t} \tag{7}$$

$$\sin\varphi = \frac{n-2}{n+2} \tag{8}$$

$$c = \sigma_t \cdot \sqrt{\frac{n}{2}} \tag{9}$$

where q_u is the compressive strength, and σ_t is the tensile strength determined by the double punch testing method.

The shear strength of concrete refers to its capacity to resist structural failure under shear stress, which is one of the crucial indicators for evaluating the mechanical properties of concrete. According to the theory of material mechanics, the shear strength of unreinforced concrete is typically 10–20% of its compressive strength. From the calculation results presented in Table 9, it can be observed that the data obtained in this investigation generally fall within this theoretical range, thereby preliminarily validating the rationality of the computational approach. However, the actual shear strength of concrete is influenced by numerous factors, including the material composition, curing conditions, and loading patterns. Therefore, to obtain precise shear strength parameters, it is recommended that subsequent research employ direct shear tests or triaxial shear tests to acquire more reliable experimental data.

Table 9. Calculation of shear strength index of CGS concrete.

Mix Scheme	1	2	3	6	7	8	11	12	13
C(MPa)	1.42	1.43	1.42	1.38	1.36	1.30	1.24	1.28	1.20
$\varphi(°)$	42.73	41.56	40.67	31.3	38.71	28.34	22.76	32.16	29.47

3.4. Microstructure Analysis

In order to investigate the relationship between the microstructure and macroscopic mechanical properties of CGS concrete, microscopic analyses of three groups of CGS concrete with activator moduli of 1.0, 1.5, and 2.0 were carried out. The samples studied were Nos. 1, 6, and 11 in Table 6. The microstructure of concrete was studied by scanning electron microscopy (SEM).

Figure 14 shows the SEM analysis images of CGS concrete at 1000 and 2000 times. The formation and development of micro-morphology such as interfacial transition zones, unreacted materials, pores and microcracks of concrete were analyzed. It can be seen that there are obvious pores and microcracks in the structure. A large number of unreacted FA and SF particles were observed under high-power scanning electron microscopy, indicating that the polymerization reaction between the two was not sufficient. Therefore, increasing the content of FA and SF reduces the compressive strength of the geopolymer structure. However, appropriate FA and SF particles can fill the tiny pores near the interfacial transition zone, improve the pore structure of the concrete, and provide a more stable internal structure.

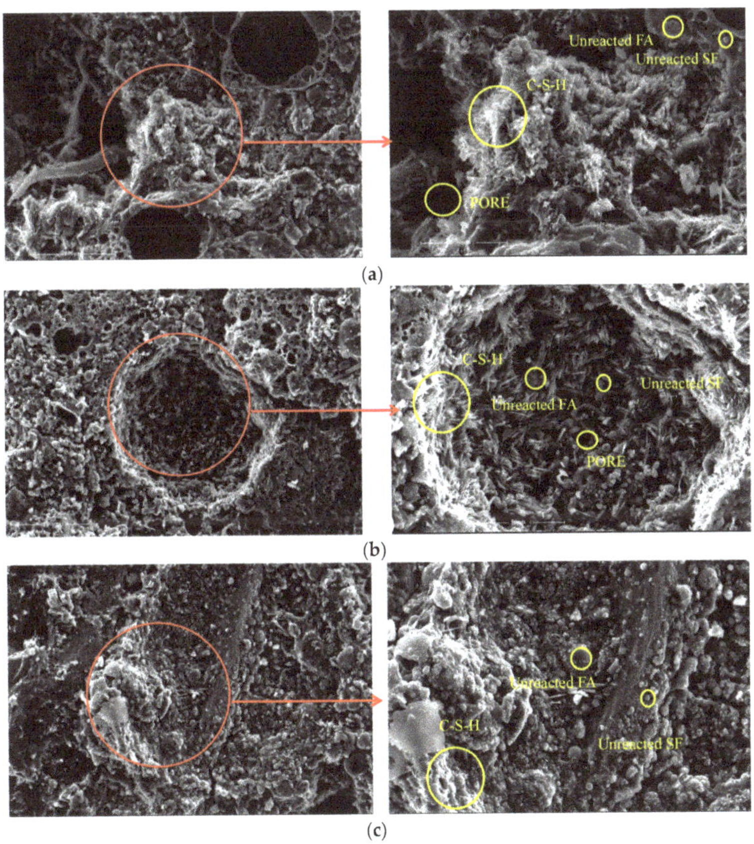

Figure 14. SEM images of samples with different sodium silicate moduli (M): (**a**) M1.0 at 1000× and 2000×; (**b**) M1.5 at 1000× and 2000×; (**c**) M2.0 at 1000× and 2000×.

4. Conclusions

In this study, CGS is innovatively utilized as the primary material for preparing geopolymer concrete, with a sodium silicate solution serving as the alkali activator. Simultaneously, a significant amount of powdered waste is employed to replace traditional cementitious materials, such as cement, which are environmentally detrimental during production. This approach aims to achieve the resource utilization of coal-based solid waste, contributing to sustainable construction practices.

(1) This study focuses on CGS as the research subject. Through testing its particle size distribution, chemical composition, and mineral constituents, it was found that CGS exhibits characteristics such as low density, uneven particle distribution, high specific surface area, weak alkalinity, and strong water absorption. Based on determining the material's Si/Al and Ca/Si ratios, a mixture proportion design was developed using a simplex centroid design method, incorporating CGS, FA, SF, and lime as cementitious materials. The ratios were set within the following ranges: Si/Al (2.8–3.5), and Ca/Si (0.3–0.5). The water-to-binder ratio was set at 0.35 and the binder-to-sand ratio was set at 1.0. Sodium silicate was selected as the alkali activator, and three different moduli (1.0, 1.5, 2.0) were chosen for experimental testing.

(2) Based on experimental results, the mechanical strength of concrete is jointly influenced by the modulus of sodium silicate (sodium silicate) and the relative proportions of raw materials. Additionally, the dry density and curing age of concrete also affect its mechanical properties. For alkali-activated coal gasification slag concrete under different activator moduli, compressive performance is primarily governed by the combined effects of SF content, CGS content, and FA content. When the modulus is 1.0 or 1.5, the flexural strength of the concrete is predominantly influenced by SF content, with SF significantly enhancing flexural strength. In contrast, CGS and FA contribute minimally to improving flexural strength.

(3) Research on compressive, flexural, and tensile strengths of specimens with different sizes revealed that the compressive failure process consists of three distinct stages: the elastic phase, crack propagation phase, and failure phase. Analysis of load-displacement data showed that the stress-strain curves at failure exhibited characteristics similar to conventional concrete. Through fitting compressive strength data from 100 mm and 20 mm cubic specimens, the size conversion coefficient was determined to be approximately 0.456 for compressive strength and 0.599 for flexural strength. After applying a size effect conversion, the 28-day compressive strength of specimens ranged from 7 MPa to 11.97 MPa, while flexural strength varied between 1.95 MPa and 5.58 MPa. Both the Brazilian splitting test and double punch test methods proved effective for indirect measurement of splitting tensile strength. The results obtained from these two methods showed good agreement, demonstrating their reliability in characterizing the tensile properties of alkali-activated CGS concrete.

Author Contributions: Conceptualization, R.S. and X.L.; methodology, R.S. and X.L.; validation, S.L.; formal analysis, X.L.; investigation, R.S.; resources, X.L.; data curation, R.S.; writing—original draft preparation, S.L.; writing—review and editing, S.L. and X.L.; visualization, R.S.; supervision, X.L.; project administration, R.S.; funding acquisition, R.S. All authors have read and agreed to the published version of the manuscript.

Funding: This research is funded by the Special Key Research Project of Integration of Industry and Innovation Chain of QINCHUANGYUAN Industrial Cluster Program (grant number 2023QCY-LL-24), the Shaanxi Key Research Program on QINCHUANGYUAN Scientist and Engineer Project (grant number 2023KXJ-134), the Research Program of Anhui Water Conservancy Development Co., Ltd. (grant number KY2022-B16), and the Research Program of Anhui Water Conservancy Development Co., Ltd. (grant number KY-2021-13).

Institutional Review Board Statement: Not applicable.

Informed Consent Statement: Not applicable.

Data Availability Statement: The original contributions presented in this study are included in the article. Further inquiries can be directed to the corresponding author(s).

Conflicts of Interest: Author Rongjian Shen was employed by the company Shaanxi Anjian Investment and Construction Co Ltd. The remaining authors declare that the research was conducted in the absence of any commercial or financial relationships that could be construed as a potential conflict of interest.

References

1. He, J.; Gu, D.; Chen, J.; Wu, X.; Kelly, T.N.; Huang, J.F.; Chen, J.C.; Chen, C.S.; Bazzano, L.A.; Reynolds, K.; et al. Enabling a rapid and just transition away from coal in China. *One Earth* **2020**, *3*, 187–194. [PubMed]
2. Wang, Q.; Song, X. Why do China and India burn 60% of the world's coal? A decomposition analysis from a global perspective. *Energy* **2021**, *227*, 120389.
3. Nikolaidis, P.; Poullikkas, A. A comparative overview of hydrogen production processes. *Renew. Sustain. Energy Rev.* **2017**, *67*, 597–611.4.
4. Du, M.; Huang, J.; Liu, Z.; Zhou, X.; Guo, S.; Wang, Z.; Fang, Y. Reaction characteristics and evolution of constituents and structure of a gasification slag during acid treatment. *Fuel* **2018**, *224*, 178–185.
5. Pomykała, R. The mechanical properties of coal gasification slag as a component of concrete and binding mixtures. *Pol. J. Environ. Stud.* **2014**, *23*, 1403–1406.
6. Blaisi, N.I.; Roessler, J.G.; Watts, B.E.; Paris, J.; Ferraro, C.C.; Townsend, T.G. Construction material properties of high temperature arc gasification slag as a portland cement replacement. *J. Clean. Prod.* **2018**, *196*, 1266–1272.
7. Xiaofu, S.; Jianli, M.; Jian, Z.; Danyu, X.; Liangyun, Z.; Jinqian, Z.; Xiaoyu, D.; Xiaomin, Z. Research status and prospects of utilization technologies of slag from coal gasification. *J. Environ. Eng. Technol.* **2017**, *7*, 712–717.
8. Li, Z.; Zhang, Y.; Zhao, H.; Chen, H.; He, R. Structure characteristics and composition of hydration products of coal gasification slag mixed cement and lime. *Constr. Build. Mater.* **2019**, *213*, 265–274.
9. Li, X.; Jia, S.; Chen, L.; Shen, R.; Liu, Y.; Mou, R. Experimental Study on Mechanical Properties of Thermally Conductive High-Strength Concrete. *Materials* **2025**, *18*, 642. [CrossRef]
10. Zhu, Z.; Lian, X.; Zhai, X.; Li, X.; Guan, M.; Wang, X. Mechanical properties of ultra-high performance concrete with coal gasification coarse slag as river sand replacement. *Materials* **2022**, *15*, 7552. [CrossRef]
11. Jiao, Y.; Zhang, Y.; Guo, M.; Zhang, L.; Ning, H.; Liu, S. Mechanical and fracture properties of ultra-high performance concrete (UHPC) containing waste glass sand as partial replacement material. *J. Clean. Prod.* **2020**, *277*, 123501.
12. Ishikawa, Y. Utilization of coal gasification slag collected from IGCC as fine aggregate for concrete. In Proceedings of the EUROCOALASH 2012 Conference, Thessaloniki, Greece, 25–27 September 2012.
13. Zhao, S.; Yao, L.; He, H.; Zou, Y.; Hu, L.; Zhai, Y.; Yu, Y.; Jia, J. Preparation and environmental toxicity of non-sintered ceramsite using coal gasification coarse slag. *Arch. Environ. Prot.* **2019**, *45*.
14. Yang, H.; Wang, H. Coal Gasification Slag as a Green Additive in Supplementary Cementitious Materials: Mechanical Properties and Microstructure. *Materials* **2024**, *18*, 86. [CrossRef]
15. Fu, B.; Cheng, Z.; Wang, D.; Li, N. Investigation on the utilization of coal gasification slag in Portland cement: Reaction kinetics and microstructure. *Constr. Build. Mater.* **2022**, *323*, 126587.
16. Luo, F.; Jiang, Y.; Wei, C. Potential of decarbonized coal gasification residues as the mineral admixture of cement-based material. *Constr. Build. Mater.* **2021**, *269*, 121259.
17. Wei, C.; Li, Y.; Liu, X.; Zhang, Z.; Wu, P.; Gu, J. Large-scale application of coal gasification slag in nonburnt bricks: Hydration characteristics and mechanism analysis. *Constr. Build. Mater.* **2024**, *421*, 135674.
18. *GB/T 21144-2007*; Solid Concrete Brick, The State Bureau of Quality and Technical Supervision. National Standard of the People's Republic of China: Beijing, China, 2007.
19. Zhang, Z.; Zhang, Z.; Yin, S.; Yu, L. Utilization of iron tailings sand as an environmentally friendly alternative to natural river sand in high-strength concrete: Shrinkage characterization and mitigation strategies. *Materials* **2020**, *13*, 5614. [CrossRef]
20. *Standard ASTM C230*; Specification for Flow Table for Use in Test of Hydraulic Cement. American society for Testing Materials Standard: New York, NY, USA, 2014.
21. *JTG 3432—2024*; Test Methods of Aggregate for Highway Engineering. China Communications Press: Beijing, China, 2024.
22. Fang, H.Y.; Chen, W.F. Further Study of Double-Punch Test 5or Tensile Strength of Soil. In Proceedings of the 3rd Southeast Asian Conference on Soil Engineering, ASCE, Hong Kong, China, 6–10 November 1972; pp. 211–215.
23. *SL/T 264—2020*; Specification for rock tests in water resources and hydropower projects. China Water & Power Press: Beijing, China, 2020.
24. Liu, Y.; Lu, C.; Hu, X.; Shi, C. Effect of silica fume on properties of high strength geopolymer cementitious materials. *J. Chin. Ceram. Soc.* **2020**, *48*, 1689–1699.
25. Li, N.; Shi, C.; Zhang, Z. Understanding the roles of activators towards setting and hardening control of alkali-activated slag cement. *Compos. Part B: Eng.* **2019**, *171*, 34–45.

26. Kumar, S.; Kumar, R.; Mehrotra, S.P. Influence of granulated blast furnace slag on the reaction, structure and properties of fly ash based geopolymer. *J. Mater. Sci.* **2010**, *45*, 607–615.
27. Tuode, L.; Chunmei, D. Determination of Shear Strength Parameters of Fissured Clay by Compression-Tension Tests. *Geotech. Investig. Surv.* **1985**, *3*, 45–49.

MDPI AG
Grosspeteranlage 5
4052 Basel
Switzerland
Tel.: +41 61 683 77 34

Materials Editorial Office
E-mail: materials@mdpi.com
www.mdpi.com/journal/materials

www.ingramcontent.com/pod-product-compliance
Lightning Source LLC
LaVergne TN
LVHW070453061126
851858LV00011B/233